Non-intrusive Load Monitoring

Hui Liu

Non-intrusive Load Monitoring

Theory, Technologies and Applications

Science Press
Beijing

Springer

Hui Liu
School of Traffic and Transportation
Engineering
Central South University
Changsha, China

ISBN 978-981-15-1859-1 ISBN 978-981-15-1860-7 (eBook)
https://doi.org/10.1007/978-981-15-1860-7

Jointly published with Science Press
The print edition is not for sale in Mainland of China. Customers from Mainland of China please order the print book from: Science Press.

This Springer imprint is published by the registered company Springer Nature Singapore Pte Ltd.
The registered company address is: 152 Beach Road, #21-01/04 Gateway East, Singapore 189721, Singapore

Preface

Electric products have been widely used in our life, and electric power has become an indispensable part of modern social development. With the advent of the energy crisis, energy-saving issues have attracted great attention among the whole society. As an important means of saving energy, improving energy efficiency by appropriate energy management becomes hotspot in recent years. In order to carry out more rational management of the energy consumption, the energy use information of appliances in a building is desired. By employing the information, the residents can restrict the operating time of high energy-consuming appliances and the manufacturers will be forced to upgrade their products to meet the energy-saving goal.

As an effective way to get fine-grained information, load monitoring techniques are of great importance for the efficiency improvement. Non-intrusive load monitoring is based on the information collected by the sensors at the entrance of the electrical circuit to realize the identification of switching details of appliances in the electrical circuit. Compared to the intrusive method, the non-intrusive scheme only needs install voltage and current sensors at the entrance of the electrical circuit, which makes it cost-saving and simple to install. In view of the broad practical prospect, the non-intrusive load monitoring has become a research hotspot in recent years.

Focusing on the non-intrusive load monitoring techniques in the field of smart grid, this book presents a thorough introduction of related basic principles, along with some possible improvement directions. For the graduate students and managers of relevant departments, this book can provide valuable information about the principles of the non-intrusive load monitoring techniques. For the researchers and doctoral students, this book can provide ideas and encourage subsequent research in non-intrusive load monitoring.

The studies in the book are supported by the National Natural Science Foundation of China, the National Key R&D Program of China, the Innovation Drive of Central South University, China. In the process of writing the book,

Zeyu Liu, Haiping Wu, Rui Yang, Chenqing Yu, Chengming Yu, Shuqing Dong, and other team members have done a lot of model verifications and other works; the author would like to express heartfelt appreciations.

Changsha, China
August 2019

Hui Liu

Contents

Nomenclature

ACE	Average Coverage Error
AR	Autoregressive
ARIMA	Autoregressive Integrated Moving Average
ARMA	Autoregressive Moving Average
AUC	Area Under Curve
BIC	Bayesian Information Criterion
BME	Light bulb
BP-ANN	Back-Propagation Artificial Neural Network
CNN	Convolutional Neural Network
CUSUM	Cumulative Sum
CVD	Continuously Variable Devices
CWC	Coverage Width-based Criterion
DAE	Denoised AutoEncoder
DBSCAN	Density-Based Spatial Clustering of Applications with Noise
DLC	Data-driven Linear Clustering
DNN	Deep Neural Network
DTW	Dynamic Time Warping
DWE	Dishwasher
ELM	Extreme Learning Machine
EM	Expectation-Maximization
EMD	Empirical Mode Decomposition
FA	Firefly Algorithm
FEEMD	Fast Ensemble Experience Mode Decomposition
FFT	Fast Fourier Algorithm
FGE	Refrigerator
FHMM	Factor Hidden Markov Model
FN	False Negative
FP	False Positive
FSM	Finite State Machines
GA	Genetic Algorithm

GLR	Generalized Likelihood Ratio
GOF	Goodness-of-Fit
GPS	Graph Signal Processing
HMM	Hidden Markov Model
IMF	Intrinsic Mode Function
KFDA	Kernel Fisher Discriminant Analysis
KNN	K-Nearest Neighbor
K-NNR	K-Nearest Neighbor Rule
LDA	Linear Discriminant Analysis
LP	Linear Programming
LS	Load Signature
LSTM	Long Short-Term Memory
MA	Moving Average
MAE	Mean Absolute Error
MAPE	Mean Absolute Percent Error
MIMO	Multi-Input Multi-Output
MLP	Multi-Layer Perceptron
MOFPA	Multi-objective Flower Pollination Algorithm
NILM	Non-Intrusive Load Monitoring
NMF	Non-negative Matrix Factorization
NSGA-II	Non-dominated Sorting Genetic Algorithm-II
PCA	Principal Component Analysis
PDRNN	Pool-based Deep Recurrent Neural Network
PICP	Prediction Interval Coverage Probability
PINAW	Prediction Interval Normalized Average Width
PINC	Prediction Interval Nominal Confidence
PLOA	Parallel Local Optimization Algorithm
PSO	Particle swarm optimization
QR	Quantile Regression
RF	Random Forest
RMSE	Root Mean Square Error
ROC	Receiver Operating Characteristic
SDE	Standard Deviation of Error
SLFN	Single Hidden Layer Feedforward Neural Network
SPRT	Sequential Probability Ratio Test
SSER	Sparse Switching Event Recovering
SVM	Support Vector Machine
SVR	Support Vector Regression
TN	True Negative
TP	True Positive
TWED	Time Warp Edit Distances

Chapter 1
Introduction

1.1 Overview of the Non-intrusive Load Monitoring

1.1.1 The Non-intrusive Load Monitoring

With the development of modern society, people are more closely connected with electric power. The electric power is not only the basis of modern industrial productions, but also provides convenience for the daily activities of residents. The last 4 decades (1971–2012) have seen a nearly fourfold increase in global electrical power consumption, and the global electric power consumption has reached 21,725 billion kWh in 2012 (Shi et al. 2016). The electric power consumption is one of the main sources of CO_2 emissions (Shi et al. 2018). Increasing energy efficiency is regarded as an important path to reduce carbon emissions and ease the energy crisis.

The smart grid is a widely discussed concept that can improve energy efficiency. An important feature of the smart grid is the information interaction between the energy generation side and the user side (Wang et al. 2018). Generally, more information is beneficial to the decision of the energy generator as well as the user. Taking the user side as an example, if the residents are provided with the electric power consumption information of each appliance, the residents tend to reduce the wastes and improper use of electric power. As a result, household electric power consumption can be reduced by 12% (Liu et al. 2019a).

Load monitoring methods are divided into two categories: the intrusive method and the non-intrusive method. The intrusive load monitoring method is carried out by installing sensors at the front end of each appliance and conducts real-time monitoring of appliances according to the results from all of the sensors. This method is simple in principle and usually provides high monitoring accuracy. However, it relies on a large number of sensors and needs to carry out the intrusive transformation of existing circuits, which is difficult to implement, costly and hard to popularize.

Each appliance has its special signature. Non-intrusive load monitoring is based on the information collected by the sensors at the entrance of electrical circuit to identify the switching details of appliances. Compared to the intrusive method, the

H. Liu, *Non-intrusive Load Monitoring*,
https://doi.org/10.1007/978-981-15-1860-7_1

non-intrusive scheme only needs sensors in the entrance of the electrical circuit, which makes it cost-saving and simple to install. In view of the broad practical prospect, the non-intrusive load monitoring has become a research hotspot of scholars at home and abroad in recent years. A typical example of the non-intrusive appliance load monitoring application is shown in Fig. 1.1. The algorithm framework of the Non-intrusive Load Monitoring (NILM) is presented in Fig. 1.2.

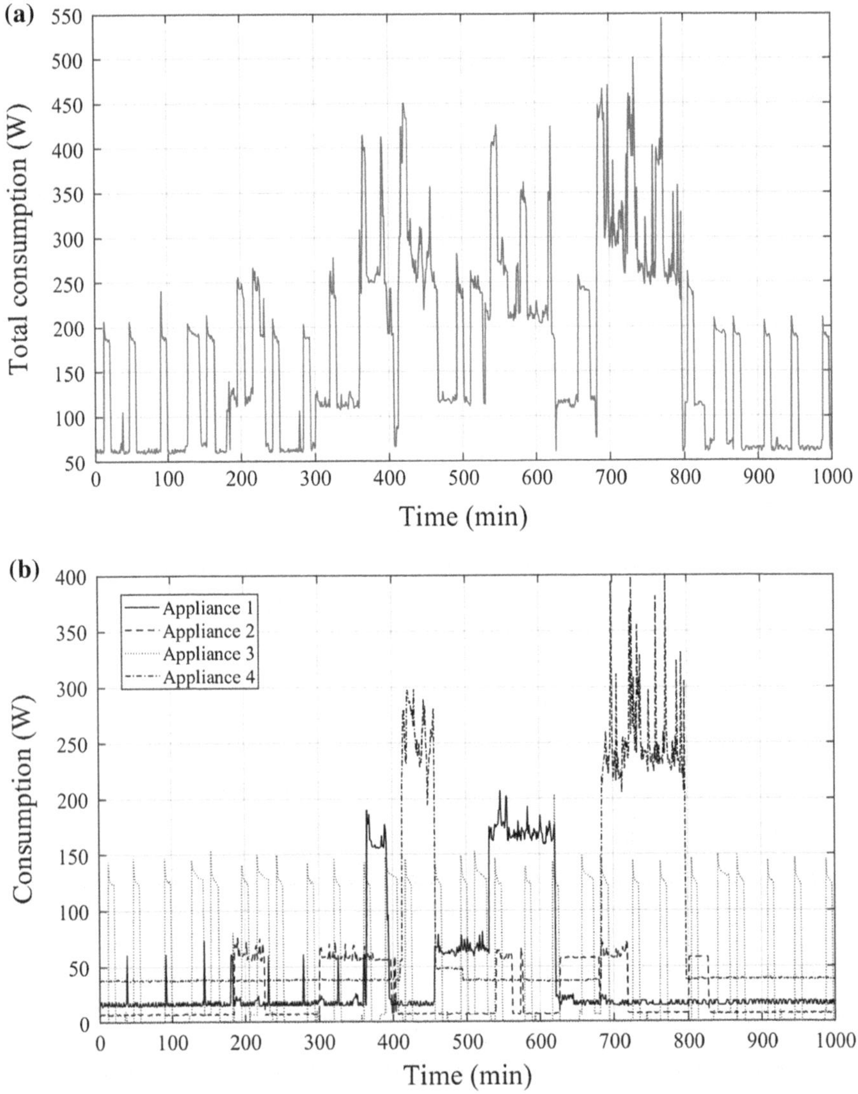

Fig. 1.1 Non-intrusive appliance load monitoring application: **a** aggregate power consumption, **b** power consumption of each appliance

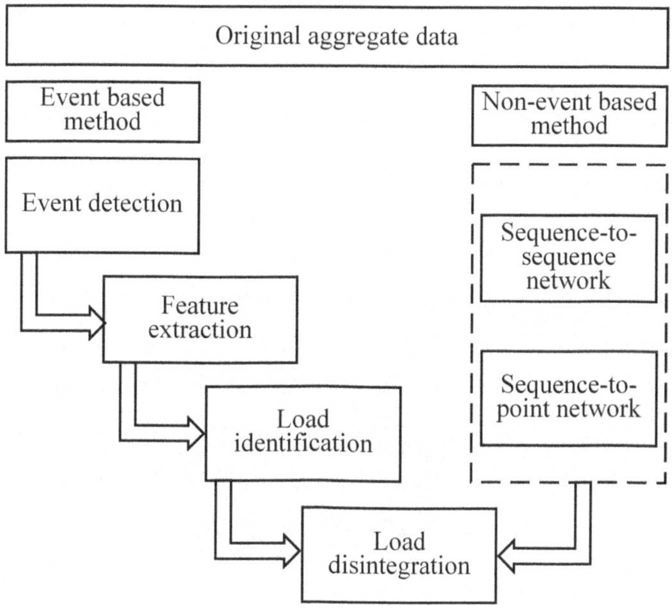

Fig. 1.2 The algorithm framework of the NILM

1.1.2 Overview of Recent Research in Non-intrusive Load Monitoring

In this section, the current state of the non-intrusive load monitoring is discussed in the view of bibliographic. The Web of Science database is utilized to retrieve information about non-intrusive load monitoring. Considering the fact that there are several commonly used expressions for non-intrusive load monitoring, the subject retrieval is expressed as "TS = (Non-intrusive load disaggregation OR Non-intrusive load monitoring OR Non-intrusive appliance load monitoring)". Applying the query by August 22nd, 2019, 617 relevant scientific documents can be found. Among which there are 2 highly cited papers. In order to compressive knowledge of all the documents, several analyses are carried out as follows:

(1) *Statistic analysis of the non-intrusive load monitoring documents*

The top 10 research directions of the 617 documents are engineering electrical electronic, energy fuels, computer science information systems, computer science artificial intelligence, computer science theory methods, telecommunications, automation control systems, construction building technology, engineering civil and green sustainable science technology (Fig. 1.3). It can be found that the non-intrusive load monitoring research is a crossover field of traditional computer science and artificial intelligence. The traditional electrical and electronic engineering provides the theoretical foundation for the non-intrusive load monitoring, the computer science artificial intelligence provides methods to deal with the non-intrusive load monitoring and the results are beneficial to energy saving of modern buildings.

Figure 1.4 provides the annual number of documents from 2004 to 2019, as shown in the figure, the year 2013 and 2015 have seen a significant increase in the number of documents. The annual number of documents in recent 2 years is over 100 while that of the years before 2012 is less than 20. The non-intrusive load monitoring has attracted much more attention in recent years. Figure 1.5 demonstrates the document

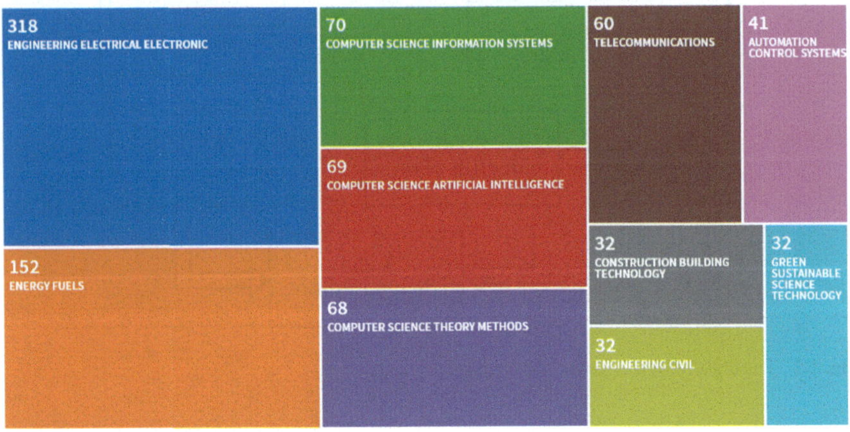

Fig. 1.3 Research direction map of documents based on the subject retrieval of "TS = (Non-intrusive load disaggregation OR Non-intrusive load monitoring OR Non-intrusive appliance load monitoring)"

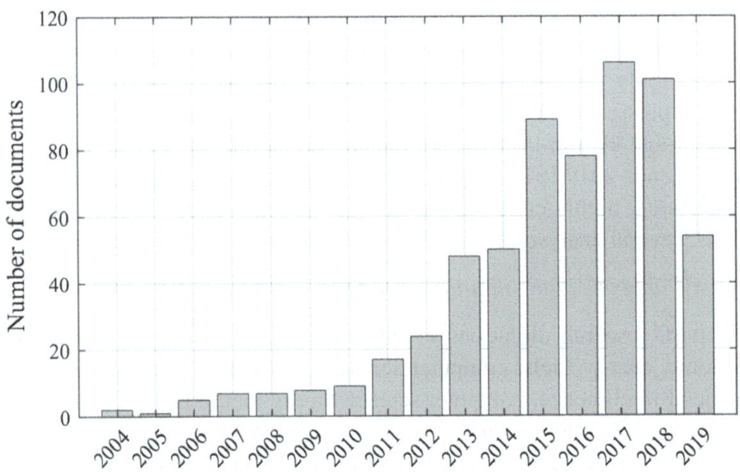

Fig. 1.4 Annual numbers of documents based on the subject retrieval of "TS = (Non-intrusive load disaggregation OR Non-intrusive load monitoring OR Non-intrusive appliance load monitoring)"

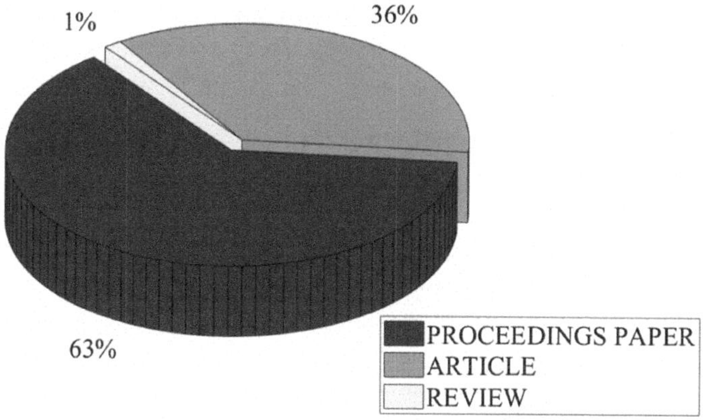

Fig. 1.5 Documents types based on the subject retrieval of "TS = (Non-intrusive load disaggregation OR Non-intrusive load monitoring OR Non-intrusive appliance load monitoring)"

types of the non-intrusive load monitoring. Among all the 617 documents, there are 393 proceeding papers, which take a proportion of 63%. The non-intrusive load monitoring was widely discussed in academic conferences.

(2) *Publication source analysis of the non-intrusive load monitoring documents*

In order to find the most relevant journals in the research of the NILM, the bibliographic coupling overlay map of publication source based on the NILM documents from the Web of Science are given in Fig. 1.4. As shown in Fig. 1.6, not only the publication source bibliographic coupling relationship but also the average publication years of the documents are explained. The color bar at the bottom of the figure indicates that the yellow color corresponds to the recent publication years while the blue color corresponds to older publication years. As shown in Fig. 1.6, several journals including the *IEEE transactions on smart grid*, *IEEE access*, *International journal of electrical power energy systems and sustainability* are colored yellow. There are more NILM documents published in these journals recently.

(3) *Geography distribution analysis of the non-intrusive load monitoring researchers*

The bibliographic coupling overlay map of organizations based on the NILM documents from the Web of Science are given in Fig. 1.7. As shown in Fig. 1.7, not only the bibliographic coupling relationship of different organizations but also the average publication years of the documents from each organization are explained. The color bar at the bottom of the figure indicates that the average publication year of documents, the years of the yellow organizations are newer than that of the blue organizations. In this way, the most active organization in current NILM study can be found. As shown in Fig. 1.7, the average publication year of documents from North China Electric Power University, China Electric Power Research Institute, University of Lisbon, City University of Hong Kong, Nanjing University of Information

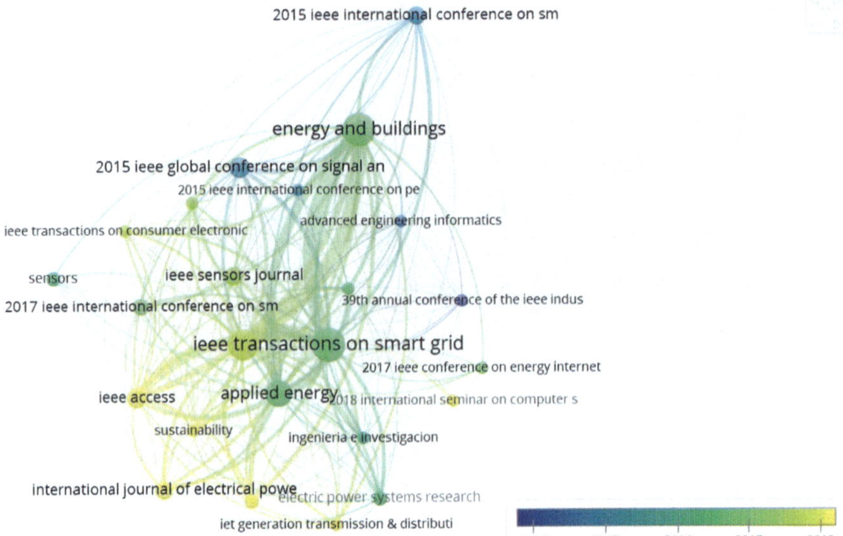

Fig. 1.6 The bibliographic coupling overlay map of publication source based on the NILM documents from the Web of Science

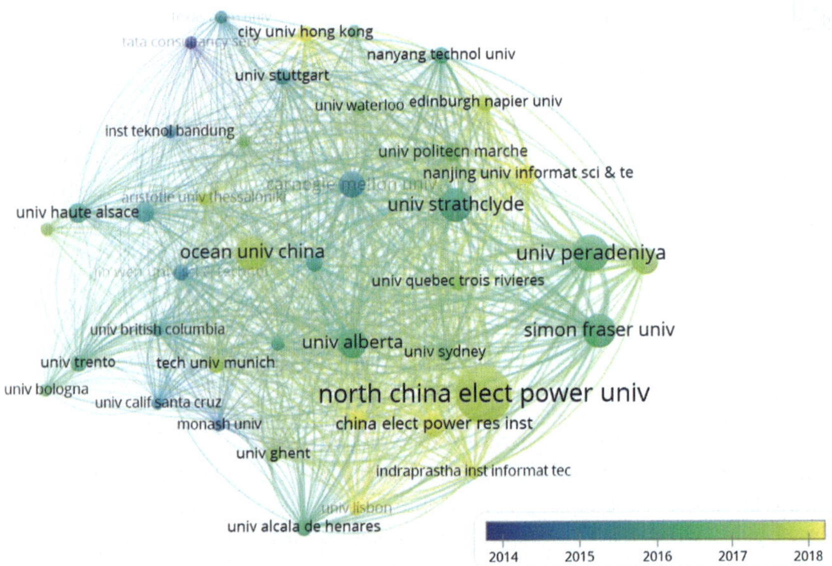

Fig. 1.7 The bibliographic coupling overlay map of organizations based on the NILM documents from the Web of Science

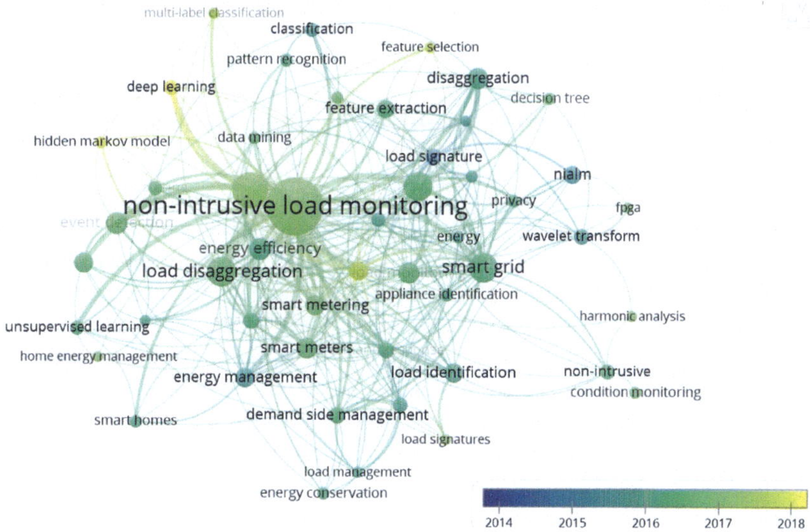

Fig. 1.8 The keyword co-occurrence overlay map of documents based on the subject retrieval of "TS = (Non-intrusive load disaggregation OR Non-intrusive load monitoring OR Non-intrusive appliance load monitoring)"

Science and Technology, and Ocean University of China are newer than that of other organizations. These researchers from these organizations are active in NILM study recently.

(4) *Keyword co-occurrence analysis of the non-intrusive load monitoring documents*

In order to have a comprehensive understanding of the NILM research, the keywords of the 617 documents are analyzed. The keyword co-occurrence item overlay map of the 617 documents is given in Fig. 1.8. As shown in Fig. 1.8, not only the keyword co-occurrence relationship but also the average published years of the documents are explained. The color bar at the bottom of the figure indicates that the yellow keywords are more frequently mentioned than the blue keywords. In this way, the hotspots in current NILM research can be found. As shown in Fig. 1.8, several keywords including the "deep learning", "hidden Markov model", "feature selection" and "machine learning" are paid more attention recently.

1.2 Fundamental Key Problems of Non-intrusive Load Monitoring

1.2.1 Event Detection in Non-intrusive Load Monitoring

The NILM technical routes can be roughly divided into the event-based method and non-event based method. In the event-based method, event detection is an important step to determine the switching time of each appliance. Generally, event detection is based on the amplitude change of active power. Many methods have been proposed to achieve the goal. According to (Anderson et al. 2012), the event detection methods can be divided into the expert heuristics models, the probabilistic models and matched-filters based models. By following the taxonomy proposed in (Anderson et al. 2012) as well as considering the recent progress in the event detection, the event detection models are categorized into four categories: expert heuristics models, probabilistic models, template matching models and hybrid models. The details of those models are given as Fig. 1.9.

(1) *Expert heuristics models*

Hart proposed a sliding window-based event detection method, in the proposed method, the normalized wind power time series was divided into a series of segments (Hart 1992). The segments were classified into steady or changing states according to the standard deviation and a predefined threshold. The segments whose standard deviation value is smaller than the threshold were regarded as the steady states and the segments whose standard deviation value is smaller than the threshold were regarded as the steady states. The events were determined between two steady-states. Alcalá et al. provided an envelope extraction based method (Alcalá et al. 2017). In the proposed method, the root means square calculation was utilized to get the envelope

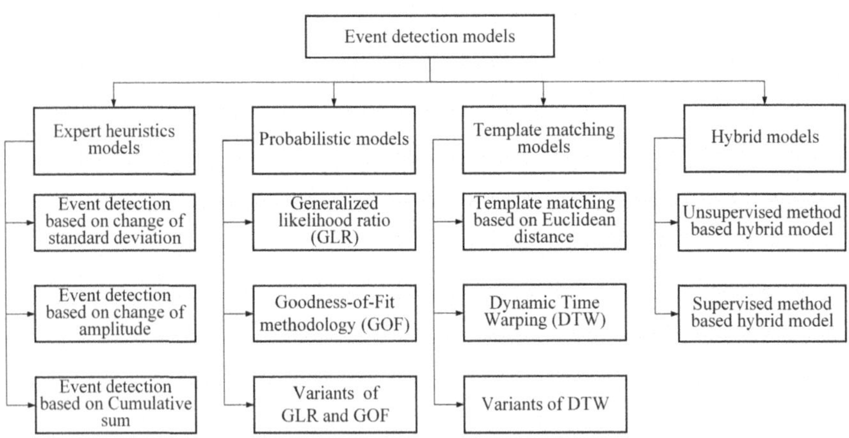

Fig. 1.9 Taxonomy of event detection models

of the current. A derivation was performed to get the change of the envelope. In this way the changes were converted into several peaks. Finally, peaks whose amplitudes were bigger than the threshold were regarded as the events. Experimental results based on real datasets indicated that the proposed method could achieve similar performance as the probabilistic models without any training process.

(2) *Probabilistic models*

The events are defined as the operation state change of appliance (i.e. turn on, turn off), in which the total power consumption changes at the same time. The most direct way to detect the event is to compare the power change deviates of adjacent samples. When the power change deviates exceed a predefined threshold, an event is detected. However, the power time series is influenced by many factors and fluctuates randomly in the real application. It becomes hard to distinguish the events from the noise by a single threshold. In this condition, the probabilistic models were put forward. The Generalized Likelihood Ratio (GLR) is a typical probabilistic model that was first proposed to solve the abrupt change problem. It was involved in the event detection study by (Luo et al. 2002).

In the GLR, pre-event averaging window, post-event averaging window and detection window are defined. The pre-event averaging window and post-event averaging window are utilized to define two Gaussian distributions. The probability density functions about the two Gaussian distributions are utilized to determine the possibility that there is an event in the detection window.

Mario et al. proposed an improved version of the GLR model in which a voting scheme was utilized to improve the performance (Mario et al. 2011). Similar to the GLR based model, the event detection models based on the Goodness-of-Fit (GOF) methodology are also widely investigated (Yang et al. 2014). The event is detected based on the difference between distributions before and after the event. The experimental results indicated that the false-positive performance of the GOF based event detection model was better than that of the expert heuristics method.

(3) *Template matching models*

Template matching based event detection method is carried out by comparing the total detected power signal with the template power signals (Leeb et al. 1995). The basic steps of the template matching based event detection methods are as follows: (a) calculating the spectral envelope of the original times series, the original voltage and current time series with high frequency are converted to signals with obvious characteristic in transients; (b) setting up the template library, the typical instances of the turn on and turn off process is extracted and normalized during the training process; (c) detecting event based on the template matching, a change-of-mean mechanism is utilized to triggers the template matching process, the distance between the total detected power signal and the template power signals is calculated, if the distance is smaller than the threshold, an event is detected.

It can be found that the distance calculation is a key step in the process of template matching based event detection method. The Euclidean distance is the most widely used distance and it was utilized in the template matching process in (Shaw et al.

2008). However, the Euclidean distance is sensitive to the phase of the signal, i.e., two similar signals may get large Euclidean distance if there is a phase difference between them. In this condition, time-series similarity measure methods such as the dynamic time warping method (Basu et al. 2015) and its variant (Liu et al. 2017) are proposed to carry out the template matching.

It is worth noting that template matching can not only be regarded as an event detection method but also be regarded as an appliance detection method. If the instances in the library are connected with different appliances, the appliance can be detected by template matching.

(4) *Hybrid models*

Zheng et al. proposed a Density-Based Spatial Clustering of Applications with Noise (DBSCAN) based event detection in which not only the timestamp but also the two adjacent steady states were detected (Zheng et al. 2018). Afzalan et al. investigated a hybrid framework in which the parameters of the model in different situations can be determined automatically (Afzalan et al. 2019). The clustering method and a neuron network-based classification method were involved in the study. The effectiveness of the proposed framework was validated on a date from three houses. Zhao et al. provided an adaptive event detection method based on the Graph Signal Processing (GPS) in which the threshold is determined according to an iterative clustering scheme (Zhao et al. 2016). Wild et al. proposed a concept named active section which was an extension of conventional event (Wild et al. 2015). The active sections were defined to include the complex situations (e.g. peaks, short pulses and variable load) into consideration. And the Kernel Fisher Discriminant Analysis (KFDA), which is an unsupervised method is utilized to carry out the event detection task. The effectiveness of the proposed method is validated based on the BLUED dataset. Variation of the GOF based method is proposed in (De Baets et al. 2017). In the proposed hybrid GOF based model, a median filter is utilized to remove impulsive noise of the original power time series, a voting mechanism is adopted to improve the conventional GOF algorithm, and the surrogate-based optimization method is proposed to optimize the parameters of the hybrid model. Besides, the cepstrum method is also involved as an event detector in the frequency domain in the hybrid event detection framework. The experimental results indicated that the proposed hybrid models are more robust against base load than the conventional GOF model.

1.2.2 Feature Extraction in Non-intrusive Load Monitoring

The idea of the NILM derives from the fact than each appliance has unique signatures or features. The performance of the NILM is highly depended on the uniqueness of features from appliances. So the features extraction methods are critical to the performance of the NILM. Generally, the feature extraction process can be regarded as signal processing to get important information from the original voltage signal and current signal. Various methods have been proposed to get features to improve

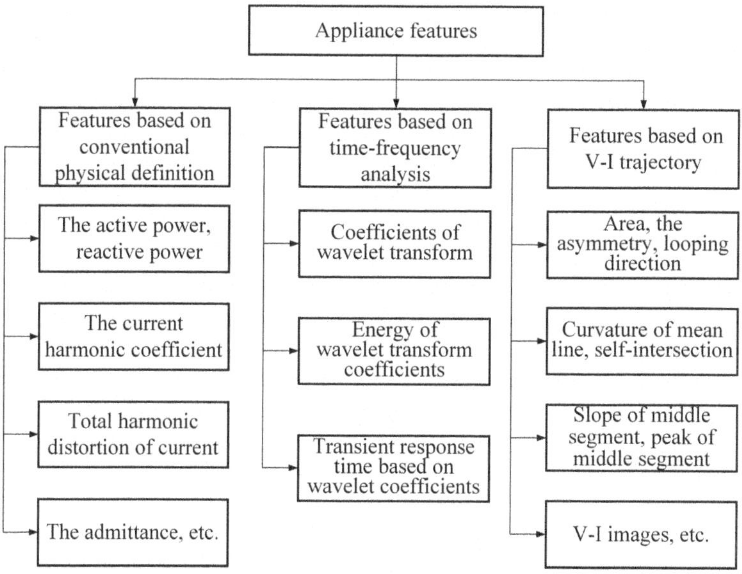

Fig. 1.10 Taxonomy of appliance features

the performance of the NILM. Reviews of the features can be found in (Zoha et al. 2012) and (Sadeghianpourhamami et al. 2017). This section will analyze the current researches of feature extraction in the following aspects as given in Fig. 1.10.

(1) *Features based on conventional physical definition*

In the early study of the NILM, the features utilized to describe the appliance are mostly based on the conventional physical definition. For instance, the change of active power during an event is utilized in (Powers et al. 1991). The combination of active power and reactive power is given in (Hart 1992). Besides, the current harmonic coefficient (Bouhouras et al. 2019), the total harmonic distortion of current (Dong et al. 2013) and the admittance (Liu et al. 2018) are defined as features in the non-intrusive load monitoring. It is worth noting that the features based on the conventional physical definition can also be used in the form of shape. For instance, the start-up transients of real power are utilized as features to match with known patterns (Norford and Leeb 1996). Similar ideas can be found in (Liu et al. 2017), where both the transients real power and reactive power are adopted to match with known patterns.

(2) *Features based on time-frequency analysis*

The current and voltage signals with higher frequency can bring more information about the appliances, but this information is redundant. To guarantee the performance of NILM, it's important to convert the original signals to less-redundant form (Tabatabaei et al. 2017). In this condition, the features based on time-frequency

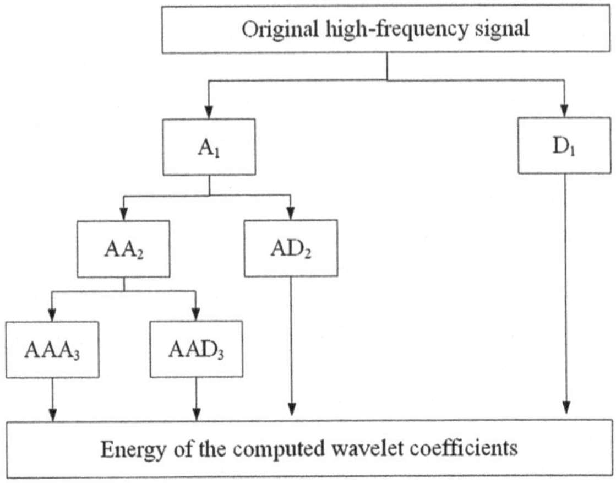

Fig. 1.11 The binary tree of wavelet transform. A represents the approximate subseries, while D represents the detailed subseries

analysis are proposed. Among the time-frequency analysis tools, the wavelet transform is the most widely used. In the study of (Su et al. 2011), both the short-time Fourier transform and the wavelet transform are utilized to extract important information from the current waveform of turn on and turn off transient. The binary tree of wavelet transform is shown in Fig. 1.11. The experimental results indicated that the performance of the wavelet transform was much better than that of the short-time Fourier transform. The wavelet function in the wavelet transform is an important factor that can affect the performance of feature extraction. Gray et al. compared the performance of Daubechies wavelets with different orders in the appliance classification (Gray and Morsi 2015). The experimental results showed that the higher-order Daubechies wavelets tended to provide better performance. Gillis et al. designed a wavelet to extract the most important information from the high-frequency current signal (Gillis and Morsi 2017).

(3) *Features based on V-I trajectory*

The V-I trajectory was proposed in (Lam et al. 2007), in which the features of the V-I trajectory including the asymmetry, the looping direction, the area, the curvature of mean line, the self-intersection, the slope of middle segment and the peak of middle segment are defined to describe the characteristic of the appliance. The experimental results indicated that the appliances can be well-separated by the V-I trajectory. Hassan et al. carried out an empirical investigation of the V-I trajectory based features and found that the V-I trajectory based features are effective for appliance classification (Hassan et al. 2014). Some variants of the V-I trajectory based features were proposed to simplify the feature extraction process. Binary V-I trajectory was proposed in (Hassan et al. 2014). The binary V-I trajectory is generated by converting the original V-I trajectory into a grid of cells colored with black and white (i.e. the cells that the V-I trajectory passes is black while others are white). In this way, the computation burden is reduced significantly. However, the information loss occurs. In order to make a tradeoff between the computation burden and information

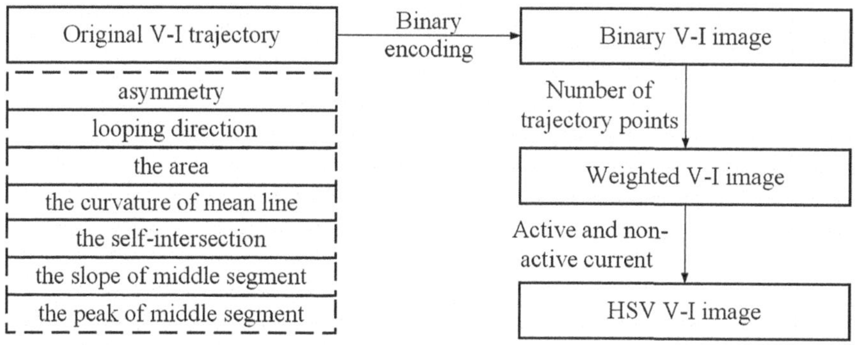

Fig. 1.12 The development and improvement process of the V-I trajectory based features

loss, De Baets et al. proposed a weighted V-I trajectory (De Baets et al. 2018). In the weighted V-I trajectory, the cells in the grid were weighted according to contained trajectory points. Besides, a new type of the V-I trajectory that contains more information was put forward in (Liu et al. 2019b). In the proposed V-I trajectory, the active and non-active components of the current as well as the repeatability of the trajectory were encoded in the HSV color space. The development and improvement process of the V-I trajectory based features are shown in Fig. 1.12.

1.2.3 Load Identification in Non-intrusive Load Monitoring

The load identification is necessary to achieve non-intrusive load monitoring. In the event based NILM methods, the load identification is carried out to classify the event according to the features. In this way, the operation state of each appliance is monitored, and the power consumption of each appliance is calculated. In the non-event based NILM methods, the load identification is generalized to load disaggregation. For instance, in the end to end deep learning based NILM methods, the load disaggregation can be regarded as a special kind of load identification. In this section, the load identification is classified and described in details as follows:

(1) *Machine learning based load identification*

In the event based NILM methods, when the features of the appliances are extracted, the appliance identification task is converted to a classification task. The machine learning methods are widely used in the classification task. In order to classify the events of the electrical appliances, Zheng et al. built a model based on the multilayer perceptron using current characteristics (Zheng et al. 2018). The experimental results showed that the Multi-Layer Perceptron (MLP) classifier was suitable for the classification of the non-linear loads. *Hassan* et al. compared the accuracy and robustness of Support Vector Machine (SVM), the Artificial Neural Network (ANN) and the adaptive boost (Hassan et al. 2014). The experimental results indicated that

Table 1.1 The comparison of two types of template matching based load identification method

Type	Direct template matching	Combinatorial search based template matching
Algorithm	DTW, shapelets	GA, linear programming
Typical reference	Basu et al. (2015), Liu et al. (2017)	Bouhouras et al. (2019)
Input signal	Transient signal	Steady state signal
Output	Type of appliance	Combination of appliances

the adaptive boost provided the best performance among the involved methods. Liu et al. put forward a hybrid load identification framework based on the time series features (Liu et al. 2019a). In the proposed framework, the Random Forest (RF) was utilized as the base classifier. Besides, machine learning methods including the K-nearest neighbor (KNN) (Mario et al. 2011), Naïve Bayes model (Mathis et al. 2014) and decision trees (Gillis et al. 2016) were also adapted for the load identification task.

(2) *Template matching based load identification*

Template matching based load identification is carried out by matching the detected signal with the template signals. The template matching based load identification can be divided into two classes: the direct template matching and the combinatorial search based template matching. The direct template matching is carried out by calculating the distance between the target signal and the template signals directly (Basu et al. 2015), while the combination based template matching is carried out by minimizing the distance between the target signal and the combination of template signals (Bouhouras et al. 2019). The difference between the two types of load identification methods is given in Table 1.1. The direct template matching methods mainly focus on the shape characteristics of the signal, so the shape similarity evaluation algorithms like the DTW and Shapelets are utilized, and the transient signal that have significant shape features are adopted. The combinatorial search based template matching method is mainly focused on the steady-state features (e.g. the current harmonics) of the signal, the appliance can be determined by searching the best combination of appliances. So the optimization algorithms like the Genetic Algorithm (GA) and the Linear Programming (LP) are utilized.

(3) *Deep learning based load identification*

Deep learning was first involved in the NILM by (Kelly and Knottenbelt 2015). There are four basic scheme of application of deep learning in NILM: (a) sequence-to-sequence, (b) sequence-to-point, (c) sequence-to-regression, and (d) sequence-to-type. The details of the four schemes are given in Table 1.2. As shown in Table 1.2, the scheme is classified according to their input and output structure. Generally, the input of the deep neuron network is a one-dimensional aggregate power data or two-dimensional V-I trajectory figure, both of them are regarded as a sequence. If the output is single point, the scheme is named as sequence-to-point. If the output

Table 1.2 Typical scheme of deep learning applied in NILM

Scheme	Typical models	Input	Output	Typical reference
Sequence-to-point	LSTM	Aggregate power data	Single sample of power data for target appliance	Kelly and Knottenbelt (2015)
Sequence-to-sequence	DAE/LSTM/CNN	Aggregate power data	Power data for the target appliance	Wang et al. (2018)
Sequence-to-regression	DNN	Aggregate power data	Start time, end time and average power of the target appliance	Kelly and Knottenbelt (2015)
Sequence-to-type	CNN	V-I trajectory figure	Type of appliance	De Baets et al. (2018), Liu et al. (2019)

is a series of power data for target appliance, the scheme is named as sequence-to-sequence. Kelly et al. provided a scheme whose output is regression of the start time, end time and average power of the target appliance, which is a sequence-to-regression method (Kelly and Knottenbelt 2015). Besides, the Convolutional Neural Network (CNN) is utilized to carry out the appliance classification task in (De Baets et al. 2018). In the work, the input is V-I trajectory figure and the output is the type of appliance, which is a sequence-to-type method.

1.2.4 Energy Forecasting in Smart Buildings

Energy forecasting in the smart building is a hot topic in the research of smart grid and smart buildings. In this section, the current state of load forecasting is reviewed taxonomically. As shown in Fig. 1.13, the forecasting horizon, the input features, the forecasting results and application scenarios of the current load forecasting research are investigated.

(1) *Taxonomy based on the forecasting horizon*

The electric load forecasting models are categorized into different types by their forecasting horizon. As explained in (Friedrich and Afshari 2015), the long-term load forecasting is utilized to achieve years ahead load forecasting, the medium-term load forecasting predicts the load months to a year ahead, the short-term load forecasting

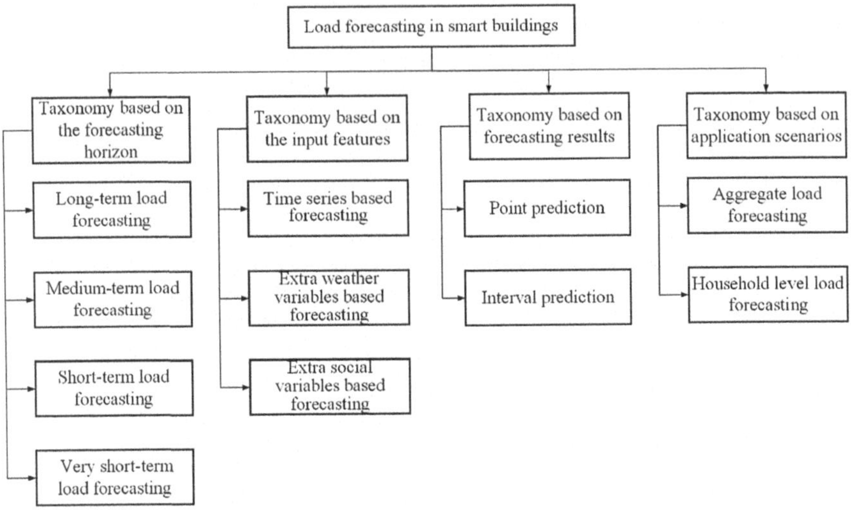

Fig. 1.13 The taxonomy of load forecasting in smart buildings

provides the load information a day to weeks ahead and the very short-term load forecasting refers to minutes to hours ahead predictions.

(2) *Taxonomy based on the input features*

The electric load forecasting models can be divided into different types according to the input features. The time series based forecasting is the most widely used framework. In the framework, only the historical data utilized as the input of the model (Hamzacebi and Es 2014). Considering the fact that the electric power consumption is closely related to the weather conditions, in (Friedrich and Afshari 2015), the weather variables are utilized as the extra input features. In order to further improve the forecasting performance, some extra social factors were also taken into consideration. For instance, *Son* et al. provided a short-term electricity demand forecasting model in which 19 features including 14 weather variables and 5 social variables are taken into consideration (Son and Kim 2017). The experimental results indicated that the mean temperature, the maximum temperature, the wind speed, the daylight time, the global solar radiation, the cooling degree days, the total number of cooling degree days, the vapor pressure, the consumer price index, and the real electricity price are the top 10 variables that affect the monthly electricity demand.

(3) *Taxonomy based on forecasting results*

The point prediction and interval prediction are two typical forms of electric load forecasting. The point prediction which provides determinate prediction results is the most commonly investigated form in electric load forecasting (Fan et al. 2018). It usually established by minimizing the difference between the real points and the forecasting points. Considering the fact that the load data is non-stationary and stochastic in some cases, it is hard to get an accurate point prediction. The interval prediction which can provide the upper and lower bounds is proposed. The points are expected to lie in the upper and lower bounds with a certain prescribed probability. The interval prediction can prove more information about the forecasting results and is regarded as an effective way to model the uncertainty in the electric load (Rana et al. 2013).

(4) *Taxonomy based on application scenarios*

The electric load forecasting can be categorized into different types by their forecasting scenarios. The electric load forecasting was investigated based on data at national level, region level, city level and household level, etc. (Dedinec et al. 2016). Electric load forecasting based on data at national level, region level and city level is regarded as aggregate load forecasting. The aggregate load forecasting is widely studied in smart energy management. Dudek proposed a model to forecast the hourly national load in Poland (Dudek 2016). Taylor et al. proposed short-term load forecasting models based on dataset from ten European countries (Taylor and McSharry 2007). Khwaja et al. established a bagged neuron network based short term load forecasting model, the effectiveness of the proposed model was validated on dataset from New England Pool region (Khwaja et al. 2015).

With the installation of the smart meters, it is becoming much easier to get more information from the household sector. The household level load forecasting is put forward recently. Hsiao put forward a household forecasting load forecasting method by analyzing the daily schedule of the residents (Hsiao 2015). Yu et al. proposed an individual household load forecasting model based on the sparse coding technique (Yu et al. 2016). Alobaidi et al. constructed a household energy consumption forecasting framework based on ensemble learning techniques (Alobaidi et al. 2018). In the proposed framework, the ANN models were established based on the two-stage resampling techniques. Besides, deep learning methods such as the deep recurrent neural network were also utilized to model the uncertainty in the household load time series (Shi et al. 2018a).

1.3 Scope of This Book

As the base of demand-side energy management, the non-intrusive load monitoring techniques are promising in energy saving and carbon emission reduction. Focusing on the non-intrusive load monitoring techniques in the field of smart grid, this book presents a thorough introduction of related basic principles, along with some possible improvement directions. The book includes 10 chapters, the organization of each chapter is given as follows:

This chapter: Introduction

This chapter first gives an overview of the non-intrusive load monitoring, and then the fundamental key problems of non-intrusive load monitoring which include event detection, feature extraction, load identification and energy forecasting in smart buildings are discussed.

Chapter 2: Detection of Transient Events in Time Series

This chapter first gives an introduction of the current researches of event detection, and then three typical event detection algorithms including Cumulative Sum (CUSUM) based transient event detection algorithm, the generalized likelihood ratio, and the Sequential Probability Ratio Test (SPRT) are investigated.

Chapter 3: Appliance Signature Extraction

This chapter first gives an introduction of the current researches of appliance signature extraction, and then several experiments are carried out to give comprehensive evaluations of different features.

Chapter 4: Appliance Identification Based on Template Matching

This chapter first gives an introduction of the current researches of template matching based appliance identification, and then the application of three typical template matching based appliance identification methods are introduced. Several experiments are carried out to evaluate the real performance of the appliance identification models based on the decision tree, KNN algorithm and Dynamic Time Warping (DTW) algorithm.

Chapter 5: Steady State Current Decomposition Based Appliance Identification

The steady state current decomposition based appliance identification methods are reviewed, the application of the classical steady state current decomposition model, harmonic phasor based current decomposition model and the non-negative matrix factor based model are discussed in detail.

Chapter 6: Machine Learning Based Appliance Identification

The machine learning based appliance identification methods are reviewed, the application of three typical machine learning based appliance identification methods are introduced. Several experiments are carried out to evaluate the real performance of the appliance identification models based on the extreme learning machine, support vector machine, and random forest.

Chapter 7: Hidden Markov Models Based Appliance Identification

This chapter first gives an introduction of the hidden Markov models based appliance identification. Then the factorial hidden Markov models and hidden semi-Markov models are investigated.

Chapter 8: Deep Learning Based Appliance Identification

The deep learning based appliance identification researches are reviewed, different schemes of deep learning application are investigated, and then several experiments are carried out to give comprehensive evaluations of deep learning based appliance identification methods.

Chapter 9: Deterministic Prediction of Electric Load Time Series

This chapter first gives an introduction of the current researches of deterministic prediction of electric load time series, and then the application of the Autoregressive Integrated Moving Average (ARIMA) based models and the Elman neural network based models are investigated. Several experiments are carried out to give comprehensive evaluations of different forecasting models.

Chapter 10: Interval Prediction of Electric Load Time Series

The interval prediction of electric load time series are reviewed, the applications of two typical interval methods are introduced. Several experiments are carried out to evaluate the real performance of the quantile regression based models and the Gaussian process regression based models.

References

Afzalan M, Jazizadeh F, Wang J (2019) Self-configuring event detection in electricity monitoring for human-building interaction. Energy Build 187:95–109. https://doi.org/10.1016/j.enbuild.2019.01.036

Alcalá J, Ureña J, Hernández Á, Gualda D (2017) Event-based energy disaggregation algorithm for activity monitoring from a single-point sensor. IEEE Trans Instrum Meas 66(10):2615–2626. https://doi.org/10.1109/TIM.2017.2700987

Alobaidi MH, Chebana F, Meguid MA (2018) Robust ensemble learning framework for day-ahead forecasting of household based energy consumption. Appl Energy 212:997–1012. https://doi.org/10.1016/j.apenergy.2017.12.054

Anderson KD, Berges ME, Ocneanu A, Benitez D, Moura JMF (2012) Event detection for non intrusive load monitoring. In: IECON 2012—38th annual conference of IEEE industrial electronics, 10/2012. IEEE, Montreal, QC, Canada, pp 3312–3317. https://doi.org/10.1109/iecon.2012.6389367

Basu K, Debusschere V, Douzal-Chouakria A, Bacha S (2015) Time series distance-based methods for non-intrusive load monitoring in residential buildings. Energy Build 96:109–117. https://doi.org/10.1016/j.enbuild.2015.03.021

Bouhouras AS, Gkaidatzis PA, Panagiotou E, Poulakis N, Christoforidis GC (2019) A NILM algorithm with enhanced disaggregation scheme under harmonic current vectors. Energy Build 183:392–407. https://doi.org/10.1016/j.enbuild.2018.11.013

De Baets L, Ruyssinck J, Develder C, Dhaene T, Deschrijver D (2017) On the Bayesian optimization and robustness of event detection methods in NILM. Energy Build 145:57–66. https://doi.org/10.1016/j.enbuild.2017.03.061

De Baets L, Ruyssinck J, Develder C, Dhaene T, Deschrijver D (2018) Appliance classification using VI trajectories and convolutional neural networks. Energy Build 158:32–36

Dedinec A, Filiposka S, Dedinec A, Kocarev L (2016) Deep belief network based electricity load forecasting: an analysis of Macedonian case. Energy 115:1688–1700. https://doi.org/10.1016/j.energy.2016.07.090

Dong M, Meira XuW, Chung CY (2013) Non-intrusive signature extraction for major residential loads. IEEE Trans Smart Grid 4(3):1421–1430. https://doi.org/10.1109/TSG.2013.2245926

Dudek G (2016) Pattern-based local linear regression models for short-term load forecasting. Electr Power Syst Res 130:139–147. https://doi.org/10.1016/j.epsr.2015.09.001

Fan G-F, Peng L-L, Hong W-C (2018) Short term load forecasting based on phase space reconstruction algorithm and bi-square kernel regression model. Appl Energy 224:13–33. https://doi.org/10.1016/j.apenergy.2018.04.075

Friedrich L, Afshari A (2015) Short-term forecasting of the Abu Dhabi electricity load using multiple weather variables. Energy Procedia 75:3014–3026. https://doi.org/10.1016/j.egypro.2015.07.616

Gillis JM, Morsi WG (2017) Non-intrusive load monitoring using semi-supervised machine learning and wavelet design. IEEE Trans Smart Grid 8(6):2648–2655. https://doi.org/10.1109/TSG.2016.2532885

Gillis JM, Alshareef SM, Morsi WG (2016) Nonintrusive load monitoring using wavelet design and machine learning. IEEE Trans Smart Grid 7(1):320–328. https://doi.org/10.1109/TSG.2015.2428706

Gray M, Morsi WG (2015) Application of wavelet-based classification in non-intrusive load monitoring. In: 2015 IEEE 28th Canadian conference on electrical and computer engineering (CCECE), May 2015, pp 41–45. https://doi.org/10.1109/ccece.2015.7129157
Hamzacebi C, Es HA (2014) Forecasting the annual electricity consumption of Turkey using an optimized grey model. Energy 70:165–171. https://doi.org/10.1016/j.energy.2014.03.105
Hart GW (1992) Nonintrusive appliance load monitoring. Proc IEEE 80(12):1870–1891
Hassan T, Javed F, Arshad N (2014) An empirical investigation of V-I trajectory based load signatures for non-intrusive load monitoring. IEEE Trans Smart Grid 5(2):870–878
Hsiao Y-H (2015) Household electricity demand forecast based on context information and user daily schedule analysis from meter data. IEEE Trans Industr Inf 11(1):33–43. https://doi.org/10.1109/TII.2014.2363584
Kelly J, Knottenbelt W (2015) Neural nilm: deep neural networks applied to energy disaggregation. In: Proceedings of the 2nd ACM international conference on embedded systems for energy-efficient built environments. ACM, pp 55–64
Khwaja AS, Naeem M, Anpalagan A, Venetsanopoulos A, Venkatesh B (2015) Improved short-term load forecasting using bagged neural networks. Electr Power Syst Res Complet 125:109–115. https://doi.org/10.1016/j.epsr.2015.03.027
Lam HY, Fung GSK, Lee WK (2007) A novel method to construct taxonomy electrical appliances based on load signaturesof. IEEE Trans Consum Electron 53(2):653–660. https://doi.org/10.1109/TCE.2007.381742
Leeb SB, Shaw SR, Kirtley JL (1995) Transient event detection in spectral envelope estimates for nonintrusive load monitoring. IEEE Trans Power Delivery 10(3):1200–1210. https://doi.org/10.1109/61.400897
Liu B, Luan W, Yu Y (2017) Dynamic time warping based non-intrusive load transient identification. Appl Energy 195:634–645. https://doi.org/10.1016/j.apenergy.2017.03.010
Liu Y, Wang X, Zhao L, Liu YJE, Buildings (2018) Admittance-based load signature construction for non-intrusive appliance load monitoring. Energy Build 171:209–219
Liu H, Wu H, Yu C (2019a) A hybrid model for appliance classification based on time series features. Energy Build 196:112–123. https://doi.org/10.1016/j.enbuild.2019.05.028
Liu Y, Wang X, You W (2019b) Non-Intrusive Load Monitoring by Voltage-Current Trajectory Enabled Transfer Learning. IEEE Trans Smart Grid 10(5):5609–5619. https://doi.org/10.1109/TSG.2018.2888581
Luo D, Norford LK, Shaw SR, Leeb SB (2002) Monitoring HVAC equipment electrical loads from a centralized location–methods and field test results/Discussion. ASHRAE Trans; Atlanta 108:841
Mario B, Ethan G, Scott MH, Lucio S, Kyle A (2011) User-centered nonintrusive electricity load monitoring for residential buildings. J Comput Civ Eng 25(6):471–480. https://doi.org/10.1061/(ASCE)CP.1943-5487.0000108
Mathis M, Rumsch A, Kistler R, Andrushevich A, Klapproth A (2014) Improving the recognition performance of NIALM algorithms through technical labeling. In: 2014 12th IEEE international conference on embedded and ubiquitous computing, August 2014, pp 227–233. https://doi.org/10.1109/euc.2014.41
Norford LK, Leeb SB (1996) Non-intrusive electrical load monitoring in commercial buildings based on steady-state and transient load-detection algorithms. Energy Build 24(1):51–64. https://doi.org/10.1016/0378-7788(95)00958-2
Powers JT, Margossian B, Smith BA (1991) Using a rule-based algorithm to disaggregate end-use load profiles from premise-level data. IEEE Comput Appl Power 4(2):42–47. https://doi.org/10.1109/67.75875
Rana M, Koprinska I, Khosravi A, Agelidis VG (2013) Prediction intervals for electricity load forecasting using neural networks. In: The 2013 international joint conference on neural networks (IJCNN), August 2013, pp 1–8. https://doi.org/10.1109/ijcnn.2013.6706839
Sadeghianpourhamami N, Ruyssinck J, Deschrijver D, Dhaene T, Develder C (2017) Comprehensive feature selection for appliance classification in NILM. Energy Build 151:98–106. https://doi.org/10.1016/j.enbuild.2017.06.042

Shaw SR, Leeb SB, Norford LK, Cox RW (2008) Nonintrusive load monitoring and diagnostics in power systems. IEEE Trans Instrum Meas 57(7):1445–1454. https://doi.org/10.1109/TIM.2008.917179

Shi K, Chen Y, Yu B, Xu T, Yang C, Li L, Huang C, Chen Z, Liu R, Wu J (2016) Detecting spatiotemporal dynamics of global electric power consumption using DMSP-OLS nighttime stable light data. Appl Energy 184:450–463. https://doi.org/10.1016/j.apenergy.2016.10.032

Shi H, Xu M, Li R (2018a) Deep learning for household load forecasting—a novel pooling deep RNN. IEEE Trans Smart Grid 9(5):5271–5280. https://doi.org/10.1109/TSG.2017.2686012

Shi K, Yu B, Huang C, Wu J, Sun X (2018b) Exploring spatiotemporal patterns of electric power consumption in countries along the Belt and Road. Energy 150:847–859. https://doi.org/10.1016/j.energy.2018.03.020

Son H, Kim C (2017) Short-term forecasting of electricity demand for the residential sector using weather and social variables. Resour Conserv Recycl 123:200–207. https://doi.org/10.1016/j.resconrec.2016.01.016

Su Y, Lian K, Chang H (2011) Feature selection of non-intrusive load monitoring system using STFT and wavelet transform. In: 2011 IEEE 8th international conference on e-business engineering, October 2011, pp 293–298. https://doi.org/10.1109/icebe.2011.49

Tabatabaei SM, Dick S, Xu W (2017) Toward non-intrusive load monitoring via multi-label classification. IEEE Trans Smart Grid 8(1):26–40. https://doi.org/10.1109/TSG.2016.2584581

Taylor JW, McSharry PE (2007) Short-term load forecasting methods: an evaluation based on European data. IEEE Trans Power Syst 22(4):2213–2219. https://doi.org/10.1109/TPWRS.2007.907583

Wang J, Li H, Lu H (2018) Application of a novel early warning system based on fuzzy time series in urban air quality forecasting in China. Appl Soft Comput 71:783–799. https://doi.org/10.1016/j.asoc.2018.07.030

Wild B, Barsim KS, Yang B (2015) A new unsupervised event detector for non-intrusive load monitoring. In: 2015 IEEE global conference on signal and information processing (GlobalSIP), December 2015, pp 73–77. https://doi.org/10.1109/globalsip.2015.7418159

Yang CC, Soh CS, Yap VV (2014) Comparative study of event detection methods for non-intrusive appliance load monitoring. Energy Procedia 61:1840–1843. https://doi.org/10.1016/j.egypro.2014.12.225

Yu C-N, Mirowski P, Ho TK (2016) A sparse coding approach to household electricity demand forecasting in smart grids. IEEE Trans Smart Grid 1–11. https://doi.org/10.1109/tsg.2015.2513900

Zhao B, Stankovic L, Stankovic V (2016) On a training-less solution for non-intrusive appliance load monitoring using graph signal processing. IEEE Access 4:1784–1799. https://doi.org/10.1109/ACCESS.2016.2557460

Zheng Z, Chen H, Luo X (2018) A supervised event-based non-intrusive load monitoring for non-linear appliances. Sustainability 10(4):1001. https://doi.org/10.3390/su10041001

Zoha A, Gluhak A, Imran M, Rajasegarar S (2012) Non-intrusive load monitoring approaches for disaggregated energy sensing: a survey. Sensors 12(12):16838–16866. https://doi.org/10.3390/s121216838

Chapter 2
Detection of Transient Events in Time Series

2.1 Introduction

As for Non-intrusive Load Monitoring (NILM), transient event detection is estimating whether and when the change point occurs. The change point refers to the change of some statistical characteristics in a power time series, which is caused by load switching.

The change point detection theory developes rapidly. Leeb et al. demonstrated the theoretical basis and prototype implementation of power system transient event detection in the NILM, and developed a transient event detection algorithm (Leeb et al. 1995). Cox et al. studied a simple detection system which can identify the operation of each load with the voltage transient mode (Cox et al. 2006). Norford et al. presented a non-intrusive power load monitoring method. It used the steady-state and transient load detection algorithms (Norford and Leeb 1996). De Oca et al. proposed a change point detection algorithm using cumulative sum. It is used for non-stationary sequences (De Oca et al. 2010). Niu et al. proposed a new sliding window bilateral CUSUM algorithm for detecting transient signals (Niu and Jia 2011).

An experiment of load transient event detection is carried out on the AMPds dataset and the BLUED dataset. Three loads of the appliances from AMPds dataset are selected, including Light bulbs (BME), Dishwashers (DWE) and Refrigerators (FGE). The multiple-load data in the BLUED dataset is selected as the event detection objects. Based on the principle of the traditional event detection algorithm, the models containing the traditional CUSUM methods, the generalized likelihood ratio (GLR) and the sliding windows based on a two-sided CUSUM algorithm, are respectively used for experimental comparative analysis.

H. Liu, *Non-intrusive Load Monitoring*,
https://doi.org/10.1007/978-981-15-1860-7_2

2.2 Cumulative Sum Based Transient Event Detection Algorithm

The CUSUM algorithm is a common algorithm in the statistical process, which was first proposed by Page in 1954 (Page 1954). Subsequently, many scholars have conducted in-depth research on this algorithm. The CUSUM method is accumulating the small offsets in the process to achieve the amplification effect, so as to improve the sensitivity for small offsets in the detection process.

2.2.1 Mathematical Description of Change Point Detection

Assuming that the time series is $T = \{t_n\}_{n=1}^{\infty}$, the change time λ can be defined as follows:

$$\lambda = \inf\{i : f_i(x_1, \ldots, x_i) \geq h\} \tag{2.1}$$

where h is a preset value to judge whether the time-series change is beyond the limit and f_i is a function of the observed values $(x_1, \ldots, x_k)$.

2.2.2 Parametric CUSUM Algorithm

In Eq. (2.1), f_i is a function of the observed values $(x_1, x_2, \ldots, x_n)$. In parametric CUSUM algorithm, the statistical function f is defined as follows:

$$\begin{aligned} f_i &= \max(0, (f_{i-1} + r_i)) \\ f_0 &= 0 \end{aligned} \tag{2.2}$$

where r_i is the likelihood logarithm ratio function:

$$r_i = \ln \frac{p_{\theta_1}(x_i)}{p_{\theta_0}(x_i)} \tag{2.3}$$

Before the abrupt change point, $p_{\theta_1}(x_i) < p_{\theta_0}(x_i)$, $s_k < 0$ and $f_i = 0$. After changing, $p_{\theta_1}(x_k) > p_{\theta_0}(x_k)$, $s_k > 0$ and f_i increases.

Define decision function $Q_i(f_i)$:

$$Q_i(f_i) = \begin{cases} 1 \; if \; f_i \geq h \\ 0 \; if \; f_i < h \end{cases} \tag{2.4}$$

If f_k reaches to the threshold value h, $Q_k(f_k)$ is equal to 1, which means the change point is detected. Moreover, the occurrence time of change point is defined as follows:

$$T = \min\{i : f_i \geq h\} \tag{2.5}$$

The parametric CUSUM algorithm needs to pre-know the probability density function in the time series $x_1, x_2, \ldots, x_n$. However, in the scenario of non-intrusive load identification, it is difficult to obtain the probability distribution of random variables such as voltage and current. Therefore, it is necessary to propose a non-parametric CUSUM algorithm.

2.2.3 Non-parametric CUSUM Algorithm

The core principle of the non-parametric CUSUM algorithm is that when the cumulative sum under the monitoring part is obviously higher than the average cumulative sum under the near-by normal smooth part, it means that the change point has been detected.

The main difference between the non-parametric and parametric CUMSUM is the definition of iteration function. The function f in the nonparametric CUSUM algorithm is defined as:

$$\begin{cases} f_0 = 0 \\ f_i = \max(0, f_{i-1} + r_i) \end{cases} \tag{2.6}$$

where $r_i = x_i - \upsilon_0 - c$, and υ_0 is the average value under the whole time series $x_1, x_2, \ldots, x_n$; c is the signal noise.

The bilateral CUSUM algorithm is shown as follows.

$$\text{If } r_i > 0, \quad \begin{cases} f_0^+ = 0 \\ f_i^+ = \max(0, f_{i-1}^+ + x_i - (\upsilon_0 + c_i)) \end{cases} \tag{2.7}$$

$$\text{If } r_i < 0, \quad \begin{cases} f_0^- = 0 \\ f_i^- = \max(0, f_{i-1}^- + x_i - (\upsilon_0 + c_i)) \end{cases} \tag{2.8}$$

When the monitoring time series $x_1, x_2, \ldots, x_p$ is in the steady state, f_i^+ and f_i^- fluctuate is around 0. When the change point x_τ exits in monitoring time series $x_1, x_2, \ldots, x_p$ and $f_k^- > h$, $\tau = k - d^-$ (d^- is the iteration time of the function f^-); when the change point x_τ exits in monitoring time series $x_1, x_2, \ldots, x_p$ and $f_k^+ > h$ > h, $\tau = k - d^+$ (d^+ is the iteration time of the function f^+).

2.2.4 Sliding Windows Based on Two-Sided CUSUM Algorithm

Actually, the load of the power system is time-varying. Thus, the υ_0 in Eq. (2.7) and Eq. (2.8) is not a stable value and it decreases the accuracy of transient events detection when applying general non-parametric CUSUM algorithm (Trivedi and Chandramouli 2005). In order to improve the algorithm, a sliding window based on the two-sided CUSUM algorithm is proposed.

The main idea of the sliding windows based on the two-sided CUSUM algorithm concluding the following steps: (a) dividing time series in a certain length before monitoring time series; (b) calculating the average value under the divided time series which are supposed to be relatively stable; (c) calculating the function f_k^+ or f_k^- with the average value under the partial time series.

2.2.5 Original Dataset

(1) *AMPds dataset*

The AMPds dataset was measured from a residual house build in the 1950s in Canada. The energy consumption which includes water, gas, and electricity were monitored for 2 years in the house. The study pays special attention to electricity. The house is provided with the electricity service via a 240 V, 200A power supply. Figure 2.1 shows the relationships between all the monitored loads. It can be found that the main house power consumption consists of consumption of the bedroom, fridge, bedrooms, furnace, basement, heat pump, clothes dryer, hot water, clothes washer, home office, dishwasher, workbench, utility, equipment, wall oven and TV etc. (Makonin et al. 2016). Table 2.1 gives detailed information on the monitored loads in the house, it can be found that the heat pump, clothes dryer and the wall oven are powered with the 240 V voltage while rest of the loads are powered with

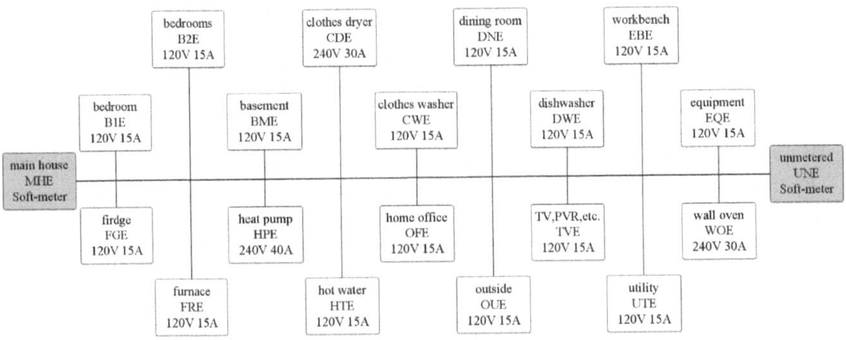

Fig. 2.1 Schematic diagram of the relationships between all the monitored loads

Table 2.1 The basic information of the metered loads

Abbreviation	Description	Supply connection	Consumption (kWh)
B1E	Bedroom	120 V 15 A	93
FGE	Fridge	120 V 15 A	505
B2E	Bedrooms	120 V 15 A	282
FRE	Furnace	120 V 15 A	1496
BME	Basement	120 V 15 A	363
HPE	Heat pump	240 V 40 A	1638
CDE	Clothes dryer	240 V 30 A	482
HTE	Hot water	120 V 15 A	117
CWE	Clothes washer	120 V 15 A	59
OFE	Home office	120 V 15 A	421
DNE	Dinner room	120 V 15 A	13
OUE	Outside	120 V 15 A	8
DWE	Dishwasher	120 V15 A	133
TVE	TV, PVR, etc.	120 V 15 A	476
EBE	Workbench	120 V 15 A	168
UTE	Utility	120 V 15 A	619
EQE	Equipment	120 V 15 A	470
WOE	Wall oven	240 V 30 A	96

the 120 V voltage. The yearly power consumption of the heat pump is the highest among all the loads. This may due to two reasons: (a) the power of the heat pump is higher than that of other loads; (b) the operation time of the heat pump is longer than that of other loads.

In this chapter, the active power is selected as load characteristics. Three loads are selected as the research objects, including the BME, DWE and FGE. According to the running states of the loads, the loads are divided into the ON/OFF load, the Finite State Machines (FSM) load, and the Continuously Variable Devices (CVD) load. The three selected loads correspond to different types of loads. The BME belongs to ON/OFF load, the DWE belongs to the FSM load, and the FGE belongs to the CVD load. Among the three types of loads, there are not only simple ON/OFF load, but also complex CVD load and FSM load, which makes the selected research objects more in line with the use of loads in real life. Original active power reporting data for BME, DWE and FGE are shown in Figs. 2.2, 2.3 and 2.4.

(2) *BLUED dataset*

In this chapter, the event detection models are performed on the BLUED dataset. The dataset was collected by measuring the total voltage and current data from a single house for a week (Anderson et al. 2012). The data were measured at 12 kHz

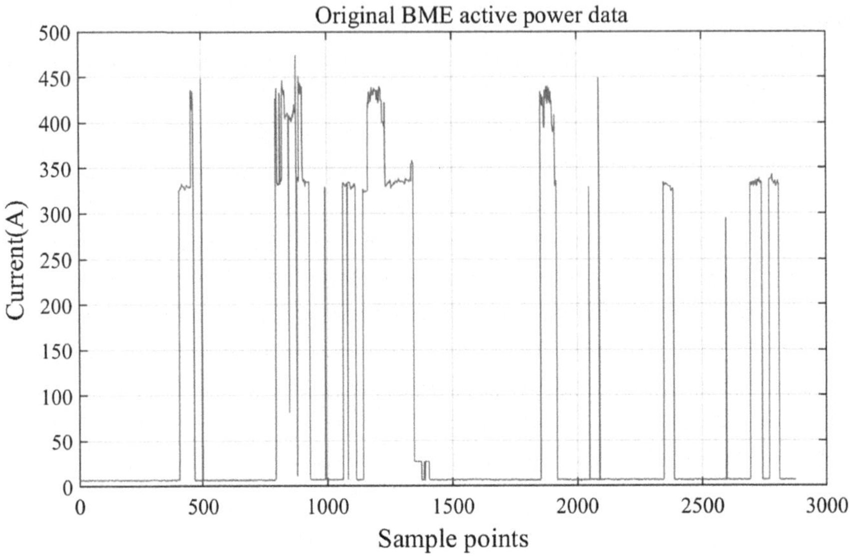

Fig. 2.2 Original BME active power data

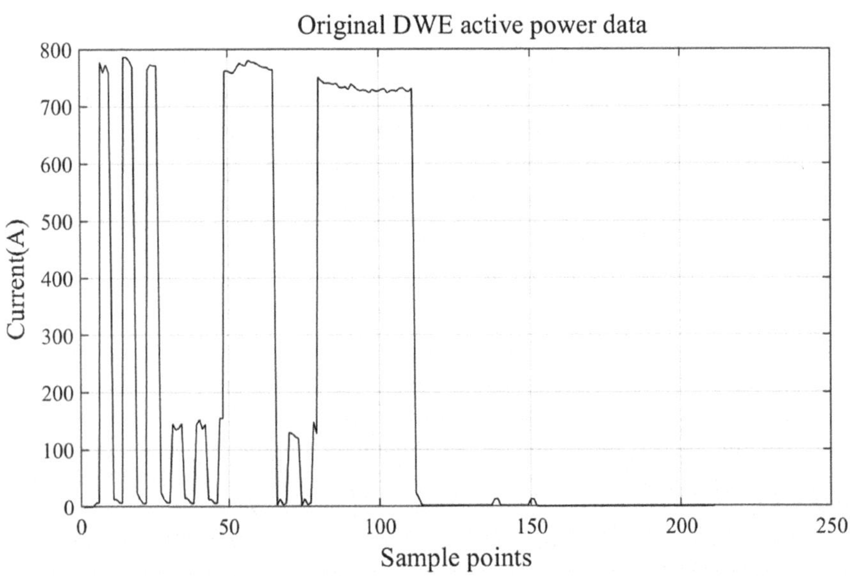

Fig. 2.3 Original DWE active power data

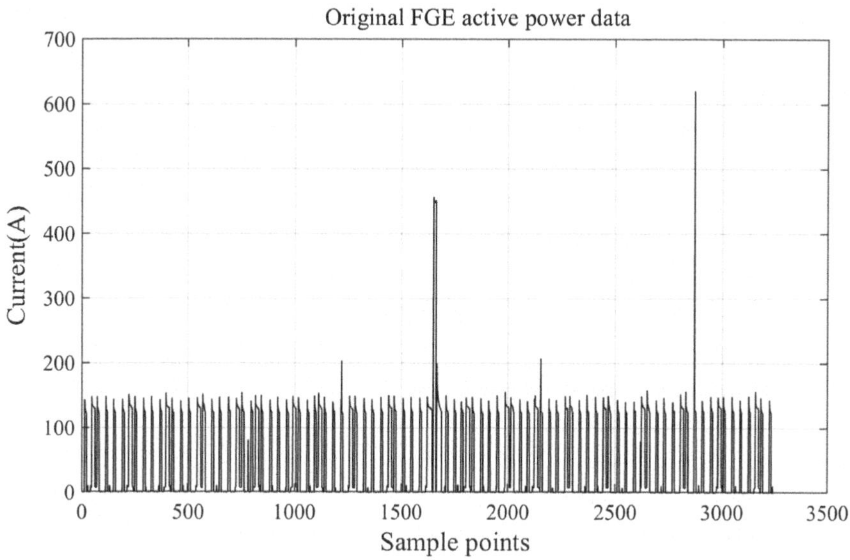

Fig. 2.4 Original FGE active power data

and the switching events of nearly all appliances were well labeled. There are 2482 events of 43 types of appliances in the collected data and the detailed information of the appliances is explained in Table 2.2.

Additionally, the BLUED dataset also provides some supplementary information, such as the environmental data and the data of each appliance sampled at a lower frequency. But it should be noticed that the harmonic of the current must be lower than the 6th harmonic because the higher harmonics are unreliable, due to the errors caused by current sensors. Original voltage and current data are shown as Figs. 2.5 and 2.6.

In this chapter, active power is selected as the characteristic parameter for event detection. The active power can be calculated from the voltage and current. The active power data after processing is shown in Fig. 2.7.

2.2.6 Evaluation Criteria and Results Analysis

(1) *Accuracy, Recall and F1-Score*

The event detection can be regarded as a dichotomous problem. For the binary classification problem, the sample can be divided into four cases: True Positive (TP), False Positive (FP), True Negative (TN) and False Negative (FN) (Junker et al. 1999). The confusion matrix of the classification results is shown in Table 2.3.

Table 2.2 Equipment types and load information

Index	Name	Average power consumption (W)	Number of events
1	Sub-woofer living room (SWLR)	0	
2	Computer A (CA)	60	45
3	Laptop B (LB)	40	14
4	Dehumidifier		0
5	Vacuum cleaner (VC)		0
6	Receive blueray playe basement	55	34
7	Sub-woofer basement (SWB)	0	
8	Apple TV basement (ATB)	0	
9	Air compressor (AC)	1130	20
10	LCD monitor A (LMA)	35	77
11	Desktop lamp (DL)	30	26
12	Tall desk lamp (TDL)	30	25
13	Garage door (GD)	530	24
14	Washing machine (WM)	130–700	95
15	Kitchen music (KM)	0	
16	Kitchen aid chopper (KAC)	1500	16
17	Tea kettle (TK)		0
18	Toaster oven (TO)		0
19	Fridge	120	616
20	Living room (LR)	45	8
21	Closet lights	20	22
22	Upstairs hallway light (UHL)	25	17
23	Hallways stairs lights (HSL)	110	58
24	Kitchen hallway light (KHL)	15	6
25	Kitchen overhead light (KOL)	65	56
26	Bathroom upstairs lights (BUL)	65	98
27	Dining room overhead light (DROL)	65	32
28	Bedroom lights (BEL)	190	19
29	Basement light (BAL)	35	39
30	Microwave	1550	70
31	TV basement (TB)	190	54
32	Harddrive B (HB)		0
33	Printer	930	150

(continued)

Table 2.2 (continued)

Index	Name	Average power consumption (W)	Number of events
34	Hair dryer (HD)	1600	8
35	Iron	1400	40
36	Empty living room socket (ELRS)	60	2
37	Monitor B (MB)	40	150
38	Backyard lights (BL)	60	16
39	Washroom light (WL)	110	6
40	Office lights (OL)	30	54
41	Air conditioner (ACO)		0
42	Dryer		0

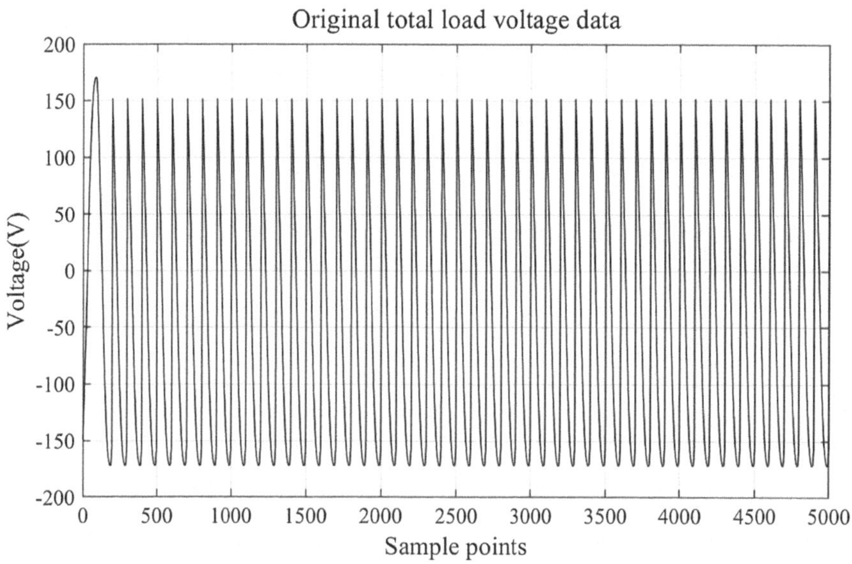

Fig. 2.5 Original total load voltage data

Accuracy is the percentage of the total sample that predictes the correct result. Accuracy is regarded as the most original evaluation index and its calculation method is given as the following formula:

$$Accuracy = \frac{TP + TN}{TP + TN + FP + FN} \tag{2.9}$$

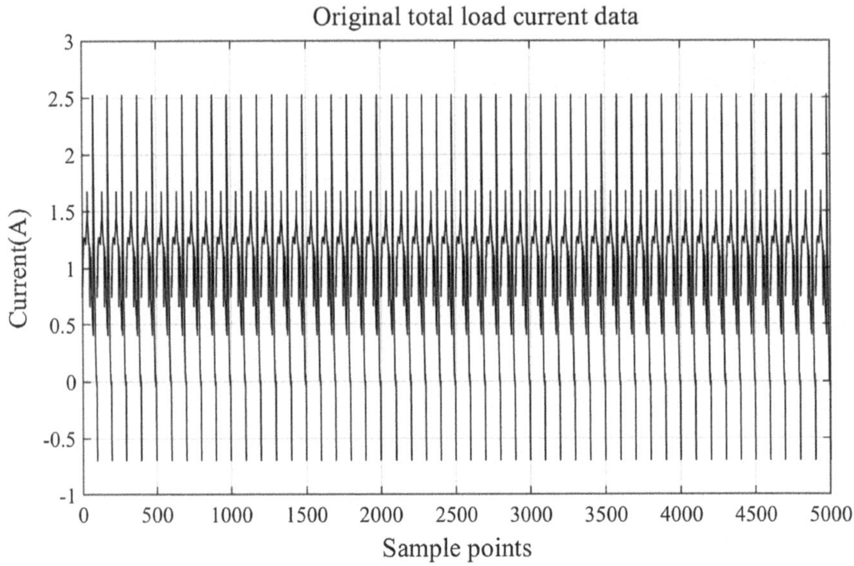

Fig. 2.6 Original total load current data

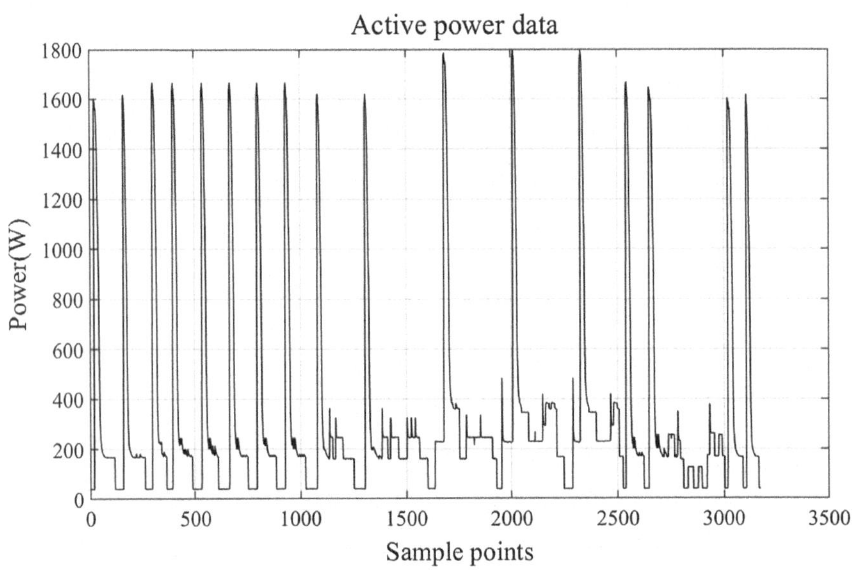

Fig. 2.7 Active power data

Table 2.3 Confusion matrix of classification results

Real situation	Predicted results	
	Positive	Negative
Positive	TP (true positive)	FN (false negative)
Negative	FP (false positive)	TN (true negative)

The Precision is for the prediction results. The definition of precision is the probability that all the samples predicted to be positive are actually positive. The calculation method of precision is given below:

$$Precision = \frac{TP}{TP + FP} \tag{2.10}$$

The recall is for the original sample. The recall refers to the probability that a sample that is actually positive is predicted to be a positive sample. The calculation method of recall is given as follows:

$$Recall = \frac{TP}{TP + FN} \tag{2.11}$$

F1-Score is the weighted harmonic mean of Precision and Recall. The Precision and Recall values are usually not consistent. When one is higher, the other is usually lower. The most common method to combine the two indexes is to use F1-Score. The calculation method of F1-Score has been given below:

$$F1{-}Score = \frac{2 \times Precision \times Recall}{Precision + Recall} \tag{2.12}$$

(2) *Comparison of event detection results*

To verify the effectiveness of the CUSUM algorithm, event detection experiments are carried out. The event detection results of the BME, DWE and FGE are shown in Figs. 2.8, 2.9 and 2.10. The results of the multi-load event detection experiments are shown in Fig. 2.11. It can be found that the CUSUM algorithm is so sensitive that it often judges normal power fluctuation as event. The results were worst when event detection was performed on DWE load. Where, the dotted line represents the event detection result.

These experimental results of electrical event detection are analyzed and evaluated as Table 2.4.

The conclusions can be drawn as follows:

(a) The detection results of three groups of single load events show that the BME has the best effect, the FGE has the second-best effect, and the DWE has the worst effect. This is because the DWE is an FSM type load and has multiple running states. Therefore, it is easy to make errors in event detection. The BME and FGE belong to ON/OFF load and the CVD load respectively. The active

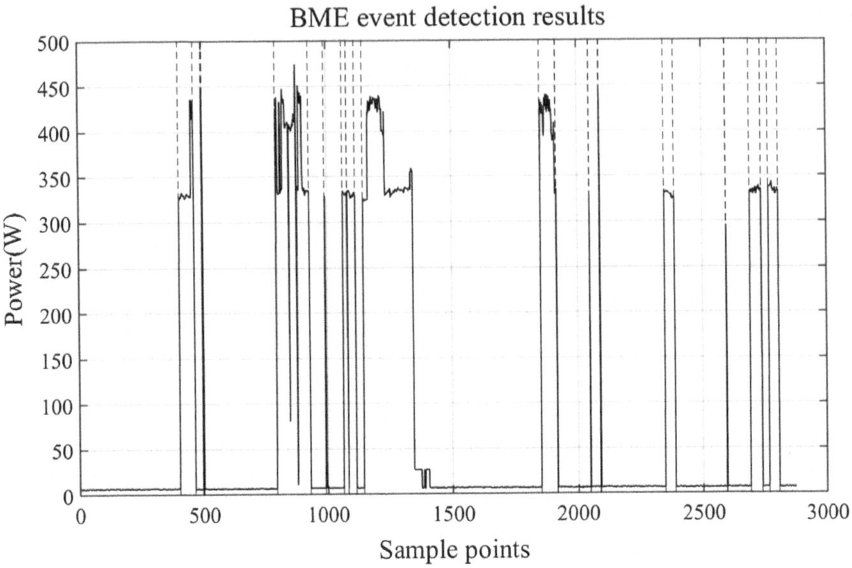

Fig. 2.8 The results of the BME event detection experiments

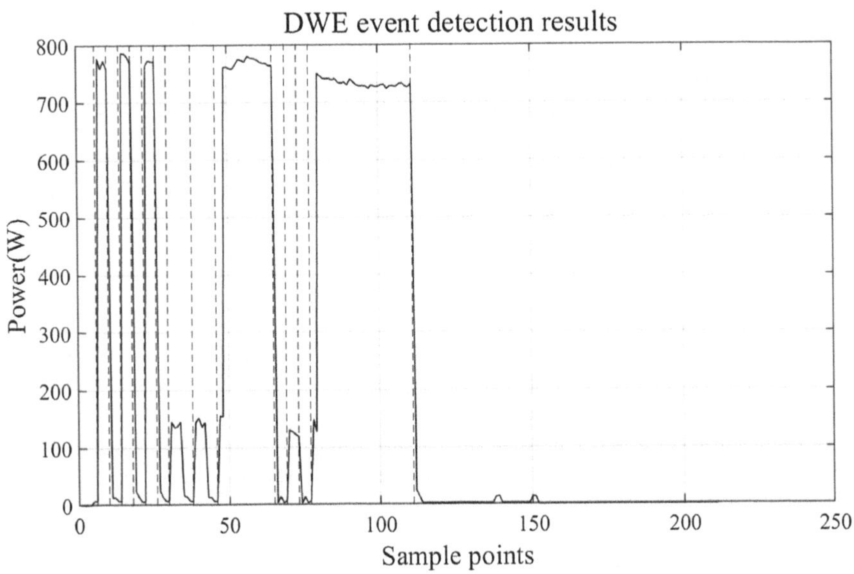

Fig. 2.9 The results of the DWE event detection experiments

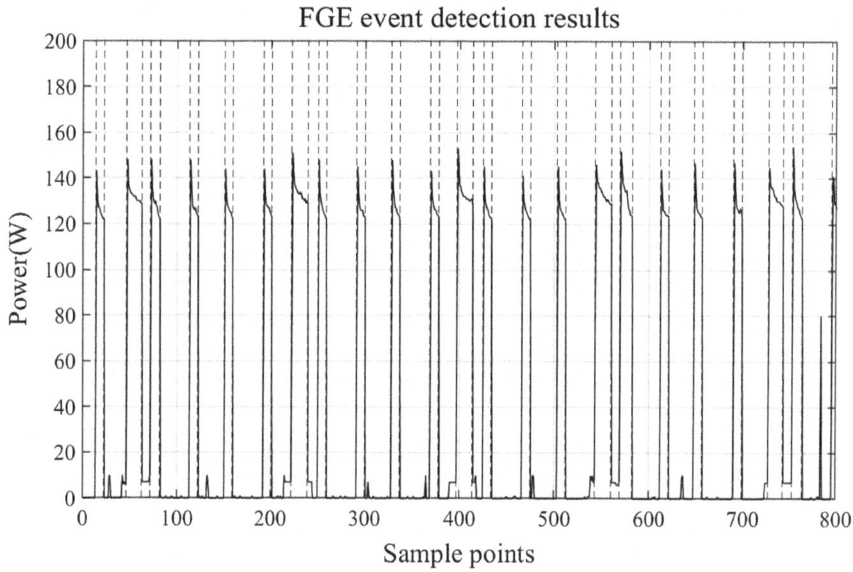

Fig. 2.10 The results of the FGE event detection experiments

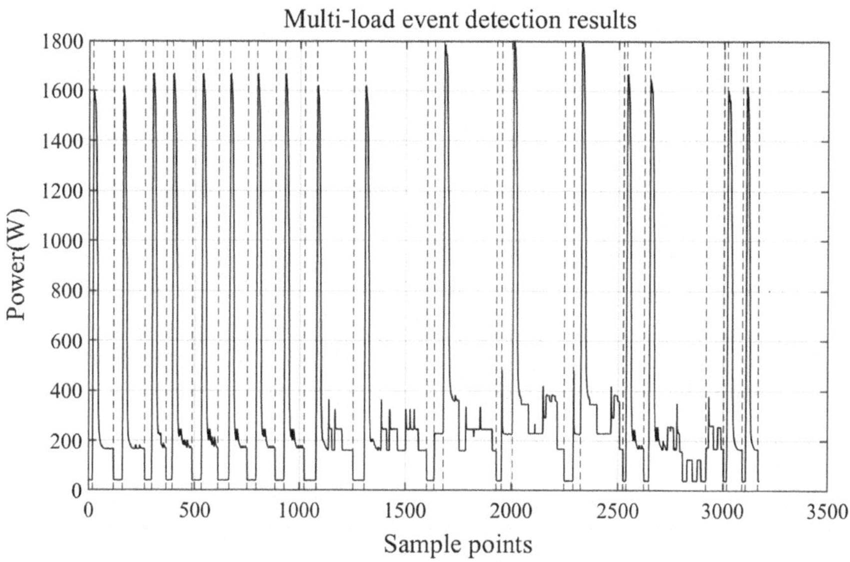

Fig. 2.11 The results of the multi-load event detection experiments

Table 2.4 Event detection results of the three groups

	BME	DWE	FGE	Multi-load
Precision	0.8021	0.7368	0.7653	0.6900
Recall	0.7700	0.7000	0.7500	0.6765
F1-score	0.7857	0.7179	0.7576	0.6832

power characteristic of open and close events is obvious. So the identification accuracy is relatively high.

(b) The detection effect of multi-load operation events is significantly lower than that of single load events. This is due to the increase in the number of loads, power changes are more complex. So the performance of event monitoring is poor.

(c) The results of four groups of experiments show that the traditional CUSUM algorithm has a higher accuracy, but a lower recall. This is due to the high detection sensitivity of the traditional CUSUM algorithm, which makes it easy to judge power fluctuations as open events. Therefore, the recall is low.

2.3 Generalized Likelihood Ratio

The generalized likelihood ratio (GLR) test is one of the most important methods for solving composite hypotheses testing problems (Siegmund and Venkatraman 1995).

2.3.1 The Theoretical Basis of GLR

Let the density function of samples $\hat{X} = (X_1, X_2, \ldots, X_n)$ be $p(\tilde{x}, \theta), \theta \in \Theta$. Considering the hypothesis test problem: $H_0 : \theta \in \Theta_0$ versus $H_1 : \theta \in \Theta_1$, the statistic called generalized likelihood ratio statistic is defined as follows:

$$\lambda(\tilde{x}) = \frac{\sup\limits_{\theta\in\Theta_0} p(\tilde{x};\theta)}{\sup\limits_{\theta\in\Theta} p(\tilde{x};\theta)} = \frac{p(\tilde{x};\hat{\hat{\theta}})}{p(\tilde{x};\hat{\theta})} \tag{2.13}$$

where $\hat{\theta}$ and $\hat{\hat{\theta}}$ are the maximum likelihood estimates of parameter θ in the whole parameter space Θ and in the parameter subspace Θ_0, respectively. And the maximum likelihood estimates are independent of parameter θ. Owing to $\Theta_0 \in \Theta$, $0 \leq \lambda(\tilde{x}) \leq 1$.

2.3.2 Comparison of Event Detection Results

To verify the effectiveness of the GLR, event detection experiments are carried out. The event detection results of the BME, DWE and FGE are shown in Figs. 2.12, 2.13 and 2.14. The results of the multi-load event detection experiments are shown in Fig. 2.15. It can be found that the GLR algorithm has low sensitivity to closing events, which results in high miss rate.

These experimental results of electrical event detection are analyzed and evaluated as Table 2.5.

The conclusions can be drawn as follows:

(a) The DWE has the worst detection performance in the single load detection, and the multi-load detection has worse performance than any single load detection. It is because the active power of the DWE and multiple loads is more complex than others.
(b) The results of four groups of experiments show that the GLR algorithm has good recall, but its accuracy is low. This is because the traditional GLR algorithm is not sensitive to the shutdown event detection.

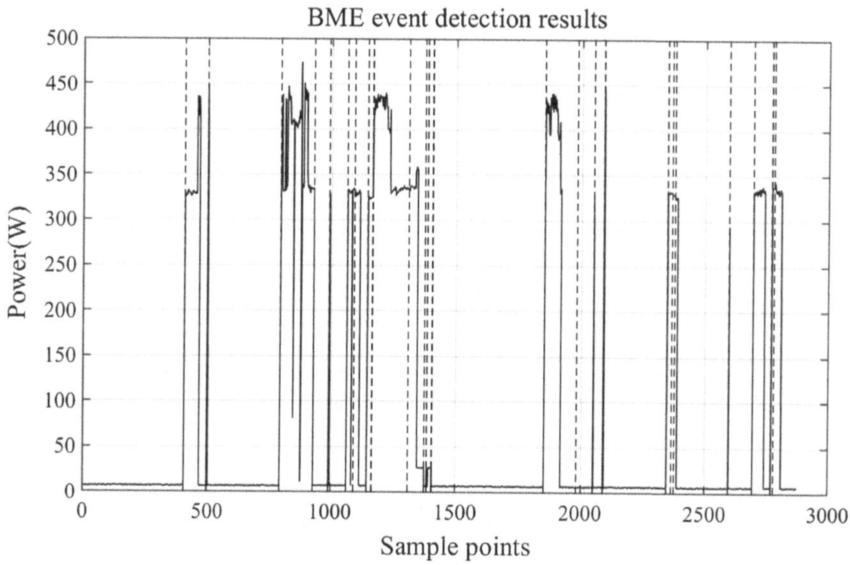

Fig. 2.12 The results of the BME event detection experiments

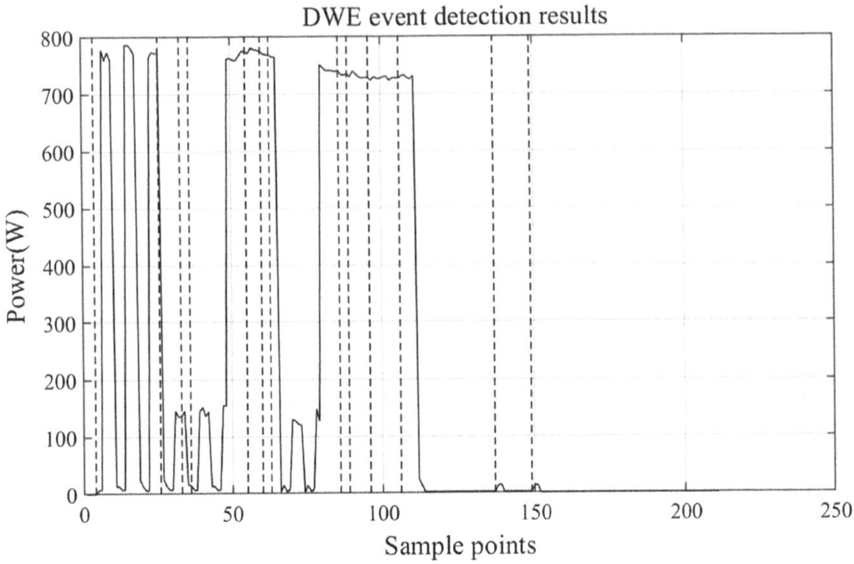

Fig. 2.13 The results of the DWE event detection experiments

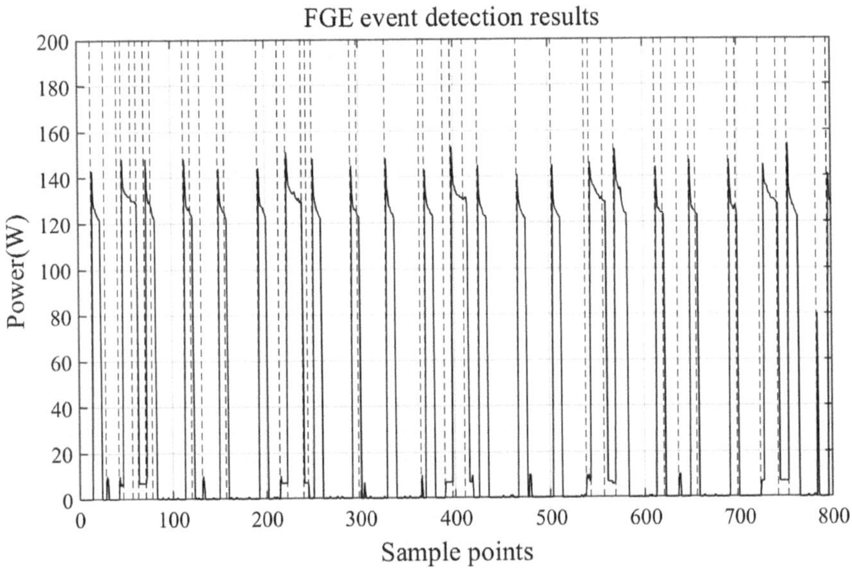

Fig. 2.14 The results of the FGE event detection experiments

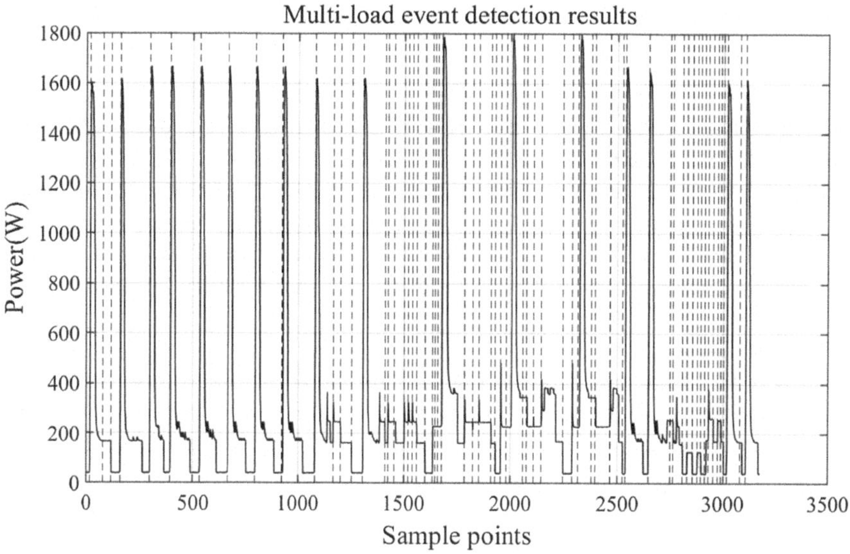

Fig. 2.15 The results of the multi-load event detection experiments

Table 2.5 Event detection results of the three groups

	BME	DWE	FGE	Multi-load
Precision	0.6827	0.6505	0.6863	0.6238
Recall	0.7100	0.6700	0.7000	0.6300
F1-score	0.6961	0.6601	0.6931	0.6269

2.4 Sequential Probability Ratio Test

2.4.1 The Theoretical Basis of SPRT

Sequential Probability Ratio Test (SPRT) is a branch of mathematical statistics. The method needs to make statistical inference by sampling samples. It does not specify the total number of samples in advance when sampling is in progress. This method takes a small number of samples first and determines the sampling stop condition according to the results. On the other hand, a sampling scheme in which the number of samples is determined in advance is called a fixed sampling scheme.

2.4.2 Comparison of Event Detection Results

To verify the effectiveness of the sliding windows based on two-sided CUSUM, event detection experiments are carried out. The event detection results of the BME, DWE and FGE are shown in Figs. 2.16, 2.17 and 2.18. The results of the multi-load event detection experiments are shown in Fig. 2.19. It can be found that the sliding windows based on the two-sided CUSUM algorithm improves the sensitivity of the CUSUM algorithm, which contributes to the highest accuracy of event detection.

These experimental results of electrical event detection are analyzed and evaluated as Table 2.6.

The conclusion can be drawn as follows:

The experimental results of the four groups show that the improved CUSUM algorithm has a high recall while a low accuracy, when judging the DWE and multiple loads. When detecting the BME and FGE, the accuracy is higher, but the recall is lower. For the DWE and multi-load loads, it is easy to detect ordinary power fluctuations as open events, due to their operational characteristics. For the BME and FGE loads, the power reduction fluctuation is easy to be identified as the off event, due to their load fluctuates.

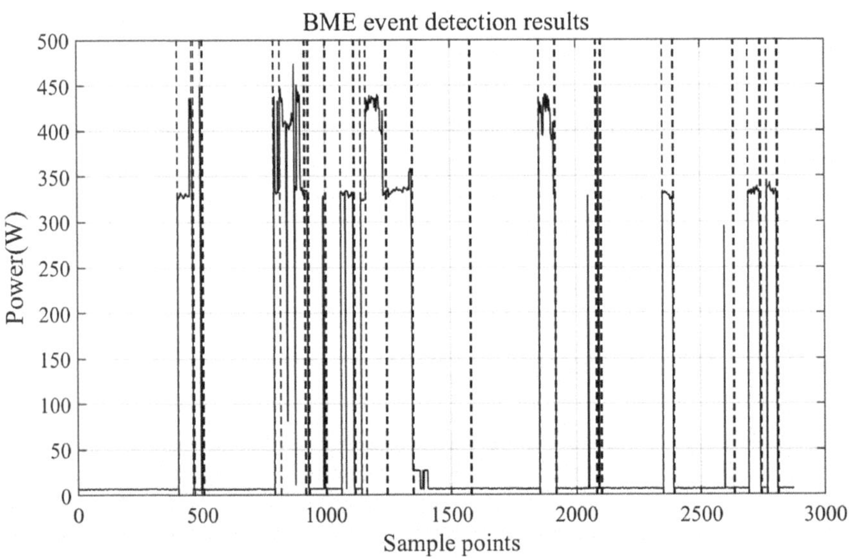

Fig. 2.16 The results of the BME event detection experiments

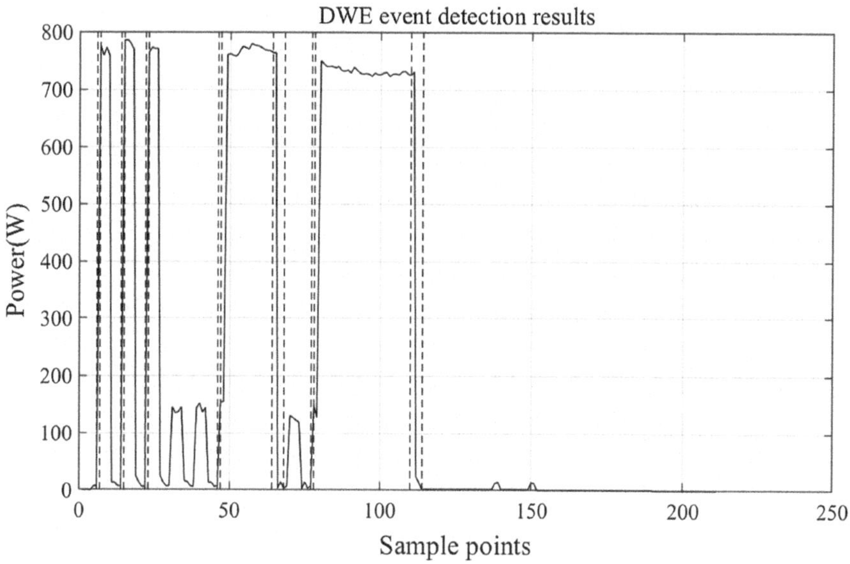

Fig. 2.17 The results of the DWE event detection experiments

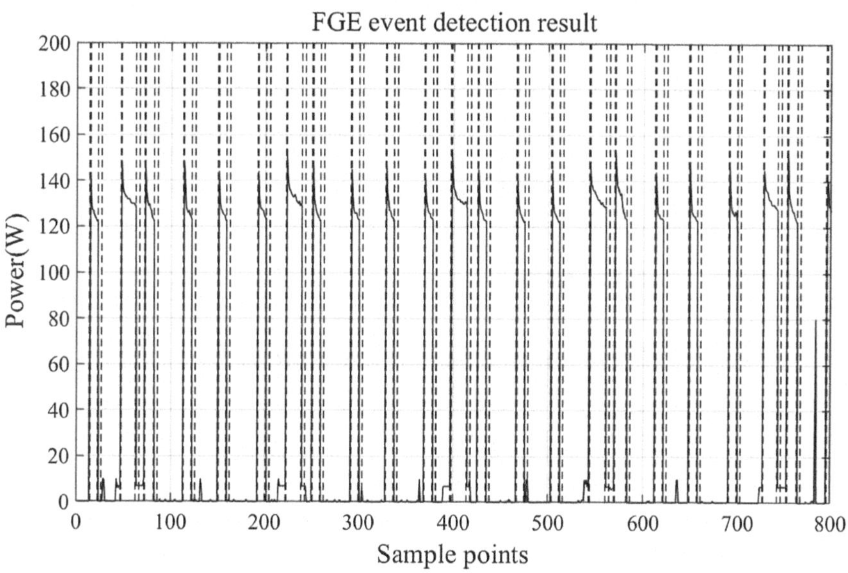

Fig. 2.18 The results of the FGE event detection experiments

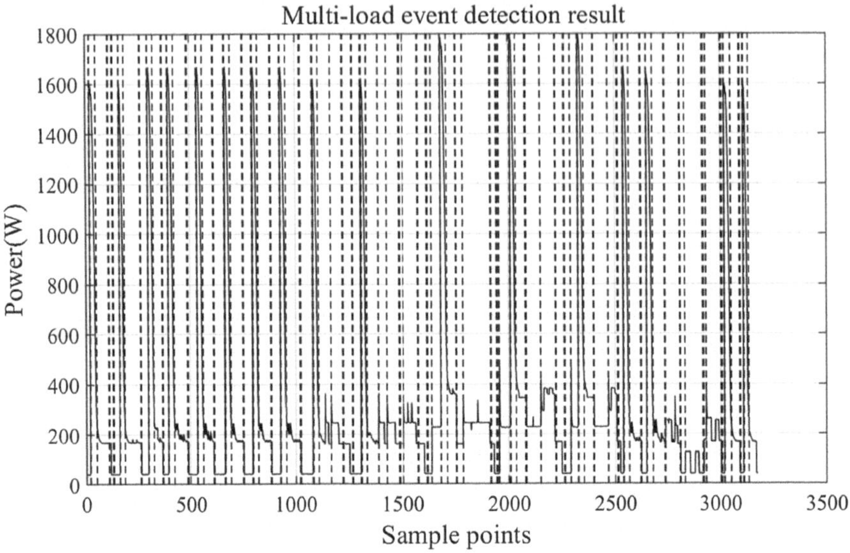

Fig. 2.19 The results of the multi-load event detection experiments

Table 2.6 Event detection results of the three groups

	BME	DWE	FGE	Multi-load
Precision	0.8200	0.7596	0.8300	0.7476
Recall	0.7810	0.7900	0.8137	0.7700
F1-score	0.8000	0.7745	0.8218	0.7586

2.5 Experiment Analysis

2.5.1 *The Results of Three Kinds of Algorithms*

The evaluation results of the event detection results are summarized in Table 2.7.

2.5.2 *Conclusion*

Based on the above experimental results, the following conclusions can be drawn.

(a) Based on the F1-score evaluation index results, it can be seen that the event detection performance of GLR algorithm is the worst, followed by the traditional CUSUM algorithm, and the improved CUSUM algorithm has the best

Table 2.7 Event detection results of the three methods

Methods	Indices	BME	DWE	FGE	Multi-load
CUSUM	Precision	0.8021	0.7368	0.7653	0.6900
	Recall	0.7700	0.7000	0.7500	0.6765
	F1-score	0.7857	0.7179	0.7576	0.6832
GLR	Precision	0.6827	0.6505	0.6863	0.6238
	Recall	0.7100	0.6700	0.7000	0.6300
	F1-score	0.6961	0.6601	0.6931	0.6269
Sliding windows based on two-sided CUSUM	Precision	0.8200	0.7596	0.8300	0.7476
	Recall	0.7810	0.7900	0.8137	0.7700
	F1-score	0.8000	0.7745	0.8218	0.7586

performance. This is because the GLR event detection algorithm based on transient power characteristics provides a poor detection performance on shutdown. As a result, the accuracy is low, so its performance is the worst. Traditional CUSUM algorithm has a too high accuracy in detecting power fluctuations, but the recall is quite low. The improved CUSUM algorithm has a high Accuracy in detecting events and also improves the recall of traditional CUSUM algorithm, so it provides the best performance.

(b) The BME and FGE have better detection performance within the single load detection, whereas the DWE is the worst. The opening and closing events of BME and FGE are relatively easy to be detected, and the accuracy is relatively high.

(c) Compared with the single load event detection effect, multi-load event monitoring performance is poor. It is because the multi-load situation is more complex than the single load.

(d) The improved CUSUM algorithm, on the basis of the advantages of traditional CUSUM algorithm, can improve the detection accuracy.

References

Anderson K, Ocneanu A, Benitez D, Carlson D, Rowe A, Berges M (2012) BLUED: A fully labeled public dataset for event-based non-intrusive load monitoring research. In: Proceedings of the 2nd KDD workshop on data mining applications in sustainability (SustKDD), pp 1–5

Cox R, Leeb SB, Shaw SR, Norford LK (2006) Transient event detection for nonintrusive load monitoring and demand side management using voltage distortion. In: Twenty-first annual IEEE applied power electronics conference and exposition, 2006. APEC '06. IEEE, p 7

De Oca VM, Jeske DR, Zhang Q, Rendon C, Marvasti M (2010) A cusum change-point detection algorithm for non-stationary sequences with application to data network surveillance. J Syst Softw 83(7):1288–1297

Junker M, Hoch R, Dengel A (1999) On the evaluation of document analysis components by recall, precision, and accuracy. In: Proceedings of the fifth international conference on document analysis and recognition. ICDAR '99 (Cat. No. PR00318). IEEE, pp 713–716

Leeb SB, Shaw SR, Kirtley JL (1995) Transient event detection in spectral envelope estimates for nonintrusive load monitoring. IEEE Trans Power Delivery 10(3):1200–1210. https://doi.org/10.1109/61.400897

Makonin S, Ellert B, Bajić IV, Popowich F (2016) Electricity, water, and natural gas consumption of a residential house in Canada from 2012 to 2014. Scientific Data 3(1):160037. https://doi.org/10.1038/sdata.2016.37

Niu L, Jia H (2011) Transient event detection algorithm for non-intrusive load monitoring. Dianli Xitong Zidonghua (Automation of Electric Power Systems) 35(9):30–35

Norford LK, Leeb SB (1996) Non-intrusive electrical load monitoring in commercial buildings based on steady-state and transient load-detection algorithms. Energy Build 24(1):51–64. https://doi.org/10.1016/0378-7788(95)00958-2

Page ES (1954) Continuous inspection schemes. Biometrika 41(1/2):100–115

Siegmund D, Venkatraman E (1995) Using the generalized likelihood ratio statistic for sequential detection of a change-point. Ann Stat 255–271

Trivedi S, Chandramouli R (2005) Secret key estimation in sequential steganography. IEEE Trans Signal Process 53(2):746–757

Chapter 3
Appliance Signature Extraction

3.1 Introduction

3.1.1 Background

Non-intrusive Load Monitoring (NILM) requires only one sensor to be installed at the residential meter to collect the total current and voltage signals. Since the type of data collected is extremely limited, extracting effective Load Signature (LS) information based on current and voltage signals has become one of the key factors to improve the identification accuracy of the NILM (Chui et al. 2014). In the NILM, load signatures are mainly divided into steady-state and transient features. The steady-state load signature is extracted from the characteristics of the physical features, harmonic features or voltage-current relationship of the load signal. It is usually defined as 'snapshot' form signature (Jian et al. 2010). The transient load signature is mainly used to analyze the load change of the electrical switching process and describe the transition of different load signatures before and after the change. It is defined as 'delta' form signature. It contains many types of features, such as the amount of power change, the amount of impedance change, the amount of harmonic change, the power peak stability ratio and etc. The transient load signature requires that the extracted features can be additive.

The physical features of the load are the most common load features, including active power, reactive power, impedance, admittance and etc (Baranski and Voss 2005). Physical features are generally superimposable and have a good degree of discrimination. However, for complex power usage situations, the use of physical features alone provides poor identification performance. The current waveform (CW) is one of the LS that used frequently (Figueiredo et al. 2011). The features for CW include current harmonics, current intensity and information entropy of the current (Duarte et al. 2012; Tsai and Lin 2012). Some CW features are additive. However, single CW features do not have strong discrimination and are susceptible to environmental noise. As a relatively novel feature extraction method, the V-I

H. Liu, *Non-intrusive Load Monitoring*,
https://doi.org/10.1007/978-981-15-1860-7_3

trajectory features have received extensive attention (Hassan et al. 2014). The V-I trajectory features are utilized for analyzing the relationship between current and voltage signals over a duty cycle by plotting the VI image. The VI image can be combined with computer graphics and deep learning for further feature extraction to receive a higher identification accuracy. However, the V-I trajectory features can not be additive and only be used for the extraction of steady-state features.

In general, the core of the appliance signature extraction is the analysis and extraction of different kinds of load feature. The chapter analyzes three common types of steady-state load features in detail, including physical definition features, current harmonic features, and VI image features. And the adaptive features based on transfer learning are evolved from VI image features. By evaluating the extracted features using the evaluation indices and calculating the classification performance of each type of feature in combination with actual data.

3.1.2 *Feature Evaluation Indices*

In order to better evaluate each feature of the extracted load, the chapter uses the knowledge of the feature engineering field to introduce three different evaluation indices. These feature evaluation indices calculate the correlation between the feature vector of each type of feature and the label vector. And then the correlation coefficient and weight value of the feature are obtained.

(1) *Spearman correlation coefficient*

The three major correlation coefficients of statistics, which include the Pearson, Spearman, and Kendall, reflect the correlation between the direction and extent of the trend of the two variables (Maravi et al. 2017). The Spearman correlation coefficient, which also called the rank correlation coefficient, is used to make a linear correlation analysis between two variables (Manurung et al. 2017). The advantage of it is that it need no distribution of the original variables. It is a non-parametric statistical method and has a wider range of applications than Pearson. When two variables do not meet the assumption of a bivariate normal distribution, Spearman rank correlation can describe the mutual relationship between them. The Spearman correlation coefficient is calculated as follows.

$$S = \frac{\sum_{i=1}^{n} (x_i - \bar{x}) \cdot (y_i - \bar{y})}{\sqrt{\sum_{i=1}^{n} (x_i - \bar{x})^2 \cdot \sum_{j=1}^{n} (y_j - \bar{y})^2}} \tag{3.1}$$

(2) *Kendall correlation coefficient*

The Kendall correlation coefficient was proposed by Kendall in 1938. It checks whether there is a correlation between the two variables by judging whether the two variables are coordinated (Doorn et al. 2018). The Kendall correlation coefficient can judge the correlation between categorical variables. It first introduces the concept of concordant. Supposing there are n observations $\{(x_i, y_i)|i = 1 \sim n\}$, if the product satisfies $(x_i - x_j) \cdot (y_i - y_j) > 0 \forall i > j$, the observations (x_i, y_i) and (x_j, y_j) will satisfy the concordant, indicating that the directions of change between the two observation sequences are the same. Conversely, the observations (x_i, y_i) and (x_j, y_j) do not satisfy the concordant. N_I is used to indicate the number of concordant observations and N_2 represents the number of non-concordant observations. The formula of Kendall correlation coefficient is drawn as follows:

$$K = 2 \times \frac{N_1 - N_d}{n(n-1)} \tag{3.2}$$

The Kendall correlation coefficient is similar to the Spearman correlation coefficient, and the variable range is between -1 and 1. When the two correlation coefficients are used for feature evaluation, the closer the calculated value is to 0, the lower the correlation between the feature vector and the label vector is. When the value is closer to 1, the positive correlation between the two is stronger. When the value is close to -1, there is a strong negative correlation between the two.

(3) *ReliefF feature selection algorithm*

The ReliefF feature evaluation algorithm was born out of the Relief algorithm which was first proposed by Kira and Rendell in 1992 (Kira and Rendell 1992). The Relief algorithm assigns a weight value to each feature to form a weight vector. For the randomly selected sample R, the Relief algorithm implements the update of the feature weight vector by finding the m nearest neighbors of the R's homogeneous and heterogeneous samples (called Near Hit) (Spolaor et al. 2014). The Relief algorithm can only analyze and process the samples of the two classifications. In order to perform feature evaluation for multi-class samples, Kononenko proposed the ReliefF feature evaluation algorithm (Kononenko 1994). The process framework of the ReliefF algorithm is shown as follows (Fig. 3.1):

The weight update formula is expressed as follows:

$$\begin{aligned}\Omega(X) =& \Omega(X) - \sum_{i=1}^{m} \frac{diff(X, R, T_i)}{m \cdot N_P} \\ &+ \sum_{Y \notin class(R)} \left[\frac{p(Y)}{1 - p(class(R))} \sum_{j=1}^{m} \frac{diff(X, R, F_j(Y))}{m \cdot N_P}\right]\end{aligned} \tag{3.3}$$

$$diff(X, R, T_i) = \begin{cases} \frac{|R[X] - T[X]|}{\max(X) - \min(X)} & \text{When } X \text{ is continuous} \\ 0 & \text{When } X \text{ is discrete and } R[X] = T[X] \\ 1 & \text{When } X \text{ is discrete and } R[X] \neq T[X] \end{cases} \tag{3.4}$$

where $p(Y)$ is the proportion of the category in class Y. $F_j(Y)$ represents the jth nearest neighbor sample in class Y.

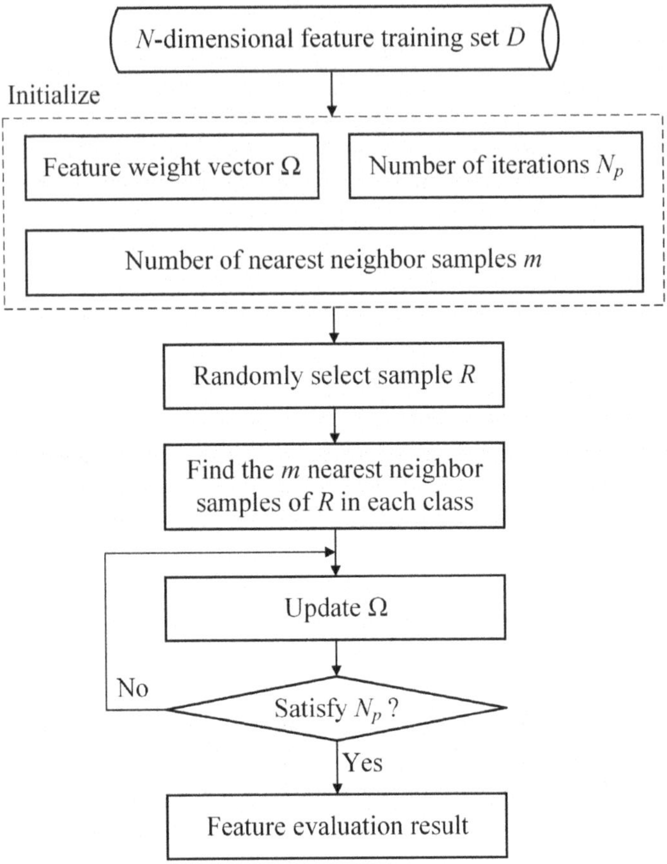

Fig. 3.1 The process of ReliefF feature evaluation algorithm

3.1.3 Classification Evaluation Indices

The commonly used two-classification evaluation indices mainly include Accuracy, Recall, Precision, and F1-Score. For a multi-classification system with multiple label values, the value of Accuracy can be obtained same as the two-classification system, which is based on the relationship between the classification result and the actual label value. However, the multi-classification system cannot obtain three types of indices: Recall, Precision, and F1-Score. In order to better evaluate the multi-classification results and evaluate the features, the chapter introduces three macro-multi-classification evaluation indices, which are Macro Recall, Macro Precision, and Macro F1_Score. The calculating formulas are expressed as follows:

$$marco\_R = \frac{1}{n}\sum_{i=1}^{n} Recall(i) \tag{3.5}$$

$$marco\_P = \frac{1}{n}\sum_{i=1}^{n} Precision(i) \tag{3.6}$$

$$marco\_F1 = \frac{2 \times macro\_R \times macro\_P}{macro\_R + macro\_P} \tag{3.7}$$

where n is the total number of categories.

3.1.4 Data Selection

In the chapter, the load feature extraction experiments are performed on a high-frequency public dataset, the WHITED dataset (Kahl et al. 2016). The WHITED dataset contains currents and voltages of all appliances operating independently. And the data were measured at 44.1 kHz. The instances of the appliances were collected from 8 areas, and there were three power supply standards among these areas. The dataset contains 1100 samples of 110 appliances and the appliances can be classified into 47 various types. Each appliance has 10 running instances and each instance records 5-second-long data. The study selects the data of six types of electrical appliances in the WHITED dataset as the processing object. They are LED light, washing machine, fan heater, compact fluorescent lamp, air conditioning, and hairdryer. The label values and abbreviations corresponding to them are shown in Table 3.1.

In order to ensure the consistency of each feature extraction process and the SVM load identification model construction process, the chapter stipulates that the length of training set for SVM load recognition model is 900, and the length of testing set is 300. At the same time, the sample number of different electrical appliances in the training set and testing set is evenly distributed.

Table 3.1 The labels and abbreviations of appliances

Label value	Appliance	Abbreviation
1	LED light	LED
2	Washing machine	Washing
3	Fan heater	Fan
4	Compact fluorescent lamp	Lamp
5	Air conditioning	AC
6	Hairdryer	Hairdryer

3.2 Features Based on Conventional Physical Definition

3.2.1 The Theoretical Basis of Physical Definition Features

In the typical framework of non-intrusive load monitoring and decomposition, load identification usually includes the following five parts: data collection, data preprocessing, event detection, feature extraction, and load identification (Hosseini et al. 2017). The emphasis of the chapter is the analysis of feature extraction. The experimental objects are the original $u(t)$ and $i(t)$ signals of the WHITED dataset. The operating voltage and current are calculated as follows.

$$U = \sqrt{\frac{1}{T}\int_0^T u^2(t)dt} \tag{3.8}$$

$$I = \sqrt{\frac{1}{T}\int_0^T i^2(t)dt} \tag{3.9}$$

The physical quantities in the power system are diverse. Common physical definition features mainly include active power, reactive power, apparent power, admittance, impedance and so on (He et al. 2012). The specific calculation method can be shown as follows:

(1) Power

In a general AC circuit, there is an active component and a reactive component in the delivered electric power. The instantaneous active power is different and constantly changing. The average active power (average of power in one cycle) is generally used to measure the energy consumption in the circuit. Reactive power is used for the exchange of electric and magnetic fields inside the circuit. It is also used to supply the electrical power to maintain the magnetic field. It only performs energy conversion inside the circuit. In order to measure the exchange energy, the maximum power of the energy exchange process is used to represent the reactive power. Apparent power refers to the total power on the circuit, which is expressed as the product of the voltage rms value and the current rms value.

The calculation formulas of active power and reactive power are given as follows:

$$P = \frac{1}{T}\int_0^T u(t)i(t)dt \tag{3.10}$$

$$Q = \sqrt{S^2 - P^2} \tag{3.11}$$

where S is the apparent power, $S = UI$.

(2) *Power factor*

As one of important power indicators (Jordan 1994), the power factor is the ratio of the active power of the AC circuit to the apparent power. The power factor is related to the physical properties of the circuit. It is usually expressed as λ. Under a certain voltage and power, when the λ of the electrical equipment is lower, the reactive power of the circuit is greater. Thereby the utilization of the equipment is reduced and the line power loss is increased. The higher the value is, the better the electrical efficiency is, and the more fully utilized the power generation equipment is. According to active and reactive power, the power factor can be calculated as follows:

$$\lambda = \frac{P}{\sqrt{P^2 + Q^2}} \tag{3.12}$$

(3) *Impedance*

In circuits with resistance, inductance, and capacitance, the obstruction to alternating current is called impedance. The impedance consists of resistance and reactance. The resistance represents the resistance of the pure resistance to the current, and the reactance consists of the inductive reactance generated by the inductor and the capacitive reactance generated by the capacitor.

The calculation formula of impedance is given as follows:

$$Z = \frac{U}{I} \tag{3.13}$$

The calculation formula of resistance is given as follows:

$$R = Z \cos \varphi_Z \tag{3.14}$$

where $\varphi_Z = \arcsin(\frac{Q}{S})$;

The calculation formula of reactance is given as follows:

$$X = Z \sin \varphi_Z \tag{3.15}$$

(4) *Admittance*

The calculation formula of admittance is given as follows:

$$Y = \frac{I}{U} \tag{3.16}$$

The calculation formula of conductance is given as follows:

$$G = Y \cos \varphi_Z \tag{3.17}$$

The calculation formula of the susceptance is given as follows:

$$B = -Y \sin \varphi_Z \tag{3.18}$$

The chapter explores the above ten physical definition features, including active power P, reactive power Q, apparent power S, power factor λ, impedance Z, resistance R, reactance X, admittance Y, conductance G, and susceptance B.

3.2.2 Feature Extraction

Since the operating frequency of the voltage and current signals is 50 Hz, when extracting the original current and voltage signals with a sampling frequency of 44,100 Hz, it is necessary to extract the current and voltage sequences of one of the working cycle lengths as sample data. Each sample dataset contains 882 points. The overall process of feature extraction is shown in Fig. 3.2.

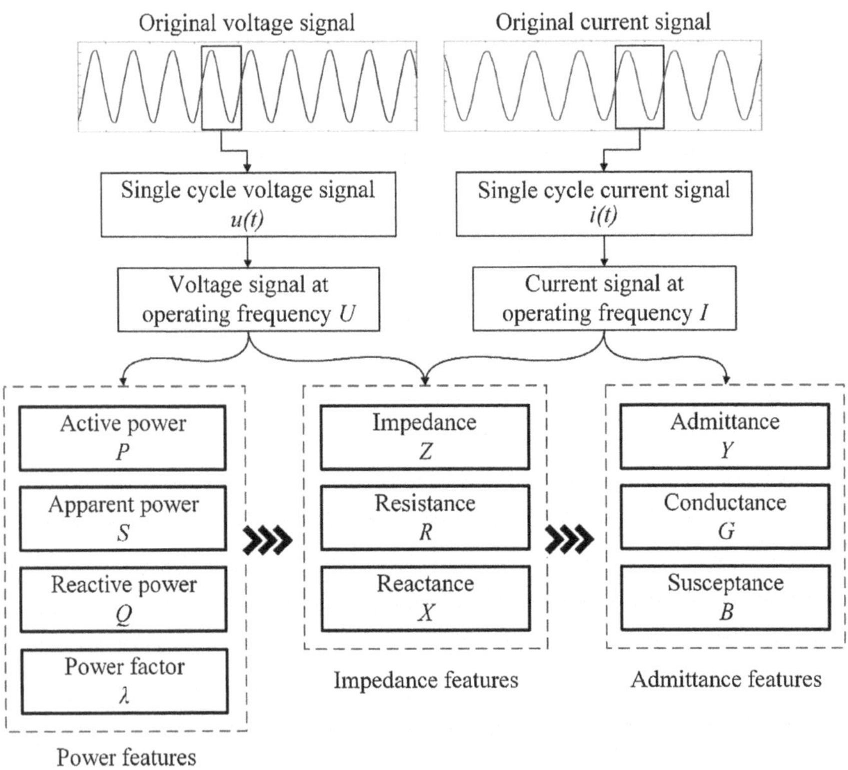

Fig. 3.2 The process of physical definition features extraction

3.2.3 Feature Evaluation

Every physical definition feature is evaluated with three types of evaluation indices. The calculations are shown in Fig. 3.3. The x-axis represents the type of feature, and the y-axis represents the calculation results of the corresponding evaluation index.

By analyzing the results of the evaluation indices, the following conclusions can be drawn:

(a) For most physical definition features, the absolute values of the Spearman correlation coefficient and Kendall correlation coefficient are lower. This indicates that the linear correlation between the feature vector and the label vector is low. However, the weights of ReliefF are high, indicating that the intra-class tightness and inter-class discrimination of the feature distribution are high.

(b) In all physical definition features, the reactive power Q and apparent power S have the highest weight values. This indicates that the values of reactive power and apparent power are basically stable for the same appliance. For different electrical appliances, the values of the two indices show significant specificity.

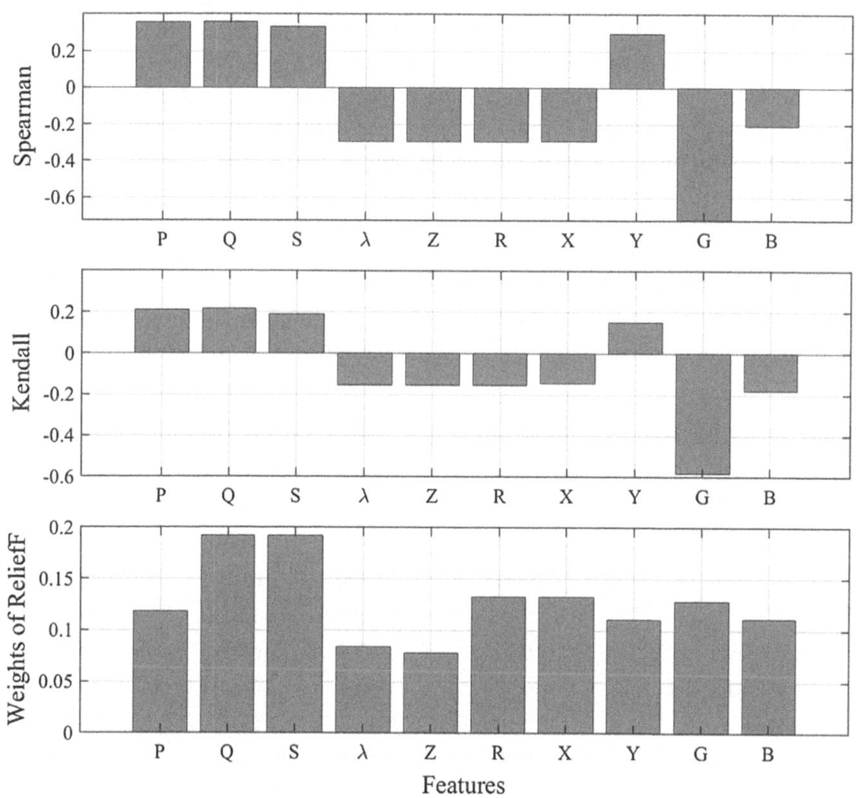

Fig. 3.3 Calculation of evaluation indices for physical definition features

The power factor λ and impedance Z not only have weak linear correlation but also are not sufficiently specific for different electrical appliances. And for different appliances, there is a certain aliasing phenomenon between the values of the two indices.

3.2.4 Classification Results

To further analyze the classification performance of physical definition features, this section uses the training samples constituted by 10 kinds of physical definition features. The classification performance indices of the SVM load identification model are shown in Table 3.2. The Confusion matrix is shown in Fig. 3.4.

By analyzing the above figure and table, the physical definition features show excellent classification accuracy. During the total test samples, only a very small number of samples are misclassified. The final classification accuracy exceeds 0.98. Combined with the conclusion of the Sect. 3.2.3, it can be concluded that the overall physical definition features have a high degree of discrimination for the selected samples, thus ensuring that the load identification model has an excellent recognition ability. The feature vector composed of physical definition features is synergistically analyzed from the power, resistance, capacitance, inductance and other aspects of the electrical appliance, which makes up for the deficiency of the single physical definition feature and can accurately depict the features of the specific electrical appliance.

However, physical definition features are not omnipotent. For some resistive appliances or inductive appliances, the distinction between physical definition features is low. At the same time, for some variable frequency electrical appliances or variable power electrical appliances, the value of physical definition features is difficult to maintain constant, and the degree of closeness of the numerical distribution is low.

Table 3.2 The classification performance indices of the SVM model by physical definition features

Model	Type of appliance	Accuracy	Recall	Precision	F1-score
SVM	LED	1.0000	1.0000	1.0000	1.0000
	Washing	0.9933	0.9600	1.0000	0.9796
	Fan	0.9933	0.9800	0.9800	0.9800
	Lamp	0.9967	0.9800	1.0000	0.9899
	AC	0.9933	1.0000	0.9615	0.9804
	Hairdryer	0.9967	1.0000	0.9804	0.9901
		Accuracy	Marco_R	Marco_P	Marco_F1
	Total	0.9867	0.9900	0.9900	0.9900

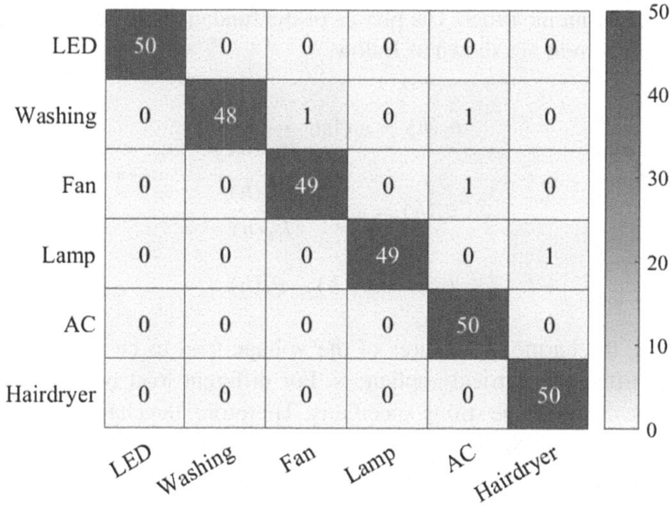

Fig. 3.4 Confusion matrix of the SVM model load identification results by physical definition features

3.3 Features Based on Time-Frequency Analysis

3.3.1 The Theoretical Basis of Harmonic Features

For the voltage and current signals of the load, in addition to obtaining the above feature information from the frequency domain, the effective information of the frequency domain can be obtained by means of time-frequency analysis (Kontos et al. 2017; Bouhouras et al. 2017). Since the voltage and current data collected by the measuring device are discrete, the data needs to be processed by the discrete Fourier transform method. It is generally analyzed using a time-sampled Fast Fourier Algorithm (FFT) (Tang et al. 2017).

The original voltage and current signals with a sample length of m and a sampling frequency of 44,100 Hz are selected, which are represented as voltage signal $U(t)$ and current signal $I(t)$, respectively. The FFT analysis is performed on the voltage signal $U(t)$ and the current signal $I(t)$, respectively.

U_{re} and U_{im} represent the real part and the imaginary part of each harmonic voltage obtained by FFT analysis, respectively. I_{re} and I_{im} represent the real part and imaginary part of each harmonic current with the FFT analysis, respectively. Therefore, the amplitude of the fundamental wave and harmonics of voltage and current are shown as follows:

$$U_h = \sqrt{U_{re}(h)^2 + U_{im}(h)^2} \qquad h = 1, 2, 3, \ldots, 20 \tag{3.19}$$

$$I_h = \sqrt{I_{re}(h)^2 + I_{im}(h)^2} \qquad h = 1, 2, 3, \ldots, 20 \tag{3.20}$$

where h is the harmonic order. The phases of the fundamental wave and harmonics of voltage and current are shown as follow.

$$\theta_1(h) = \arctan \frac{U_{im}(h)}{U_{re}(h)} \tag{3.21}$$

$$\theta_2(h) = \arctan \frac{I_{im}(h)}{I_{re}(h)} \tag{3.22}$$

$$\theta(h) = \theta_1(h) - \theta_2(h) \tag{3.23}$$

In general, the harmonic features of the voltage tend to be stable and are less affected by different electrical appliances. For different load types, the harmonic features of the current have strong specificity. Therefore, the chapter selects the real and imaginary parts of the 1st to 20th harmonics of the current as the eigenvalues and discusses the effectiveness of the harmonic features for load identification. The length of the original current signal selected for each sample is 4000.

3.3.2 Feature Extraction

For each sample signal in the original current signal, it is processed by FFT to obtain the harmonic real part feature vector I_{re} and the harmonic imaginary part feature vector I_{im} of the sample, respectively. Each feature vector contains the real or imaginary feature values of the 1st to 20th harmonics. Same as physical definition features, the experimental data is also analyzed to obtain a feature matrix. The harmonic real part is taken as the horizontal axis and the harmonic imaginary part is taken as the vertical axis. The harmonic feature distribution map of the training samples is plotted for the ith current harmonic features. Among them, Fig. 3.5 shows the distribution map of the 1st, 2nd and 20th harmonics, respectively.

By analyzing the above figures, the following conclusions can be drawn:

(a) Regardless of the real or imaginary part, the magnitude of the current harmonic features decreases significantly as the harmonic order increases. Among them, the 1st harmonic has the highest amplitude. This indicates that after the FFT analysis, the 1st harmonic retains the most information of the original signal and has the highest energy density. In contrast, the amplitude of the 20th harmonic is extremely low. This indicates that the higher harmonics retain high frequency and have lower energy density information in the original current signal.

(b) Overall, current harmonic features have high specificity for different types of electrical appliances. Obviously, for the same kind of electrical appliances, the distribution of sample points of the 1st harmonic is the densest. For different types of electrical appliances, the sample points of the 1st harmonic are the most differentiated. When the harmonic order is 2, the distribution of sample points appears to be discretized to some extent. The sample points of similar

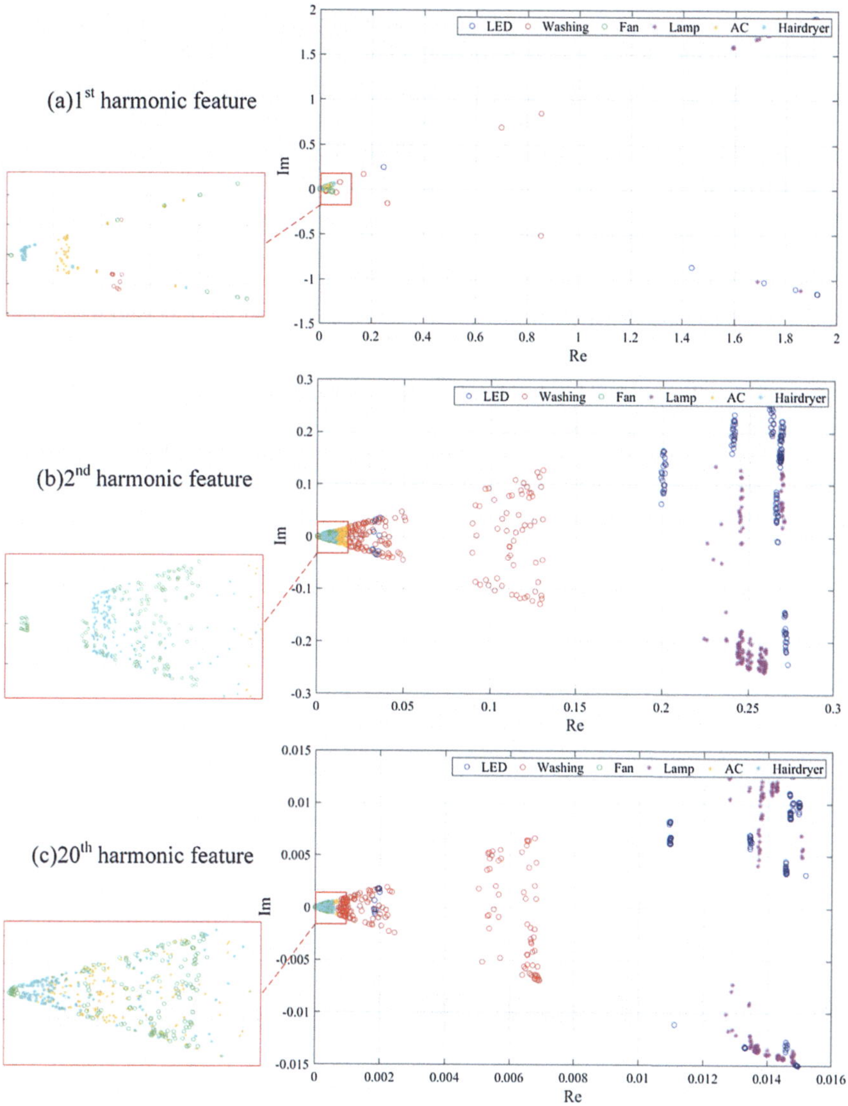

Fig. 3.5 The harmonic feature distribution map of current

electrical appliances are not closely distributed. However, there is still a high degree of discrimination among sample points of different types of electrical appliances. When the harmonic order is 20, a certain degree of aliasing occurs among the sample points of different types of electrical appliances. This situation indicates that the discrimination between different consumers is relatively low for higher-order harmonic features.

(c) In the 6 types of electrical appliances, the LED light and compact fluorescent lamp have the closest distribution to the harmonic features of different orders, and there is also a high degree of discrimination among the other electrical appliances. In contrast, when the harmonic order is higher, the feature distribution of fan heater, air conditioning and Hairdryer has a more serious aliasing phenomenon.

3.3.3 Feature Evaluation

Every feature of the current harmonics is evaluated using the three types of evaluation indices in Sect. 3.1.2. The calculated values of the three types of evaluation indices are shown in Fig. 3.6. The x-axis in the figure represents the order of the harmonics, and the y-axis represents the calculation result of the corresponding evaluation index.

By analyzing the results of the evaluation indices, the following conclusions can be drawn:

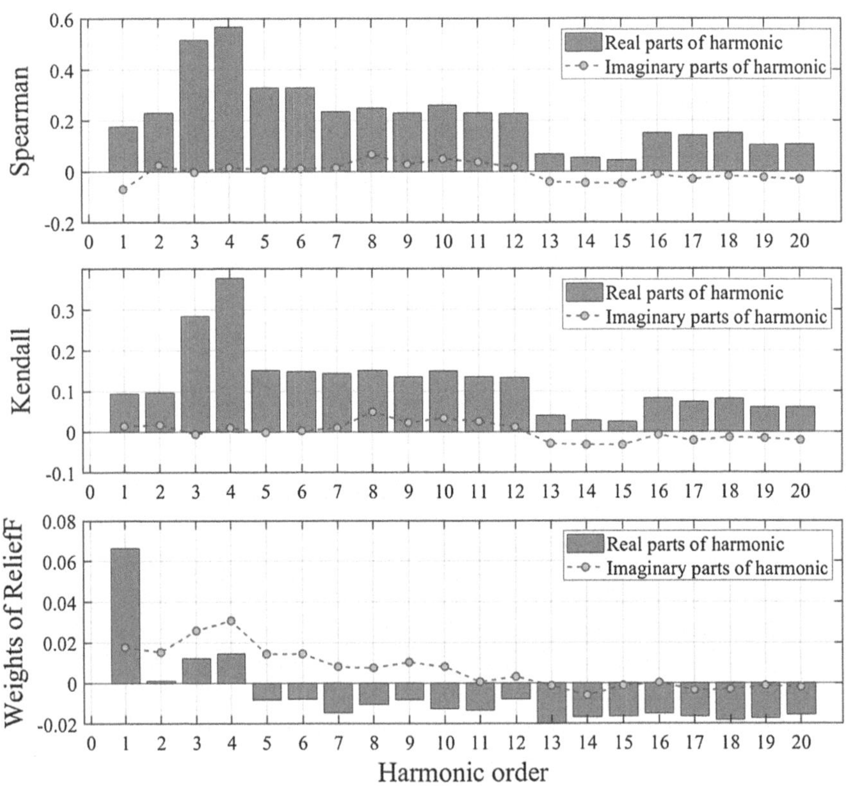

Fig. 3.6 Calculation of evaluation indices of harmonics

(a) It can be clearly seen that the calculation results of the Spearman correlation coefficient and the Kendall correlation coefficient are highly similar. Regardless of the harmonic real part or the harmonic imaginary part, the trend of the two types of correlation coefficients is consistent as the harmonic order increases. For harmonic real features, the calculated value of the correlation coefficient remains substantially positive. When the harmonic order is from 1 to 4, the correlation coefficient value gradually increases. When the harmonic order is greater than 4, the correlation coefficient value shows a gradual decreasing. In other words, there is the greatest linear correlation between the real part of the 4th harmonic feature vector and the label vector. When the harmonic order is greater than 4, the amplitude of the harmonics gradually decreases, and the linear correlation of the harmonics also shows a decreasing trend. When the harmonic order is greater than 13, the real part of the harmonic exhibits a low linear correlation. In contrast, the correlation between all imaginary parts of harmonic feature vectors and label vectors keeps at a very low level.

(b) The weight calculations of the RliefF algorithm are not exactly same as the calculation results of the two types of correlation coefficients. As the harmonic order increases, the weight value of the real part of the harmonics shows a significant downward trend. When the order is greater than 4, the weights of the real parts of the harmonics are all negative. This indicates that the high-order harmonic real part features are poorly differentiated among different consumers. The intraclass distance among the samples of the same type is high, and the distance among classes of different types of samples is low. The characteristics of the imaginary part of the harmonic also maintain a similar trend. The difference is that the weight of the imaginary part of the harmonic is higher than the real part. This shows that the linear correlation of the imaginary features is weak. However, for the features of the high-order harmonic imaginary part, the degree of discrimination is also not obvious.

(c) Among the three types of evaluation indices, the two types of correlation coefficient indices achieve the best results in the 4th harmonic real part features, and the fourth harmonic real part features also show a relatively good degree of discrimination. The weight value of the RliefF algorithm shows a significant maximum value at the real part of the 1st harmonic, indicating that the real part feature of the 1st harmonic has the highest inter-class discrimination. The above phenomena show that the real part features of the 1st harmonic real part and the 4th harmonic real part have good performance.

3.3.4 Classification Results

In real situations, odd harmonics occur more frequently than even harmonics. In a balanced three-phase system, even harmonics have been eliminated due to the symmetrical relationship, and only odd harmonics exist. Even harmonics are generated only when wiring is incorrect or under the influence of some electrical appliances, and even harmonics cannot be canceled. As long as the odd-order harmonics have nonlinear loads such as inverters and electric arc furnaces, harmonics are existing. In order to better explore the difference between the effects of odd and even harmonics on the load identification system, and to explore the difference between the harmonic real part and the harmonic imaginary part on the load identification system, the chapter constructs 9 sets of features. Each set of collections contains different kinds of features. The division of sets is shown in Table 3.3.

A multi-classification load identification model is constructed by using different feature sets as input to train the SVM model. The testing set classification result is obtained, and four multi-class evaluation indices are calculated. The final evaluation indices are calculated as shown in the following table (Table 3.4):

By analyzing the results of the evaluation indices, the following conclusions can be drawn:

Table 3.3 Harmonic feature sets division

Feature set	Odd		Even	
	Real	Imaginary	Real	Imaginary
Set_1	○	×	×	×
Set_2	×	×	○	×
Set_3	×	○	×	×
Set_4	×	×	×	○
Set_5	○	×	○	×
Set_6	×	○	×	○
Set_7	○	○	×	×
Set_8	×	×	○	○
Set_9	○	○	○	○

Table 3.4 The classification performance indices of the SVM model by different harmonic feature sets

Model	Feature set	Accuracy	Marco_R	Marco_P	Marco_F1
SVM	Set_1	0.9333	0.9333	0.9306	0.9319
	Set_2	0.9333	0.9333	0.9306	0.9319
	Set_3	0.7233	0.7233	0.7099	0.7165
	Set_4	0.6667	0.6667	0.6584	0.6625
	Set_5	0.9333	0.9333	0.9306	0.9319
	Set_6	0.7333	0.7333	0.7309	0.7321
	Set_7	0.9433	0.9433	0.9416	0.9425
	Set_8	0.9267	0.9267	0.9235	0.9251
	Set_9	0.9267	0.9267	0.9235	0.9251

(a) By comparing the calculation results of set_1~set_4, it can be seen that the harmonic real part contains more effective information than the harmonic imaginary part. Regardless of whether it is an odd harmonic real part or an even harmonic real part, the evaluation index exceeds 0.9, and the classification performance of the load recognition model is excellent. On the contrary, the classification accuracy of the imaginary features is lower than 0.8, which is not good. This proves the conclusion in the previous section that the harmonic real part has lower intraclass distance and higher interclass discrimination than the harmonic imaginary part. By comparing the calculation results of set_5 and set_9, it can be seen that adding the information of harmonic imaginary part to the real part of the harmonic does not improve the classification performance of the identification model but will reduce the classification accuracy. This indicates that the aliasing among different electrical appliances in the imaginary part of the harmonic imaginary part shows certain harmfulness to the load identification system.
(b) By comparing the calculation results of set_1~set_4 and set_7~set_8, it can be seen that the odd harmonics show better classification performance than the even harmonics. This phenomenon is largely affected by the 1st harmonic. The 1st harmonic of different electrical appliances has a very high degree of discrimination, and the 1st harmonic vector has the highest linear correlation with the label vector. For set_7, after the 1st harmonic feature is removed, model training and testing are performed again. The evaluation index values obtained at this time are 0.9367, 0.9367, 0.9347 and 0.9357, respectively. Compared with set_7, classification accuracy has decreased. Compared with set_8, classification accuracy is still higher. This shows that even if the set_1 is removed, the odd harmonics still have better prediction performance than the even harmonics. This is mainly because the odd harmonics retain more information than the even harmonics, the frequency of occurrence is higher, and the feature specificity among different electrical appliances is stronger.

For the set_7 with the highest classification accuracy, this section further analyzes its classification results. The classification performance indices of the SVM load identification model using set_7 as input are shown in Table 3.5. The Confusion matrix is shown in Fig. 3.7.

By analyzing the above figures and tables, it can be drawn that the odd-order harmonic features have a high degree of discrimination for the class 6 appliances analyzed. Especially for washing machines, fan heater, air conditioning, and hairdryer, the classification accuracy has reached 100%. The misclassifications that occur are all concentrated on the LED light and the compact fluorescent lamp. Samples with 17 compact fluorescent lamps are misidentified for LED light. By combining the conclusions in Sect. 3.3.2, this phenomenon is mainly due to the similar harmonic amplitude range between LED light and compact fluorescent lamps. There is aliasing in the distribution of some samples.

Table 3.5 The classification performance indices of the SVM model by set_7

Model	Type of appliance	Accuracy	Recall	Precision	F1-score
SVM	LED	0.9433	1.0000	0.7463	0.8547
	Washing	1.0000	1.0000	1.0000	1.0000
	Fan	1.0000	1.0000	1.0000	1.0000
	Lamp	0.9433	0.6600	1.0000	0.7952
	AC	1.0000	1.0000	1.0000	1.0000
	Hairdryer	1.0000	1.0000	1.0000	1.0000
		Accuracy	Marco_R	Marco_P	Marco_F1
	Total	0.9433	0.9433	0.9416	0.9425

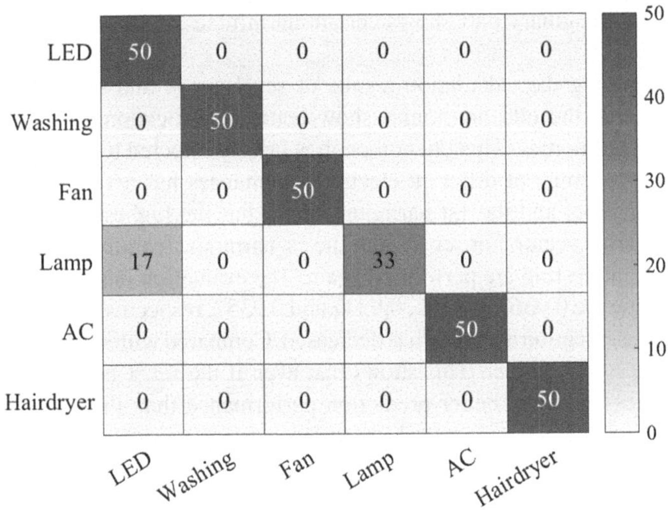

Fig. 3.7 Confusion matrix of the SVM model load identification results by set_7

3.4 Features Based on VI Image

3.4.1 *The Theoretical Basis of VI Image Features*

In order to prevent accidental interference, a total of 2000 original sample points are used as original data for more than two cycles. The VI curve is plotted with the voltage U as the abscissa and the current I as the ordinate (Lam et al. 2007). After drawing the completed VI image, the image cannot be directly used as input to the SVM. It is necessary to perform binarization processing or gradation processing on the VI image according to a certain resolution R and convert it into a feature matrix. The specific generation principle of the feature matrix is shown as follows:

After loading the original voltage and current data during the steady-state operation of the load event, t data points are selected for each group of samples to represent each data point. Then it should find the maximum value v_{max} and the minimum value v_{min} of voltage, as well as the maximum value i_{max} and the minimum value i_{min} of current in the sample. Based on the resolution of R, an $R \times R$ grid area is divided according to v_{max}, v_{min}, i_{max}, and i_{min}. The width and length of each grid are expressed as follows:

$$wide\_v = \frac{v_{\max} - v_{\min}}{R} \tag{3.24}$$

$$wide\_i = \frac{i_{\max} - i_{\min}}{R} \tag{3.25}$$

Each data point is divided into corresponding grids by area. Finally, a collection of the number of data points corresponding to the grid can be obtained. The original VI feature matrix is obtained by arranging the collections according to the position of the grid.

There are many ways to perform subsequent processing on the original VI feature matrix. The chapter mainly introduces three common methods.

(1) *Binary features matrix and binary VI image*

Binarization is to make the gray value of each pixel in the pixel matrix of the image 0 (black) or 255 (white), that is, the whole image is only black and white. The conversion threshold is set to M, the element larger than M in the feature matrix is set to 1, and the element smaller than the threshold is set to 0, thereby converting the original VI feature matrix into a feature matrix containing 0 and 1. Here, 1 is used instead of 255 to facilitate the input of subsequent identification models. The converted binarized matrix is displayed as an image in a binary image of VI containing black or white. Binarization of images is a common method of image processing. Converting the original VI image into a binary image of VI reduces the feature dimension while retaining the feature information to a large extent.

(2) *Grayscale features matrix and grayscale VI image*

The processing of the grayscale feature matrix is simple, but the grayscale image of VI holds a higher dimension than the binary image of VI. The grayscale feature matrix does not need to binarize the original VI feature matrix and only needs to set the element greater than 255 to 255 to ensure that all elements are in the gray value interval. Then all the elements are divided by 255 so that the normalization of the feature matrix is realized, and the final gray-scale feature matrix is obtained. The transformed gray matrix is displayed in the form of an image, which is a grayscale image of VI containing gray information. In theory, the grayscale image of VI holds the density information of the data points in each grid. The binary image of VI has higher recognition and avoids the trouble of setting the threshold. For the two types of VI features, there are many forms to apply to the actual classification, such as conversion to feature vectors and direct input into the deep learning network for identification. In order to match the input type of the SVM model, this section chooses to rearrange the VI feature matrix of each sample into a single-row feature vector.

(3) *Graphical features of binary VI image*

The simple binary VI image and grayscale VI image show the information in a rough way. The dimension of the feature vector increases with the increase of the resolution and is susceptible to image offset and abnormal value points. In order to better extract the information in the VI image, the corresponding graphic features of the binary VI image and the grayscale VI image can be extracted based on the knowledge of computer graphics. This section extracts 5 related features for binary VI images, which are extracted as follows:

(a) Area of image S

Since the voltage and current are periodically changed, the VI curve of each appliance will be a closed circular curve. The edge detection of the original binary VI image is performed first, and then the filling operation is performed, and the number of pixels filled in the calculation is the area S of the graphic.

(b) Number of connected areas of image N_c

The smallest unit of a image is the pixel. For every pixel, there are 8 adjacent pixels around it. Two common adjacencies can be concluded: 4 adjacencies and 8 adjacencies. If pixel points A and B were adjacent, then A and B are defined to be in communication. Visually, points that are connected to each other make up an area, and points that are not connected form different areas. Such a setting in which all the points are connected to each other is called a connected area. The area of the connected area is marked as N_c.

(c) Euler number of image E

Euler numbers belong to the category of topology. In binary image analysis, Euler number is one of the important topological features, which plays an important role in image analysis and geometric object recognition. The Euler number calculation formula for binary images is expressed as follows:

$$E = N_c - N_h \tag{3.26}$$

where N_h represents the number of holes inside the connected area.

(d) Area change rate of image P

The expansion of the image belongs to the relevant knowledge in image morphology, and the specific principle is not described in detail here. The expansion operation expands the binary image. After defining the area obtained by the expansion operation is S_e, the area change rate after the expansion of the image is expressed as follows:

$$P = \frac{S_e - S}{S} \tag{3.27}$$

(e) Perimeter of image L

The perimeter L of the image is calculated by iterating through all the pixels in the binary VI image and counting the number of elements with a value of 1.

3.4.2 Feature Extraction

The VI image features are extracted from the six types of appliances using the methods introduced in Sect. 3.4.1. The original VI image for each type of appliance is shown in Fig. 3.8. It can be seen from the figure that although the sample points of more than 4 cycles are used for drawing, the VI images of most electrical appliances are highly robust. Intuitively, the VI image features among different appliances are obvious and have a high degree of discrimination. However, the VI image of the compact fluorescent lamp is less robust and exhibits significant instability over different duty cycles. It is an important prerequisite for load identification that VI image has high discriminative features and strong period robustness.

In the chapter, the VI image features are extracted using the resolutions of 16×16, 28×28 and 50×50, respectively. For binary VI image features, the results of the extraction are shown in Fig. 3.9. For grayscale VI image features, the results of the extraction are shown in Fig. 3.10. For the convenience of observation, this section displays "0" as white and "1" as black.

By analyzing Fig. 3.9, it can be concluded that as the resolution increases, the binary VI image can more accurately depict the graphical features of the original VI image. At a resolution of 16×16, the original complex picture is simplified to a simple shape. Taking the fan heater as an example, the features of the holes in

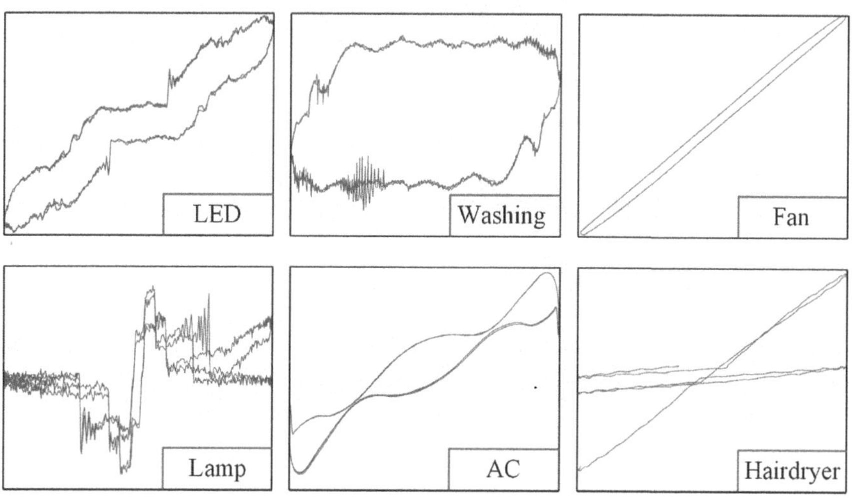

Fig. 3.8 Original VI diagram of different appliances

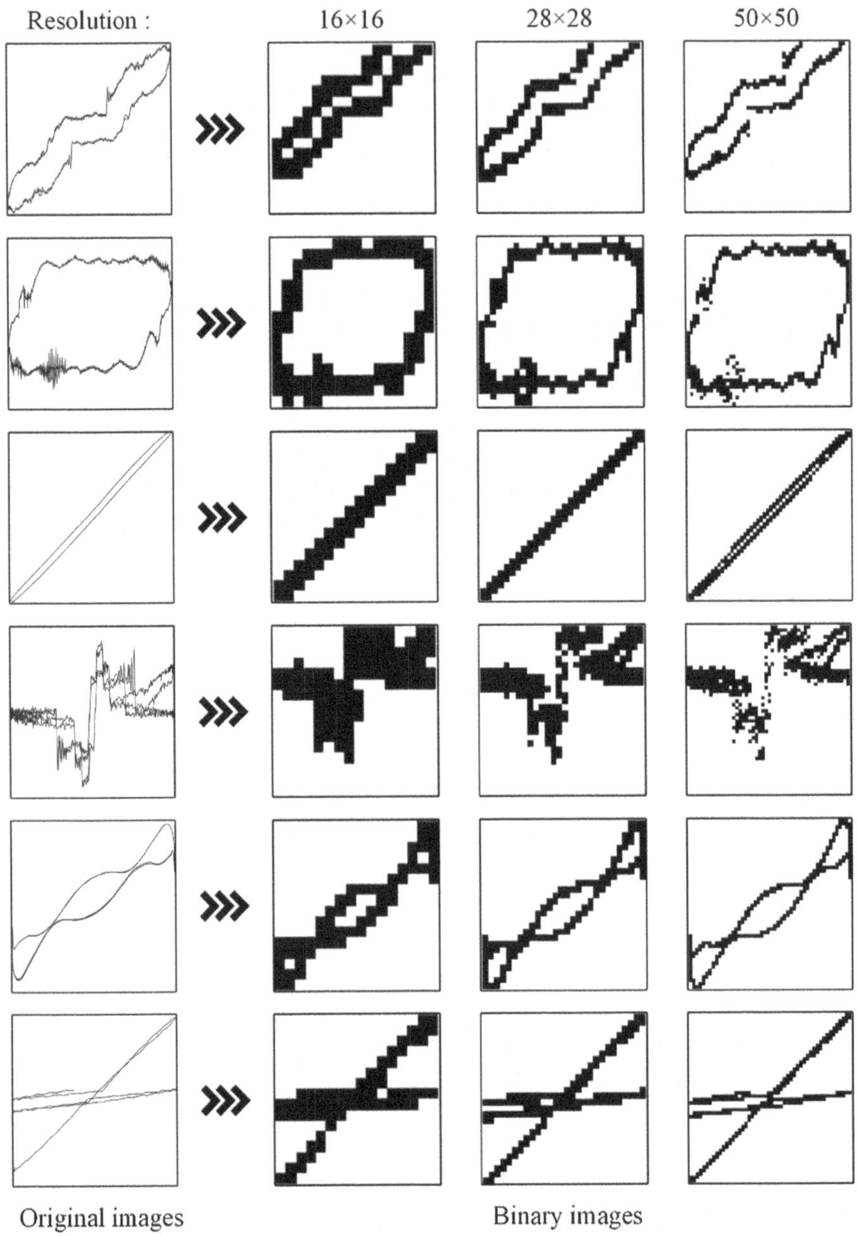

Fig. 3.9 Drawing results of binary VI images with different resolutions

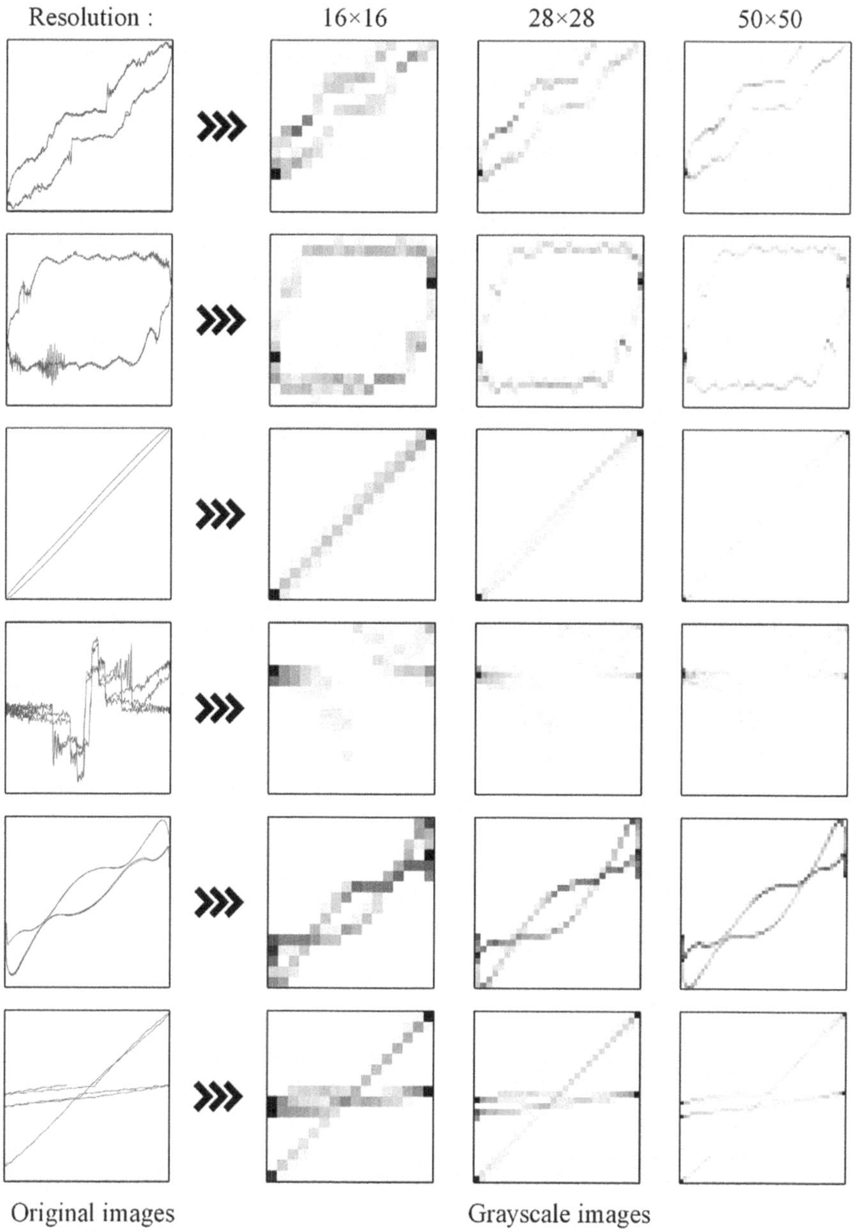

Fig. 3.10 Drawing results of grayscale VI images with different resolutions

the connected domain are expressed when the resolution is 50×50. For a compact fluorescent lamp with low robustness, the resolution of each region in the binary image features is higher when the resolution is higher.

By analyzing Fig. 3.10, it can be concluded that as the resolution increases, the grayscale VI image can also depict the graphical features of the original VI image in more detail. However, as the resolution increases, the gray value of the image gradually decreases. This is due to the fact that the increase in resolution results in a reduction in the width of each grid area, resulting in a reduction in the number of data points allocated to each grid. Among them, the compact fluorescent lamp is the most obvious. Due to the low robustness of the VI image of the device, the density of data points in each region is low. When the resolution is 50×50, its grayscale VI image is basically white, and its features cannot be clearly stated. For grayscale VI image, high resolution leads to low gray values, which makes the graphical features less noticeable?

3.4.3 Feature Evaluation

After the calculations, the values of the three types of evaluation indices for graphical features of binary VI image with different resolutions are shown in Fig. 3.11. The x-axis in the figure represents the resolutions, and the y-axis represents the calculation result of the corresponding evaluation index.

Since the dimensions of the binary VI image features and grayscale VI image features are R. As the resolution increases, the dimension grows in a series, and it is of no practical significance to analyze each of the specific features. The dimension of the feature can be reduced by principal component analysis or linear discriminant analysis to prevent the dimension explosion from affecting the performance of the classifier.

By analyzing the results of the evaluation indices for graphical features of binary VI image, the following conclusions can be drawn:

(a) The calculation results of the two types of correlation coefficients are consistent. As the resolution increases, the absolute value of the correlation coefficient of each feature increases. When the resolution is increased to a certain extent, the correlation coefficient remains substantially constant. This shows that as the resolution increases, the local features of the binary VI image are more detailed, and the graphical features can more accurately depict the original features of the VI image, to improve the linear correlation between the feature vector and the label vector. When the resolution increases to a certain extent, some of the graphical features have stabilized and the performance improvement is limited. At the same time, as the width of the grid decreases, the features are more susceptible to outliers. The linear correlation between the eigenvector and the label vector may even decrease to some extent. The calculation results of the weight

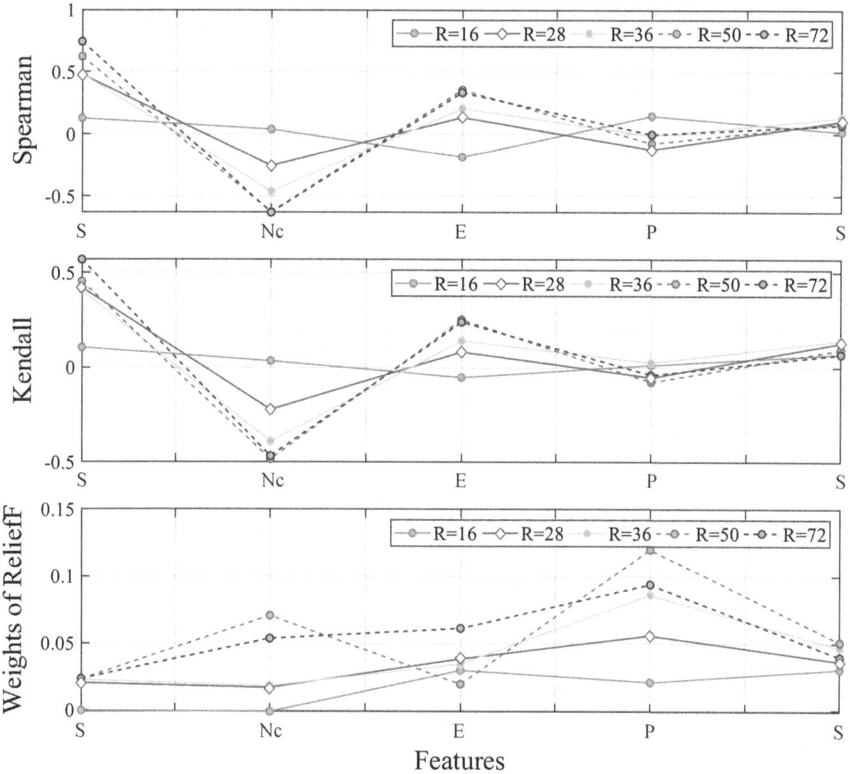

Fig. 3.11 Calculation of evaluation indices of graphical features of binary VI image with different resolutions

of ReliefF also have the above characteristics. It shows that for different electrical appliances, the degree of discrimination of various features also satisfies the above-mentioned changing trend. The difference is that when the resolution reaches a certain level, the degree of discrimination drops significantly.

(b) In the five types of graphical features, there is a high linear correlation between the area of image S and the label vector, but the feature differentiation is relatively low. In contrast, the area change rate of image P has a lower linear correlation with the label vector, but the feature discrimination is relatively higher. As the resolution increases, the feature correlation and discrimination of connected areas number of image N_c are most obvious. This is mainly due to the fact that the parts of the graphics are basically fused together at low resolution, and the connected areas cannot be accurately divided. When the resolution $R = 72 \times 72$, the degree of discrimination of the Euler number of image E showed a serious decline. This is mainly due to the fact that as the width of the grid decreases, the density of data points of each grid decreases sharply, resulting in the discretization of the graphics area and inaccurate calculation of the number of holes inside the connected area.

3.4.4 Classification Results

A multi-classification load identification model is constructed by using different VI image features as input to train the SVM model. The testing set classification result is obtained, and four multi-classification evaluation indices are calculated. The final evaluation indices are calculated as shown in the following tables (Tables 3.6, 3.7 and 3.8):

By analyzing the results of the evaluation indices, the following conclusions can be drawn:

(a) By analyzing three different VI image features, it can be seen that the SVM load identification model constructed by simply binarizing or graying the image has lower accuracy. In contrast, binary VI image features have the worst classification performance, and the classification accuracy of grayscale VI image features is higher. By extracting 5-dimensional graphical features, the classification accuracy of binary VI image has a qualitative improvement. When $R = 50 \times 50$, the classification accuracy reached 100%, indicating that all samples are completely classified correctly. The above phenomenon is mainly due to the fact that the simple VI image features are converted into feature vectors and the

Table 3.6 The classification performance indices of the SVM model by binary VI image features with different resolutions

Model	Resolution	Accuracy	Marco_R	Marco_P	Marco_F1	Time (s)
SVM	16 × 16	0.5733	0.5733	0.5313	0.5515	319.96
	28 × 28	0.4267	0.4267	NaN	NaN	374.67
	36 × 36	0.3667	0.3667	NaN	NaN	414.14
	50 × 50	0.2633	0.2633	NaN	NaN	515.74
	72 × 72	0.4467	0.4467	NaN	NaN	724.05

Table 3.7 The classification performance indices of the SVM model by grayscale VI image features with different resolutions

Model	Resolution	Accuracy	Marco_R	Marco_P	Marco_F1	Time (s)
SVM	16 × 16	0.8367	0.8367	0.8183	0.8274	322.89
	28 × 28	0.6467	0.6467	0.6241	0.6352	379.43
	36 × 36	0.5733	0.5733	0.5350	0.5535	418.53
	50 × 50	0.5133	0.5133	NaN	NaN	504.70
	72 × 72	0.4300	0.4300	NaN	NaN	710.33

Table 3.8 The classification performance indices of the SVM model by graphical features of binary VI image with different resolutions

Model	Resolution	Accuracy	Marco_R	Marco_P	Marco_F1	Time (s)
SVM	16 × 16	0.9833	0.9833	0.9833	0.9833	324.28
	28 × 28	0.9967	0.9833	0.9833	0.9833	373.30
	36 × 36	0.9733	0.9833	0.9833	0.9833	409.75
	50 × 50	1.0000	1.0000	1.0000	1.0000	489.55
	72 × 72	0.9933	0.9833	0.9833	0.9833	669.19

feature dimensions are too high. Each feature lacks sufficient robustness and is susceptible to outliers and pattern offsets. At the same time, too high a feature dimension can impose a heavy burden on a simple SVM model. The higher the feature dimension, the easier the model is to overfit, which greatly affects the classification accuracy of the classifier. Compared to binary VI image features, grayscale VI image features save the third dimension of information, so it has a stronger representation for higher VI image graphics. After extracting the graphical features of the binary VI image, the robustness of the features is greatly improved, each feature has stronger interpretability, and the feature dimension is greatly reduced. The extremely high classification accuracy indicates that the selected graphical features are very effective in expressing graphical features and have a high degree of discrimination.

(b) As the resolution increases, the fineness of the VI image increases. But the classification accuracy of the two VI image features decreases. This is mainly because the feature dimension increases exponentially as the resolution increases, which greatly affect the classification performance of the SVM model. At the same time, as can be seen from the conclusion in Sect. 3.4.2, the too low density of data points in the grid causes certain distortion of the grayscale VI image. Too high resolution also leads to the thinning of the connected domain of the image. When the proportion of “1” in the binary VI image is too small, the feature is more susceptible to pattern offset and outliers, thus affecting the robustness of the feature vector. For the graphic features of the binary VI image, the improvement of the resolution does not affect the feature dimension, and the calculation result of some features is more clear, which leads to the improvement of the classification accuracy. However, combined with the conclusions in Sect. 3.4.3, high-resolution results in the calculation costs of some features, so the classification accuracy when $R = 72 \times 72$ is lower than when $R = 50 \times 50$.

In order to analyze the classification results of the load recognition model under different features, the confusion matrix of the SVM model load identification results for binary VI image features and grayscale VI image features with $R = 16 \times 16$ are shown in Fig. 3.12. The classification performance indices are shown in Tables 3.9 and 3.10.

By analyzing the above figure and tables, it can be seen that the classification performance of the two types of VI image features is poor. As for the classification result

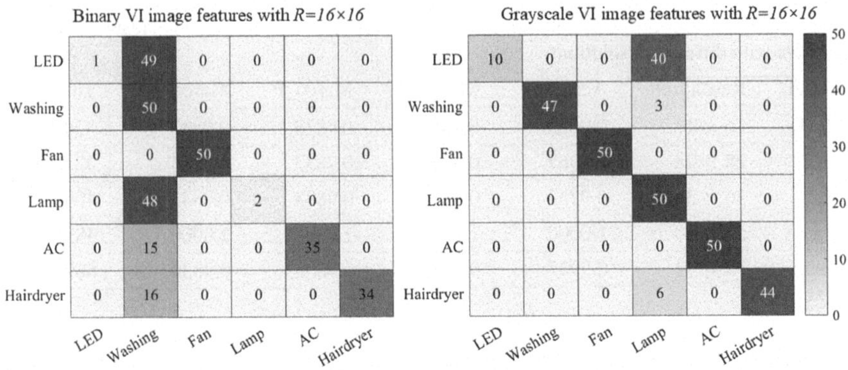

Fig. 3.12 Confusion matrix of the SVM model load identification results for binary VI and grayscale image features with $R = 16 \times 16$

Table 3.9 The classification performance indices of the SVM model for Binary VI image features with $R = 16 \times 16$

Model	Type of appliance	Accuracy	Recall	Precision	F1-score
SVM	LED	0.8367	0.0200	1.0000	0.0392
	Washing	0.5733	1.0000	0.2809	0.4386
	Fan	1.0000	1.0000	1.0000	1.0000
	Lamp	0.8400	0.0400	1.0000	0.0769
	AC	0.9500	0.7000	1.0000	0.8235
	Hairdryer	0.9467	0.6800	1.0000	0.8095
		Accuracy	Marco_R	Marco_P	Marco_F1
	Total	0.5733	0.5733	0.5313	0.5515

Table 3.10 The classification performance indices of the SVM model for grayscale VI image features with R=16 × 16

Model	Type of appliance	Accuracy	Recall	Precision	F1-score
SVM	LED	0.8667	0.2000	1.0000	0.3333
	Washing	0.9900	0.9400	1.0000	0.9691
	Fan	1.0000	1.0000	1.0000	1.0000
	Lamp	0.8367	1.0000	0.5051	0.6711
	AC	1.0000	1.0000	1.0000	1.0000
	Hairdryer	0.9800	0.8800	1.0000	0.9362
		Accuracy	Marco_R	Marco_P	Marco_F1
	Total	0.8367	0.8367	0.8183	0.8274

of binary VI image features, a large number of other samples are misclassified into the washing machine. Among them, the LED light and compact fluorescent lamp are the most serious and are basically not identified correctly. The classification results of binary VI image features are well, but there are also some misclassified samples. Among them, the vast majority of LED light is misclassified into the compact fluorescent lamp. It can be seen that the simple two types of VI image features are not suitable for directly as the features of the SVM load recognition model. Among the six types of electrical equipment, VI image features have the lowest identification ability for LED light and compact fluorescent lamps.

3.5 Features Based on Adaptive Methods

3.5.1 The Theoretical Basis of Adaptive Features

The adaptive method is worthy of further feature extraction of VI image features using transfer learning (Pan and Yang 2010). The core idea of transfer learning is using the pre-training model to retrain or extract features from the problem to be solved. The dataset of the pre-training network can be different from the training set of the problem to be solved (Ravishankar et al. 2017). For example, two training sets have inputs with the same dimension and different output types. One of the ideas of transfer learning is that the researcher needs to find a layer in the pre-training model as the feature, and then use the output of the layer as the input of the problem to be solved. Overall, the idea uses a pre-trained neural network as a feature extractor to extract features. Transfer learning can effectively solve the problem of insufficient training samples. It can help to alleviate the over-fitting.

The pre-training model used in the chapter is the LSTM deep learning network, which has 2 LSTM layers. The dataset of the pre-training network is the well-known MNIST dataset of handwritten digits (Lécun et al. 1998). The dataset is usually used to identify handwritten numbers 0–9. The dataset has a training sample size of 60,000, a test sample size of 10,000, and a total category of 10. Each MNIST handwritten digit sample is represented as a 28×28 grayscale value matrix. This is consistent with the expression of grayscale VI image features.

The LSTM model is trained using the MNIST training set to obtain the pre-training model M_p. Testing is performed using the MNIST testing set, and the testing accuracy is 0.9869. The accuracy of the classification is very high, so it meets the important premise of transfer learning. The constructed pre-training network is used as the feature extractor of VI image features. This layer is a fully connected layer with 10 neurons, defined as the L_f layer. Therefore, the feature extractor extracts a feature dimension of 10. The feature extraction process is to convert the binary VI image features or grayscale VI image features of each sample into a 28×28 input matrix for the pre-training model to obtain 10 output results of the L_f layer. Thereby a 10-dimensional adaptive feature matrix is obtained.

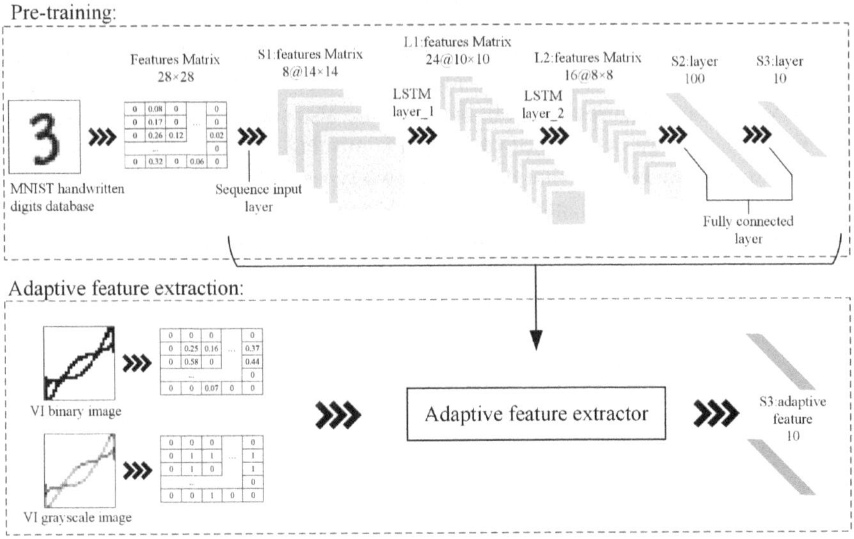

Fig. 3.13 The process of adaptive features extraction

3.5.2 Feature Extraction

The process of adaptive feature extraction and the parameter settings of the LSTM model are shown as below (Fig. 3.13).

3.5.3 Classification Results

The adaptive features extracted using transfer learning lack specific mathematical definitions, so detailed feature evaluations are not performed here. When R = 28 × 28, the adaptive features extracted by binary VI image and grayscale VI image are defined as Ad_F_b and Ad_F_g, respectively. The extracted adaptive feature is taken as input to obtain an SVM load identification model. In order to analyze the classification results of the load identification model, the confusion matrix of the SVM model load identification results for Ad_F_b and Ad_F_g are shown in Fig. 3.14. The classification performance indices are shown in Tables 3.11 and 3.12.

By analyzing the above figure and tables, it can be concluded that the adaptive features provide higher classification accuracy. Whether the binary VI image or the grayscale VI image is extracted adaptively, the classification accuracy of the final SVM model exceeds 90%. At the same time, the pre-trained LSTM model has a high classification accuracy. The above phenomenon indicates that the adaptive feature extractor can effectively mine the implicit information of the original VI feature matrix. Since the grayscale VI image has higher dimensional information and is

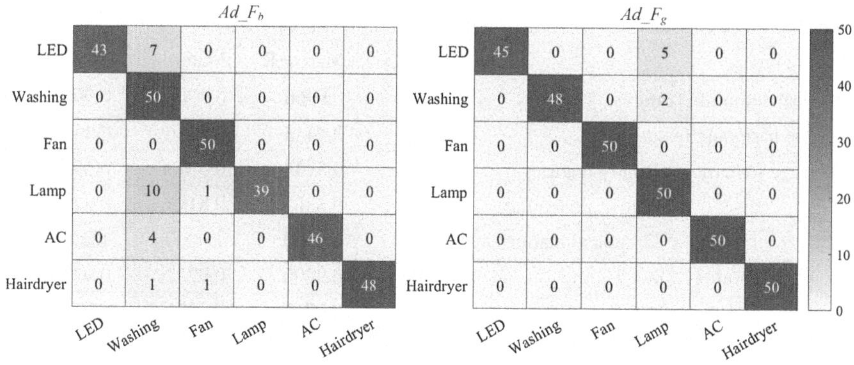

Fig. 3.14 Confusion matrix of the SVM model load identification results for adaptive features

Table 3.11 The classification performance indices of the SVM model for Ad_F_b

Model	Type of appliance	Accuracy	Recall	Precision	F1-score
SVM	LED	0.9767	0.8600	1.0000	0.9247
	Washing	0.9267	1.0000	0.6944	0.8197
	Fan	0.9933	1.0000	0.9615	0.9804
	Lamp	0.9633	0.7800	1.0000	0.8764
	AC	0.9867	0.9200	1.0000	0.9583
	Hairdryer	0.9933	0.9600	1.0000	0.9796
		Accuracy	Marco_R	Marco_P	Marco_F1
	Total	0.9200	0.9200	0.9232	0.9216

Table 3.12 The classification performance indices of the SVM model for Ad_F_g

Model	Type of appliance	Accuracy	Recall	Precision	F1-score
SVM	LED	0.9833	0.9000	1.0000	0.9474
	Washing	0.9933	0.9600	1.0000	0.9796
	Fan	1.0000	1.0000	1.0000	1.0000
	Lamp	0.9767	1.0000	0.8772	0.9346
	AC	1.0000	1.0000	1.0000	1.0000
	Hairdryer	1.0000	1.0000	1.0000	1.0000
		Accuracy	Marco_R	Marco_P	Marco_F1
	Total	0.9767	0.9767	0.9769	0.9768

more similar to the MNIST dataset, the classification accuracy is still higher than the binary VI image.

Table 3.13 The classification performance indices of different feature types

Feature types		Accuracy	Marco_R	Marco_P	Marco_F1
Physical definition features		0.9867	0.9900	0.9900	0.9900
Current harmonic features		0.9433	0.9433	0.9416	0.9425
VI image features	Binary features	0.5733	0.5733	0.5313	0.5515
	Grayscale feature	0.8367	0.8367	0.8183	0.8274
	Graphical features	1.0000	1.0000	1.0000	1.0000
Adaptive features	Ad_F_b	0.9200	0.9200	0.9232	0.9216
	Ad_F_g	0.9767	0.9767	0.9769	0.9768

3.6 Experimental Analysis

3.6.1 Comparative Analysis of Classification Performance

The best classification accuracy for each feature type is shown in the following table (Table 3.13):

By analyzing the table, it can be concluded that the graphical features of the VI image have the highest classification accuracy among all types of features introduced in the chapter. The load identification accuracy of it is 100%. Physical definition features, as well as Ad_F_g features in adaptive features, both have a excellent classification performance with over 97% accuracy. The binary features of VI images and grayscale features of VI images have lower classification accuracy. This proves that the simple VI image features are not practical for the SVM load identification model and need further processing.

3.6.2 Conclusion

The chapter focuses on the extraction of various appliance features, evaluates the extracted features, and analyzes the classification performance of each type of feature in combination with actual data. The chapter draws the following main conclusions:

(a) For most physical definition features, the linear correlation between the eigenvector and the label vector is low, while the intra-class tightness and inter-class discrimination of the feature distribution are high. Among them, for the same electrical appliances, the values of reactive power and apparent power are stable. For different electrical appliances, the values show significant specificity. The power factor λ and impedance Z not only have weak linear correlation but also are not sufficiently specific for different electrical appliances. And for different appliances, there is a certain aliasing phenomenon between the two values. Overall, physical definition features show excellent classification accuracy. The overall feature vector composed of physical definition features is synergistically analyzed from the power, resistance, capacitance, inductance and other aspects of the electrical appliance, which makes up for the deficiency of the single physical definition feature and can accurately depict the features of the specific electrical appliance.

(b) On the one hand, there is the greatest linear correlation between the 4th harmonic real part feature vector and the label vector. When the harmonic order is greater than 4, the linear correlation of harmonics shows a decreasing trend. The high-order harmonic real part features basically exhibit a low linear correlation. The correlation between all harmonic imaginary part feature vectors and the label vectors is low. On the other hand, the degree of discrimination in the real part of the harmonics shows a significant downward trend with the increase of the harmonic order. The features of the imaginary part of the harmonic show a higher degree of discrimination among the different electrical appliances than the real part. However, the degree of discrimination of the high-order harmonic imaginary features is also not obvious. The harmonic real part contains more effective information than the harmonic imaginary part with lower intraclass distance and higher interclass discrimination. The odd harmonics show better classification performance than the even harmonics.

(c) The fineness of the VI image increases as the resolution increases. The correlation and discrimination of various VI image features increase at the same time. And as the grid width decreases, the data point density of each grid decreases sharply, the graphics area is discretized, and some features are not accurately calculated, resulting in a decrease in classification accuracy when a certain resolution is reached. The dimension of the simple VI image features is too high, and each feature lacks sufficient robustness. So it is easily affected by outliers and pattern offsets, and the classification accuracy is poor. By extracting the 5-dimensional graphical features, the robustness of the features is greatly improved, the dimensions are greatly reduced, and the classification accuracy has a great improvement.

(d) The adaptive features extracted by transfer learning have higher classification accuracy. In particular, the grayscale VI image is used as the Ad_F_g feature, and the classification accuracy reaches a high level. The adaptive feature extractor can effectively mine the implicit information of the original VI feature matrix and improve the classification accuracy.

References

Baranski M, Voss J (2005) Genetic algorithm for pattern detection in NIALM systems. In: IEEE international conference on systems

Bouhouras AS, Chatzisavvas KC, Panagiotou E, Poulakis N, Parisses C, Christoforidis GC (2017) Load signatures development via harmonic current vectors. In: International Universities power engineering conference

Chui KT, Hung FH, Li BYS, Tsang KF, Chung SH (2014) Appliance signature: multi-modes electric appliances. In: IEEE international conference on consumer electronics-China
Doorn JV, Ly A, Marsman M, Wagenmakers EJ (2018) Bayesian inference for Kendall's rank correlation coefficient. Am Stat
Duarte C, Delmar P, Goossen KW, Barner K, Gomez-Luna E (2012) Non-intrusive load monitoring based on switching voltage transients and wavelet transforms. In: Future of instrumentation international workshop
Figueiredo MB, Almeida AD, Ribeiro B (2011) An experimental study on electrical signature identification of non-intrusive load monitoring (NILM) systems
Hassan T, Javed F, Arshad N (2014) An empirical investigation of V-I trajectory based load signatures for non-intrusive load monitoring. IEEE Trans Smart Grid 5(2):870–878
He D, Liang D, Yi Y, Harley RG, Habetler TG (2012) Front-End electronic circuit topology analysis for model-driven classification and monitoring of appliance loads in smart buildings. IEEE Trans Smart Grid 3(4):2286–2293
Hosseini SS, Agbossou K, Kelouwani S, Cardenas A (2017) Non-intrusive load monitoring through home energy management systems: a comprehensive review. Renew Sustain Energy Rev 79:1266–1274
Jian L, Ng SKK, Kendall G, Cheng JWM (2010) Load signature study—part I: basic concept, structure, and methodology. IEEE Trans Power Deliv 25(2):551–560
Jordan HE (1994) Power factor. In: Energy-efficient electric motors and their applications. Springer, Boston, MA
Kahl M, Haq AU, Kriechbaumer T, Jacobsen HA (2016) WHITED—a worldwide household and industry transient energy data set. In: International workshop on non-intrusive load monitoring, pp 1–5
Kira K, Rendell LA (1992) A practical approach to feature selection. In: International workshop on machine learning
Kononenko I (1994) Estimating attributes: analysis and extensions of RELIEF. In: European conference on machine learning on machine learning
Kontos E, Tsolaridis G, Teodorescu R, Bauer P (2017) High order voltage and current harmonic mitigation using the modular multilevel converter STATCOM. IEEE Access (99):1–1
Lam HY, Fung GSK, Lee WK (2007) A novel method to construct taxonomy electrical appliances based on load signaturesof. IEEE Trans Consum Electron 53(2):653–660. https://doi.org/10.1109/TCE.2007.381742
Lécun Y, Bottou L, Bengio Y, Haffner P (1998) Gradient-based learning applied to document recognition. Proc IEEE 86(11):2278–2324
Manurung RC, Perdana D, Munadi R (2017) Performance evaluation Gauss-Markov mobility model in vehicular ad-hoc network with spearman correlation coefficient. In: International seminar on intelligent technology & its applications
Maravi ME, Snyder LE, Mcewen LD, Deyoung K, Davidson AJ (2017) Using spatial analysis to inform community immunization strategies. Biomed Inform Insights 9:117822261770062-
Pan SJ, Yang Q (2010) A survey on transfer learning. IEEE Trans Knowl Data Eng 22(10):1345–1359
Ravishankar H, Sudhakar P, Venkataramani R, Thiruvenkadam S, Annangi P, Babu N, Vaidya V (2017) Understanding the mechanisms of deep transfer learning for medical images. Deep learning and data labeling for medical applications. Springer, Cham, pp 188–196
Spolaor N, Cherman EA, Monard MC, Lee HD (2014) ReliefF for multi-label feature selection. In: Proceedings of intelligent systems
Tang N, Jian-Xia SU, Wang YP, Liu Y (2017) The application of FFT algorithm in harmonic analysis of power system. J Beijing Inst Cloth Technol 37(4):83–88
Tsai MS, Lin YH (2012) Development of a non-intrusive monitoring technique for appliance' identification in electricity energy management. In: International conference on advanced power system automation & protection

Chapter 4
Appliance Identification Based on Template Matching

4.1 Introduction

4.1.1 Background

Smart grid technology has been greatly developed under the support of smart meter monitoring systems. One important reason is that smart meter monitoring systems can estimate the power consumption of each piece of equipment in a residential building. Thereby it can help consumers optimize and manage household energy consumption. Electricity information being available to the public can encourage people to save energy and achieve energy conservation. The Non-intrusive Load Monitoring (NILM) method is proposed as a more efficient way to estimate device power consumption and it is also applied in many aspects. There has been a lot of discussion in this field since Professor Hart first proposed non-intrusive load identification technology (Hart 1992).

The load-aware algorithm based on pattern recognition includes several main steps such as event detection, feature extraction, and pattern classification. The pattern recognition based load recognition depends on the load signatures acquired by feature extraction. The target of identifying the load can be achieved by learning the load features of each electrical device. The pattern recognition method can be divided into two categories: supervised learning and unsupervised learning. Commonly used pattern recognition algorithms include K-Nearest Neighbor (KNN) algorithm (Kang and Ji 2017), Artificial Neural Network (ANN) (Biansoongnern and Plangklang 2016), Support Vector Machine (SVM) (Singh et al. 2015), Ada-boost algorithm (Wang 2012) and so on.

Pattern recognition is a fundamental branch of the machine learning field and its foothold is perception. Template matching is a typical method of pattern recognition. Template matching is achieved by modeling the existing features. In the field of NILM, the template matching is to model the corresponding characteristics of the raw load data collected by the meter. When the unknown-state load is input to

H. Liu, *Non-intrusive Load Monitoring*,
https://doi.org/10.1007/978-981-15-1860-7_4

the identification system, the detected object is compared with the established template in the template library to check the similarity among them. Whether the target object is a template type appliance is determined by the maximum similarity or a certain threshold. In order to identify an electrical appliance, it is necessary to have a memory mode (also called a template) of this appliance in the past experience in the template matching method. When the current event is matched with the template in the electrical template library, the appliance is recognized.

The chapter uses the template matching method to carry out non-intrusive load identification for five household appliances. Feature extraction and preprocessing are carried out on the original voltage and current data of the PLAID dataset. Finally, more than 20 features are obtained. Two common methods, the decision tree, and the KNN are used to identify electrical appliances. In addition, the active power time series of the PLAID dataset is extracted. And the DTW template library is established to perform template matching of the power sequence. Finally, the evaluation indices such as Accuracy, F1-Score, Precision, and Recall are established to judge the performance of different algorithms in NILM.

4.1.2 Data Preprocessing of the PLAID Dataset

(1) *Basic feature extraction*

PLAID is a set of a public dataset for high-resolution electrical measurements for load identification studies. The PLAID dataset contains measurements for more than 200 different device instances, representing 11 device classes, for a total of more than one thousand records. In the chapter, five electrical equipment are selected for classification model identification.

In the typical framework of NILM and decomposition, load decomposition usually includes the following five parts: data collection, data preprocessing, event detection, feature extraction, and load identification. The emphasis of the chapter is the analysis of the classifier performance, so only the load identification part of the classification is researched. Before the experiment, it is necessary to conduct data mining on the original *U* and *I* data of the PLAID dataset, and obtain a total of 22 groups of feature vectors through the following Table 4.1.

The measured data samples in the paper include 862 sets of samples of five equipment running simultaneously, including 248 groups of the Hairdryer, 229 groups of the Microwave, 210 groups of the Fan, 90 groups of the Fridge and 85 groups of Heater.

The features extraction of electrical signals for non-intrusive load monitoring is generally divided into two categories. One is steady-state feature extraction and the other is transient feature extraction. The specific extraction classification is shown in Fig. 4.1.

Table 4.1 The variable representation of load feature vectors

Load feature vectors	Variable
Original current	I
Original voltage	U
Active power	P
Reactive power	Q
Apparent power	S
Power factor	λ
Resistance	Z
Impedance	R
Reactance	X
Admittance	Y
Conductance	G
Electrical susceptance	B
The imaginary and real parts of the odd harmonics of the current 1–9 times	H_i

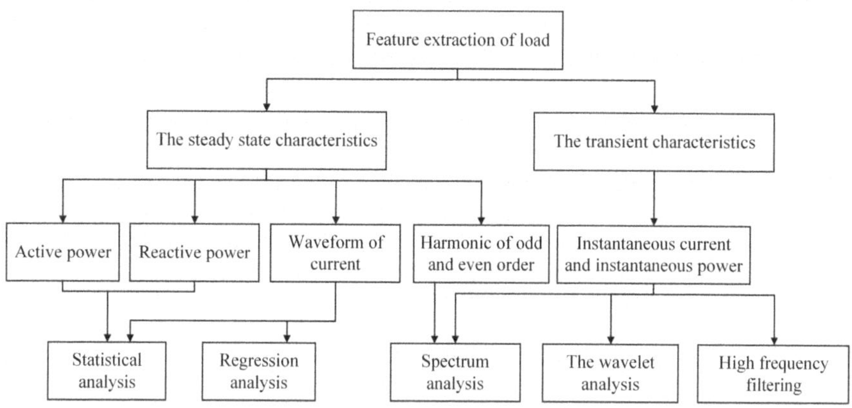

Fig. 4.1 Load characteristic types and analysis methods for non-invasive load monitoring and decomposition

(2) *Active power extraction based on time series*

Labeling the five electrical appliances in the PLAID dataset. The first 40,000 sampling points of the original current and voltage are used, and the sampling interval is 500 to obtain 80 sampling data. Finally, a total of 400 sampling points were obtained for five groups of appliances. The numerical power product (rectangle method) is used to obtain the active power value, and five sets of graphs (active power amplitude—sampling point) are drawn as follows (Fig. 4.2).

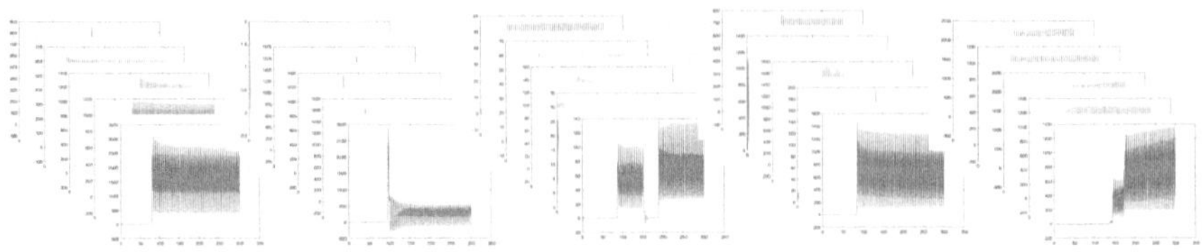

Fig. 4.2 Active power extraction of common household appliances based on time series

4.2 Appliance Identification Based on Decision Tree

4.2.1 The Theoretical Basis of Decision Tree

The algorithm for decision tree classification was first proposed by Quilan in 1986, called ID3 (Han and Kamber 2005). In the subsequent evolution, the C4.5 algorithm was generated because the ID3 algorithm could not deal with discrete attributes and is prone to over-fitting (Ruggieri 2002). The selection of the test attribute of the C4.5 algorithm is different from the ID3 algorithm, which selects the highest information gain rate as the test attribute. The C4.5 algorithm is defined as follows.

A set of data is used to form a training set, and its form can be expressed as $(V_1, V_2, \ldots, V_a, C)$, where V_i represents the attribute value of the sample and C represents the category of the sample.

Supposing the sample set T is divided into $T_1, T_2, \ldots, T_s$ subsets in turn according to s different values of the discrete attribute A. Then the information gain rate is defined as $ratio(A, T) = gain(A, T)/spilt(A, T)$ obtained by dividing the sample set T by the discrete attribute A. The *gain(A, T)* can be obtained as follows:

$$gain(A, T) = \inf(T) - \sum_{i=1}^{s} \frac{|\mathrm{T}_i|}{|T|} \times \inf(T_i) \tag{4.1}$$

where $\inf(T)$ represents the information entropy of T. The *spilt(A, T)* can be obtained as follows:

$$spilt(A, T) = -\sum_{i=1}^{s} \frac{|T_i|}{|T|} \times \log_2\left(\frac{|T_i|}{|T|}\right) \tag{4.2}$$

The structure of a decision tree can be imagined as a tree-form of the decision process. It uses the characteristics of the tree structure to classify the data, where a leaf node represents a record set under a certain constraint. Establishing a decision tree is to establish a branch of the tree based on the different values of the record fields. Then, lower nodes and new branches are repeatedly in each branch subset to form a decision tree. One of the key points in establishing a decision tree is the selection of

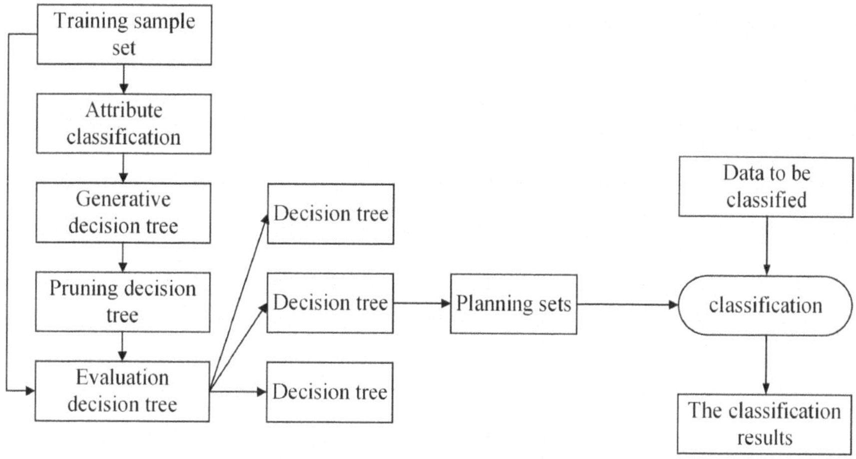

Fig. 4.3 The classification process of decision trees

different values for record fields. The quality of selection affects the growth speed and structure of the decision tree. The classification performance of the decision tree is also affected. An unsatisfactory branch value not only slows down the growth rate of the decision tree but also makes the resulting branch too thin, which may lead to the result some useful information rules are ignored. The classification process of decision trees is given as follows (Fig. 4.3).

4.2.2 *Steps of Modeling*

(a) Pre-processing the PLAID dataset. More than 20 feature quantities such as active power, reactive power, harmonics, and so on, were extracted from the original dataset;
(b) Dividing the processed sample set into the training set and testing set. The input data of the training set and testing set are normalized and equalized;
(c) Classifying the attributes of the training set to obtain the decision tree. After undergoing pruning and evaluating the decision tree, the parameters are initialized to obtain a trained decision tree model;
(d) Inputting the divided testing set samples into the trained decision tree model for electrical equipment classification. Then the prediction sample sequence relative to the real classification sequence is obtained;
(e) Calculating the corresponding evaluation indices for the output sequence in *step (d)*. The Accuracy, Recall, F1-Score, and Precision are employed to evaluate the performance of the classifier.

4.2.3 Classification Results

The classification principles of the decision tree are based on the binary classification of each feature. By constructing a decision tree, the probability that the expected value is equal to or over zero is obtained. Then the risk of appliance classification is evaluated to judge the feasibility of the decision tree decision. It can be seen from Fig. 4.4 and Table 4.2 that the decision tree shows excellent performance in the classification of electrical appliances.

From the confusion matrix, it can be seen that the identification number of the five common household appliances focus on the main diagonal of the matrix. The number of false detection cases is very low and under 5.

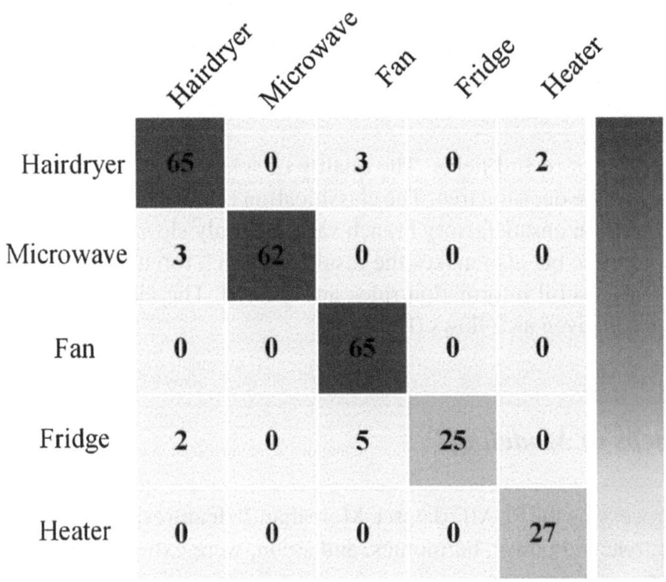

Fig. 4.4 Confusion matrix of the decision tree model load identification results

Table 4.2 The classification performance indices of the decision tree model

Model	Type of appliance	Accuracy	Recall	F1-score	Precision
Decision tree	Hairdryer	0.9614	0.9286	0.9286	0.9286
	Microwave	0.9884	0.9538	0.9764	1.0000
	Fan	0.9691	1.0000	0.9420	0.8904
	Fridge	0.9730	0.7813	0.8772	1.0000
	Heater	0.9923	1.0000	0.9643	0.9310
	Average	0.9768	0.9327	0.9377	0.9500

It can be seen from the above tables that the performance of the decision tree classification method is satisfying and stable in the identification of the five electrical appliances. The value of the F1-score also reaches 0.9377. Generally speaking, the decision tree is suitable for NILM and can achieve a better recognition effect.

However, decision tree classification is easy to over-fit. Its performance should be further discussed, which is presented in Sect. 4.5.

4.3 Appliance Identification Based on KNN Algorithm

4.3.1 The Theoretical Basis of KNN

The KNN method proposed by Cover and Hart in 1967 (Cover and Hart 1967) is an efficient algorithm for nonparametric classification. The nearest neighbor method is a passive learning method because it simply stores the known training data (Chen et al. 2013a). The basic idea of the KNN algorithm is described as follows.

In the light of traditional vector space model, the electrical information in the original PLAID dataset is formalized into weighted feature vectors, expressed as $D = D(T_1, W_1; T_2, W_2; \ldots; T_n, W_n)$. When a testing set is obtained, the model calculates its similarity to each template in the training set to find the K most similar appliances. Then the category of electrical appliances is determined according to the weighted distance.

The steps of equipment classification are shown as follows.

(a) The test appliance vector is formed based on its characteristics.
(b) Calculating the similarity between the test appliance sample and each appliance sample in the training set, and the calculation formula is given as follows:

$$Sim(d_i, d_j) = \frac{\sum_{k=1}^{M} W_{ik} \times W_{jk}}{\sqrt{\sum_{k=1}^{M} W_{ik}^2 \sqrt{\sum_{k=1}^{M} W_{jk}^2}}} \tag{4.3}$$

where d_i is the characteristic vector of the test appliance sample, d_j represents the center vector of the class j appliance, the number of dimensions of the feature vector is expressed as M, the k-th dimension of the vector is represented as W_k.

(c) According to the similarity of the electrical appliances, k electrical appliance samples that most similar to the trained electrical appliance samples are selected in the training electrical appliance.
(d) Calculating the weight of each class in the k neighbors of the test electrical sample, and the calculation formula is given as follows:

$$P(X, C_j) = \begin{cases} 1 \; if \sum\limits_{d_i \in KNN} Sim(x, d_i) y(d_i, C_j) - b \geq 0 \\ 0 \; else \end{cases} \tag{4.4}$$

(e) Comprehensively comparing the weights of various types of electrical appliances and classifying the electrical appliances of the current test samples into the category with the highest weight.

The KNN algorithm is based on analog learning and is a nonparametric classification technique. For the case of unknown state and non-normal distribution, the KNN algorithm has excellent performance in classification accuracy and robustness. But the drawbacks of the KNN algorithm are also obvious. Since the KNN algorithm is a lazy classification algorithm, the space-time overhead of it is large. In addition, when calculating similarity, feature-vectors with high dimensions have poor feature-relevance. In the calculation of sample distance, the distance between vectors is rough due to the same weight of each dimension, and the classification accuracy is not ideal.

4.3.2 Steps of Modeling

(a) Pre-processing the PLAID common dataset. More than 20 feature quantities such as active power, reactive power, harmonics, etc., were extracted from the original dataset;
(b) Dividing the processed sample set into the training set and testing set. The input data of the training set and testing set are normalized and equalized;
(c) Calculating the similarity between each electrical test sample and the template sample in the training set. According to the similarity of the electrical samples, the k samples closest to the test samples are selected in the training set, and the weights of each type of electrical appliances are calculated in turn;
(d) Comparing the weights of various types of electric appliances. Testing samples are assigned to the category of appliances with the highest weights. Thereby a predicted sample sequence is obtained;
(e) Calculating the corresponding evaluation indices for the output sequence *in step (d)*. The Accuracy, Recall, F1-Score, and Precision are employed to evaluate the performance of the classifier.

4.3.3 Classification Results

The KNN algorithm is a simple and effective non-parametric classification method. It can be seen from Fig. 4.5 and Table 4.3 that its practical effect is relatively ideal when the method is applied in the NILM.

From the results of the confusion matrix diagram, the identification numbers of the Hairdryer, Microwave, and Fan are mostly distributed on the main diagonal of the matrix. This means that the value of true positive is higher and the KNN algorithm has a high recognition accuracy for the above three appliances. However, the false detection rate of the Fridge and Heater is also high. This may be caused by the improper selection of K values, or because the samples are insufficient, making the

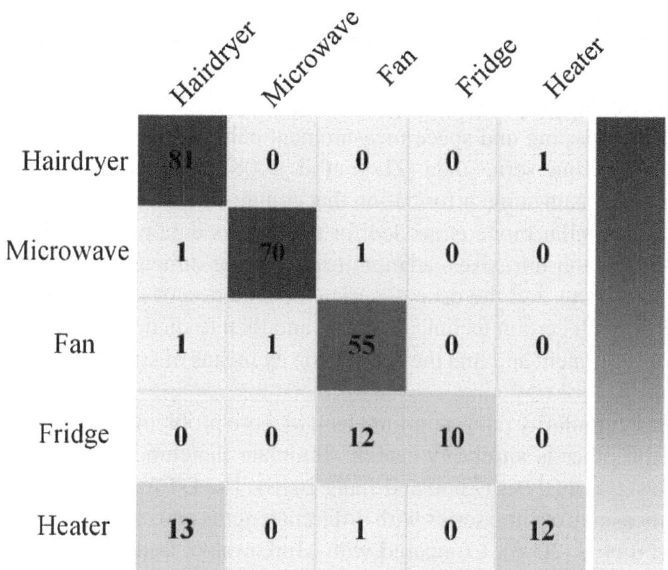

Fig. 4.5 Confusion matrix of the KNN model load identification results

Table 4.3 The classification performance indices of the KNN model

Model	Type of appliance	Accuracy	Recall	F1-score	Precision
KNN	Hairdryer	0.9382	0.9878	0.9101	0.8438
	Microwave	0.9884	0.9722	0.9790	0.9859
	Fan	0.9382	0.9649	0.8730	0.7971
	Fridge	0.9537	0.4545	0.6250	1.0000
	Heater	0.9421	0.4615	0.6154	0.9231
	Average	0.9521	0.7682	0.8005	0.9100

template libraries of these two types of electrical appliances incomplete and difficult to cope with unknown conditions.

It can be seen from the above tables that the accuracy and precision are both above 0.9 when using the KNN algorithm to identify electrical appliances. It is shown that the KNN algorithm is very appropriate for NILM. However, the value of recall is not ideal enough, and it shows the limitations of simply using this original algorithm.

In general, the KNN algorithm is suitable for NILM and can achieve better results. It has high accuracy and is not sensitive to abnormal points.

4.4 Appliance Identification Based on DTW Algorithm

4.4.1 The Theoretical Basis of DTW

Dynamic Time Warping (DTW) algorithm is a nonlinear warping technique that combines time warping and space measurement calculations. When facing a large number of multi-time series data (Zhou et al. 2008), it is necessary to dig deeper knowledge and obtain some information that contains the laws of system evolution. At this time, a similar mode is needed for time series data mining. The original U and I obtained in non-intrusive load monitoring are one-dimensional time series data. Data mining aims to find the deep electrical information of each running electrical equipment through certain technical means, and then to distinguish different pieces of electrical equipment and find the connection by means of small details (Chen et al. 2013b).

Time series similarity pattern mining has two core problems, one is time template matching, the other is similarity search. Template matching plays a fundamental role in time series analysis (Zhou and Jiang 2010). The DTW algorithm can support similarity measures of time series with different lengths and can bend and stretch the time axis (Fu et al. 2008). Compared with Minkowski's arithmetic distance, it has a better matching ability, and the DTW algorithm is superior in the performance of robustness (Bhaduri et al. 2010).

In the point-to-match relationship, the distance calculation of the DTW is actually the minimum of the sums of the point-to-base distances. At this time, the corresponding curved path is the best path, and the expression of DTW distance has been given as follows.

$$DTW(X, Y) = \min\left\{\sum_{K=1}^{K} D_{base}|(w_k)\right\} \tag{4.5}$$

To find the best path, it is necessary to construct a matrix $M_{\mathrm{m}\times\mathrm{n}}$. This matrix is called the cumulative distance matrix with m rows and n columns. Each element $\gamma_{i,j}$ in matrix M is defined as follows.

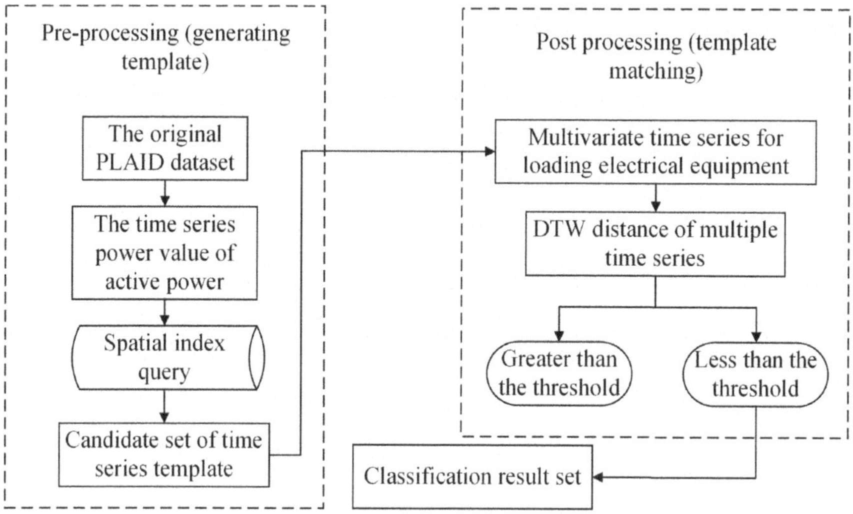

Fig. 4.6 The process of DTW template matching

$$\gamma_{i,j} = D_{base}\left(x_i, y_j\right) + \min\left\{\gamma_{i,j-1}, \gamma_{i-1}, \gamma_{i-1,j-1}\right\} \tag{4.6}$$

In addition, $\gamma_{i,j}$ is the DTW distance between sequence $X\left[1{:}j\right]$ and sequence $Y[1{:}j]$. Dynamic programming can be used to solve the equation $D_{dtw}(X, Y) = \gamma_{m,n}, \gamma_{m,n}$. The process of DTW template matching is given as follows (Fig. 4.6).

4.4.2 Steps of Modeling

(a) Pre-processing the PLAID dataset of load identification research. The active power time series is extracted from the original sequence;
(b) Dividing the processed sample set into the training set and testing set. The input data of the training set and testing set are normalized and equalized;
(c) Performing a template synthesis on the testing set samples, and sequentially generating DTW matching templates for the five types of electrical appliances selected in the chapter;
(d) Inputting the divided testing set samples into the system, calculating the DTW distance between the test samples and the template, and dividing the qualified sequence into a class of electrical properties. Finally, a sequence of predicted samples relative to the real classification sequence is obtained;
(e) Calculating the corresponding evaluation indices for the output sequence in *step (d)*. Accuracy, Recall, F1-Score, and Precision are included to evaluate the performance of the classifier.

4.4.3 Classification Results

The dynamic time warping (DTW) algorithm is a classical optimization problem. In the algorithm, if the time warping function $D(n)$ satisfies certain condition requirements, it is used to describe the time correspondence between the test template and the reference template. Thus, the structural function corresponding to the minimum cumulative distance between the two templates is solved, and the classification result of electrical matching is obtained finally. The performance of the DTW algorithm in the field of non-intrusive load monitoring is shown in Fig. 4.7 and Table 4.4.

It can be seen from the results of the confusion matrix that the DTW algorithm is not ideal for the recognition of the five types of commonly used home appliances selected in the experiment. In particular, the error rate of the Fridge and Heater is

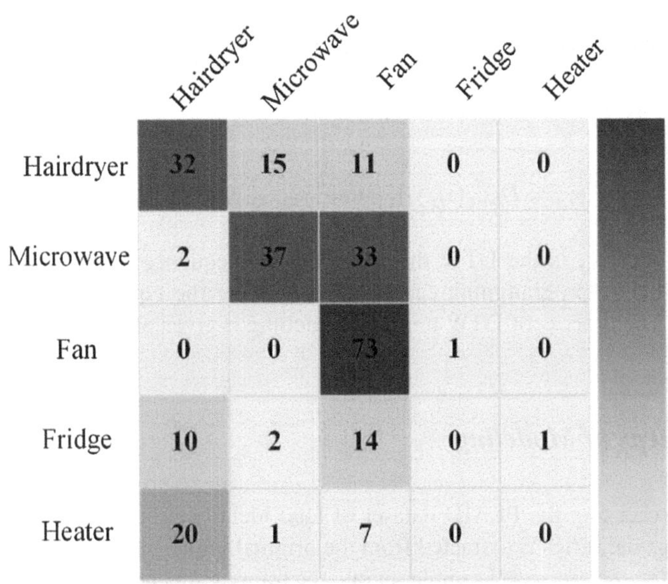

Fig. 4.7 Confusion matrix of the DTW model load identification results

Table 4.4 The classification performance indices of DTW model

Model	Type of appliance	Accuracy	Recall	F1-score	Precision
DTW	Hairdryer	0.7761	0.5517	0.5246	0.5000
	Microwave	0.7954	0.5139	0.5827	0.6727
	Fan	0.7452	0.9865	0.6887	0.5290
	Fridge	0.8919	0.0000	NaN	0.0000
	Heater	0.8880	0.0000	NaN	0.0000
	Average	0.8193	0.4104	0.5987	0.3403

very high, and both of them are mistakenly tested as Hairdryer or Fan. This may be because the algorithm is not very effective, and it is not easy to use the various knowledge of the bottom and top layers in the recognition algorithm. The recognition effect is poor when the model is dealing with a large number of samples. It may also be because the small number of samples of Fridge and Heater makes the generated templates incomplete and hard to cope with many unknown situations.

It can be seen from the results of Table 4.4 that it is difficult for the DTW to balance the accuracy and the recall, resulting in a low value of F1-Score. And NaN value appeared, mainly because TP (i.e., true positive) equals to 0.

4.5 Experiment Analysis

4.5.1 Model Framework

Flow chart of load identification and classification algorithm based on template matching is shown as follows (Fig. 4.8).

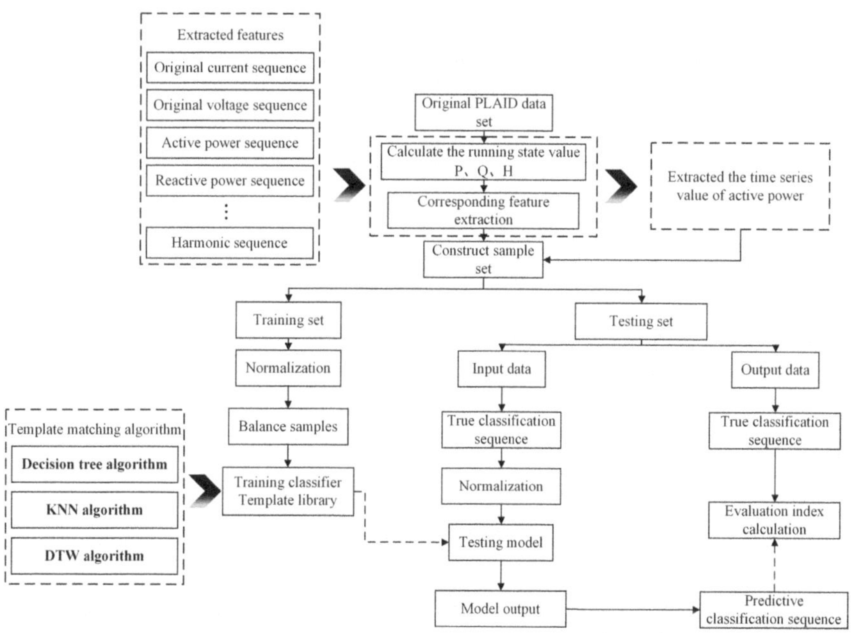

Fig. 4.8 Load identification and classification algorithm based on template matching

4.5.2 Comparative Analysis of Classification Performance

(1) *Optimal classification model for decision trees*

The decision tree classification method is easy to understand, and the classification accuracy is high. The decision tree clearly shows which fields are important and friendly to the user in understanding the rules. In addition, the decision tree algorithm requires little prior knowledge of users and hardly needs any domain knowledge, nor to assume parameters. And it can be used for high dimensional data processing.

However, there are many cases in which decision tree classification methods are difficult to deal with (Olanow and Koller 1998). For example, in the decision tree classification process, it is easy to over-fit and the classification accuracy is low. The decision tree tends to ignore the correlation between the data. For data with inconsistent sample numbers, the information from the decision tree seems to be more biased towards those features with more values. Based on the shortcomings of decision tree performance, some optimization methods can be used to improve the performance of the decision tree (Kung 1994).

(a) Analyze the impact of different leaf node positions and minimum sample numbers on the performance of decision trees.

The minimum sample number of leaf nodes iterate in the range of 10–100. The difference in cross-validation errors is analyzed to find the minimum number of samples for the optimal leaf node. The results are shown in Fig. 4.9. It can be seen from the figure that the optimal min-leaf value is 13. The resampling error and

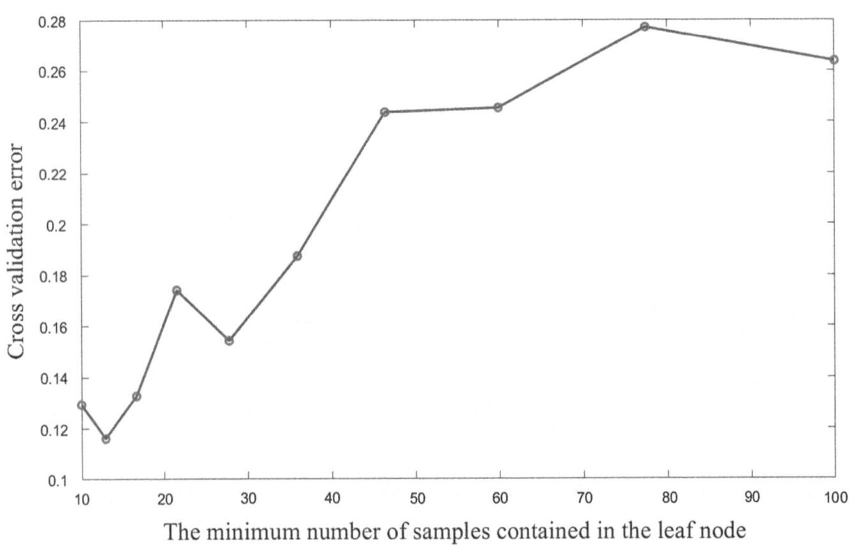

Fig. 4.9 The influence of the minimum sample number of leaf node on the performance of decision tree

Table 4.5 Resampling error and cross-validation error of decision trees in different stages

Parameter index	Decision tree	
Resub-opt	0.0896	Resampling error of optimized decision tree
Loss-opt	0.1095	Cross-validation error of optimized decision tree
Resub-default	0.0133	Resampling error of pre-optimization decision tree
Loss-default	0.0846	Cross-validation error of pre-optimization decision tree
Resub-prune	0.0133	Resampling error of decision tree after pruning
Loss-prune	0.0813	Cross-validation error of decision tree after pruning

cross-validation error after the initial optimization of the decision tree are shown in Table 4.5.

(b) Analysis of the impact of pruning on the performance of decision trees

Pruning is one of the ways to stop the decision tree from branching, and pruning is also divided into pre-pruning and post-pruning. This section uses post-pruning to cross-validate the NILM load classification performance of the decision tree. In the post-pruning, the tree should be fully grown until the leaf nodes have the smallest impurity, which is to overcome the "field limitation". The post-pruning first generates a complete decision tree from the training set, and then evaluates whether the node in the decision tree improves the accuracy. If pruning will increase, pruning will be carried out; if pruning is not improved (including no effect), it can be retained. The resampling error and cross-validation error after pruning are shown in Table 4.5. The performance of the optimized decision tree has been demonstrated in Fig. 4.10 and Table 4.6.

The results show that after changing the min-leaf of the decision tree, the cross-validation error and re-sampling error are higher than the decision tree index before the optimization. In fact, by changing the value of min-leaf, the optimization result of the decision tree does not necessarily show positive. In this experiment, only the minimum sample number of the leaf nodes in a part of the interval has been tried, and there are many unknown cases. In addition, this strategy encounters difficulties when there is noise in the data or the number of training examples is too small to produce representative samples of the objective function.

Improving the performance of decision trees by integrating algorithms is an effective method, and its analysis has been described in the chapter on random forests in Chap. 6. The higher is the classification intensity of a single decision tree, the better is the classification performance of random forests. The greater is the correlation between decision trees, the weaker is the classification ability of random forests.

In general, the decision tree classification method is suitable for load classification in NILM.

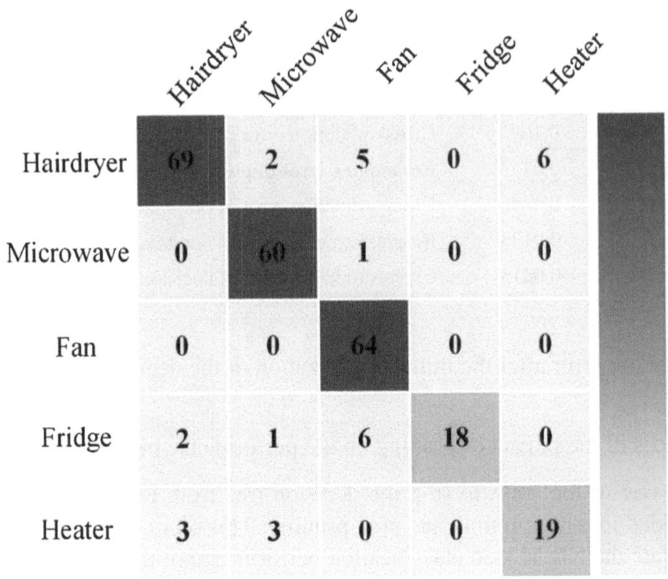

Fig. 4.10 Confusion matrix of the optimization decision tree model load identification results

Table 4.6 The classification performance indices of optimization decision tree model

Model	Type of appliance	Accuracy	Recall	F1-score	Precision
Optimization decision tree	Hairdryer	0.9305	0.8415	0.8846	0.9324
	Microwave	0.9730	0.9836	0.9449	0.9091
	Fan	0.9535	1.0000	0.9143	0.8421
	Fridge	0.9653	0.6667	0.8000	1.0000
	Heater	0.9537	0.7600	0.7600	0.7600
	Average	0.9552	0.8504	0.8608	0.8887

(2) *Optimal classification model for KNN*

The KNN voting process for five common household appliances is given as follows. The KNN algorithm is extended to the non-intrusive load monitoring field and its voting process is shown in Fig. 4.11.

U and I data of five common household appliances Hairdryer, Microwave, Fan, Fridge and heater are extracted from the PLAID data set. The extracted current and voltage data are divided into 70% data as training set and 30% data as testing set. By clustering the electrical equipment information of training samples, it can be seen in the figure that the originally chaotic electrical information has been divided into five categories, which are equivalent to the formation of electrical template library. When a new electrical power information enters the NILM system, it will be compared with various types of electrical templates in the electrical template library. It is to take

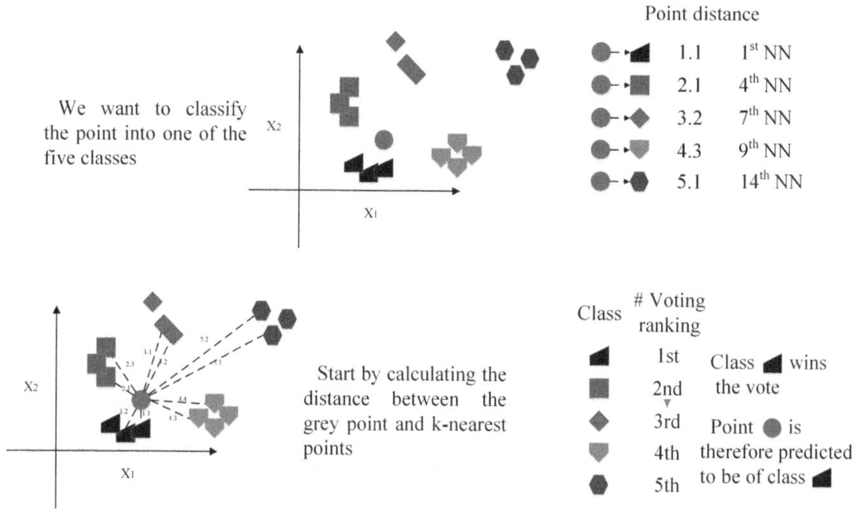

Fig. 4.11 KNN voting process for five common household appliances

the attribute of the electrical appliance A closest to the test sample as the electrical appliance classification data of the new data B. So the new electrical appliance of this class is determined to be class A.

In order to find a relatively satisfied K value, the effect of the value of K value 1 to 50 on the classification results is compared, as shown in Fig. 4.12.

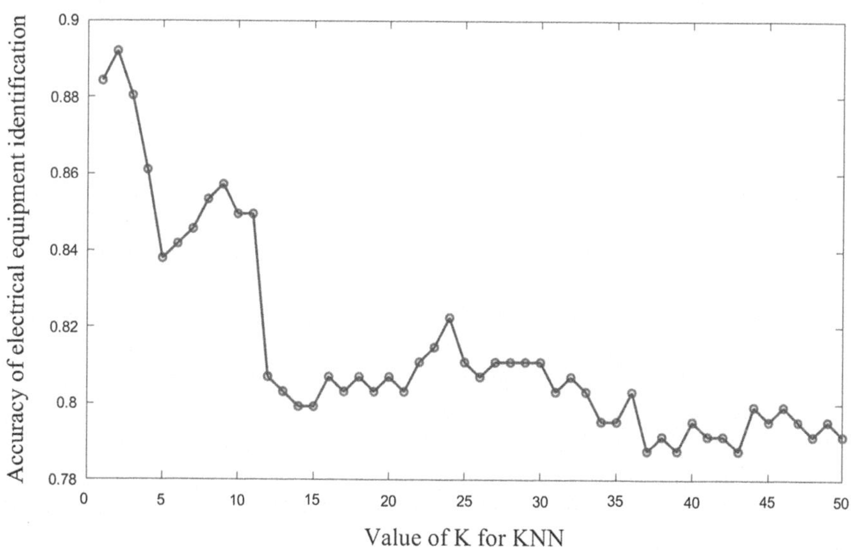

Fig. 4.12 Influence of K value change for KNN algorithm on recognition performance

The basic idea of KNN algorithm is to find k samples in the feature space that are most similar to a new sample (that is, the nearest one in the feature space). If most of the samples from the k samples have the same attributes, the new samples also belong to this attribute. The KNN algorithm is a very classic multi-classification method that is easy to implement and understand with high precision. It can be used to process numeric data and discrete data and it is not sensitive to the value of the exception.

However, the KNN algorithm also has many shortcomings that can be improved in subsequent research. If the calculated sample is unbalanced, a large number of samples that are not the nearest neighbor will cover the location of the nearest neighbor's small sample when a new sample is input. Eventually, the classification result is wrong. The sample size should not affect the classification results, so a weighted approach can be used to improve this type of problem by giving greater weight to neighbors with smaller sample distances.

Another shortcoming of the KNN algorithm is that the algorithm has higher requirements for a large amount of calculation. This is because for each electrical sample to be classified, its distance to the entire known electrical data sample will be calculated to obtain its k nearest neighbors. For this problem, the data feature preprocessing method can be used to clip the sample points which are known in advance and remove the samples which have little effect on classification.

Of course, the choice of K value will also significantly affect the appliance classification effect of the KNN algorithm. The results show that the classification accuracy is the highest when K is 2. With the increase of K, the recognition accuracy presents a gradually decreasing trend. However, when K is equal to 5–9, the load identification accuracy goes up slowly.

In general, the KNN algorithm is suitable for non-intrusive load monitoring. But there are also many places to be improved.

(3) *Optimal classification model for DTW*

(a) Active power sequence for modeling

In most disciplines, time-series is a common representation of data. For time-series processing, a common task is to compare the similarity of two sequences. Due to the switching process of similar appliances of different brands in the family is quite random, it is difficult to control the consistency of the opening time length of electrical appliances. In such a complicated situation, it is difficult to effectively obtain the traditional Euclidean distance between two time series or the similarity between two time series. The DTW algorithm can find the point that makes these two-segment time series pair of them, and then calculate the distance between them. The principle is that DTW is an effective distortion method. It can extend and distort time-series to calculate the similarity of two-segment time series. In this experiment, the original time series of current and voltage in the PLAID data set were firstly extracted. Then all the extracted time-series are classified and the electrical templates are generated successively. This sequence is compared with all the electrical templates in the electrical template library when the new electrical switch signal enters the NILM system. By calculating its Euclidean distance from other electrical templates, the system assigns it the template with the smallest distance.

The template synthesis process of several common household appliances, including Hairdryer, Microwave, Fan, Fridge and Heater, is shown in Fig. 4.13 (taking Hairdryer's template synthesis as an example).

(b) A variant of the DTW distance calculation method: Edit distance with Real Penalty

For the existing methods of calculating two types of time series distances, it is difficult to support the local time-shifting problem. In this example, a new distance function is used for calculation ("Edit distance with Real Penalty") (Lei and Ng 2004).

The ERP function can combine L1-norm and edit distance perfectly. ERP algorithm is similar to L1-norm in measuring distance. It can handle local time shifting like editing distances. Since ERP algorithm is more efficient than DTW algorithm and LCSS algorithm, it can effectively solve the distance calculation problem.

But the ERP algorithm also has certain flaws. When dealing with the distance calculation of the moving object, if the object is multi-dimensional or with a certain noise, the performance of the algorithm will be greatly reduced.

(c) A variant of the DTW distance calculation method: Time Warp Edit Distance

The Time Warp Edit Distances (TWED) algorithm (Marteau 2008) differs from the previous common time series similarity measurement algorithms in the following aspects. The TWED algorithm uses the difference between timestamps to compare the differences between the samples. In addition, the TWED distance measurement algorithm allows two paragraphs time-series of match evaluations to be linked to the down-sampled representation space. The TWED algorithm can be used when it comes to the need to match some time series with time shifting tolerance. The TWED algorithm can be used as a potential solution for metric space search methods. The evaluation indices calculation results of ERP and TWED algorithm are shown as follows.

From Fig. 4.14 and Tables 4.7 and 4.8, it can be seen that the simple and crude experiments use the ERP algorithm and the TWED algorithm for non-intrusive load monitoring is difficult to achieve the desired effect.

Accuracy of the above two algorithms barely reaches about 70% of the basic level, but the indices of recall and other two evaluation indices are quite low. When NILM experiments will be carried out, it is often necessary to consider the properties of experimental subjects in advance. If it is difficult for the experimental subjects to meet the input requirements of the algorithm, the desired effect is scarcely possible (Basu et al. 2015).

(4) *Comparison of comprehensive indices*

In the analysis of classification problems, it is necessary to use certain evaluation indices to evaluate the advantages and disadvantages of the algorithm. The evaluation index is a quantitative indicator that can be used to judge the different sets of data or different parameters of the same set of data. In the process of model evaluation, it is

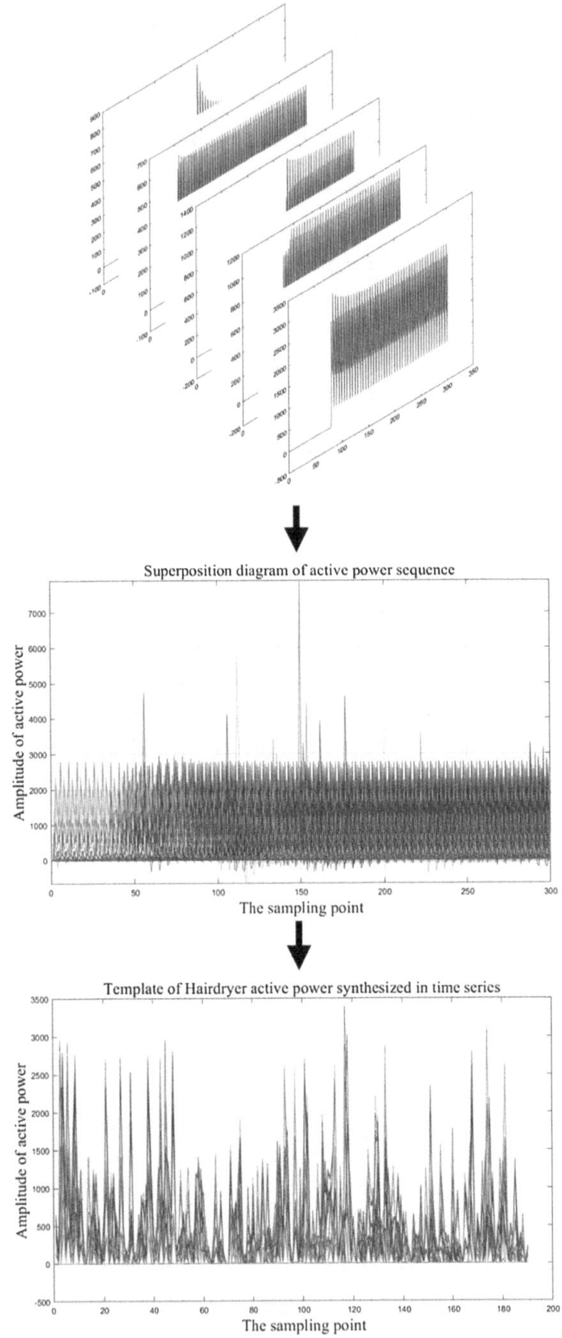

Fig. 4.13 The template synthesis process of DTW algorithm for Hairdryer

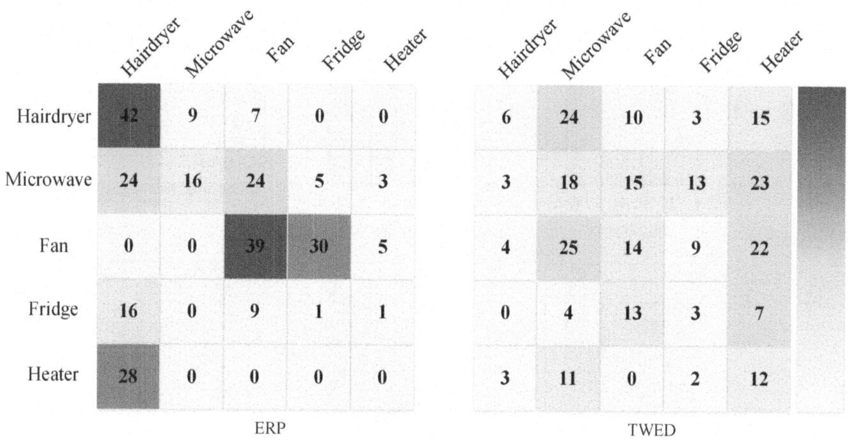

Fig. 4.14 Confusion matrix of the ERP/TWED model load identification results

Table 4.7 The classification performance indices of ERP model

Model	Type of appliance	Accuracy	Recall	F1-score	Precision
ERP	Hairdryer	0.6757	0.7241	0.5000	0.3818
	Microwave	0.7490	0.2222	0.3299	0.6400
	Fan	0.7104	0.5270	0.5098	0.4937
	Fridge	0.7645	0.0370	0.0317	0.0278
	Heater	0.8571	0.0000	NaN	0.0000
	Average	0.7513	0.3021	0.3429	0.3087

Table 4.8 The classification performance indices of TWED model

Model	Type of appliance	Accuracy	Recall	F1-Score	Precision
TWED	Hairdryer	0.7606	0.1034	0.1622	0.3750
	Microwave	0.5444	0.2500	0.2338	0.2195
	Fan	0.6216	0.1892	0.2222	0.2692
	Fridge	0.8031	0.1111	0.1053	0.1000
	Heater	0.6795	0.4286	0.2243	0.1519
	Average	0.6818	0.2165	0.1896	0.2231

often necessary to use a variety of indices for evaluation. In the process of machine learning classification, Accuracy, Recall, P-R curve, F1-score, Precision, Confusion matrix, Roc curve and Auc value are common indices that can reasonably evaluate the advantages and disadvantages of the algorithm.

The above calculation method can be handled well when dealing with the two classification problems. More issues need to be considered when training and testing

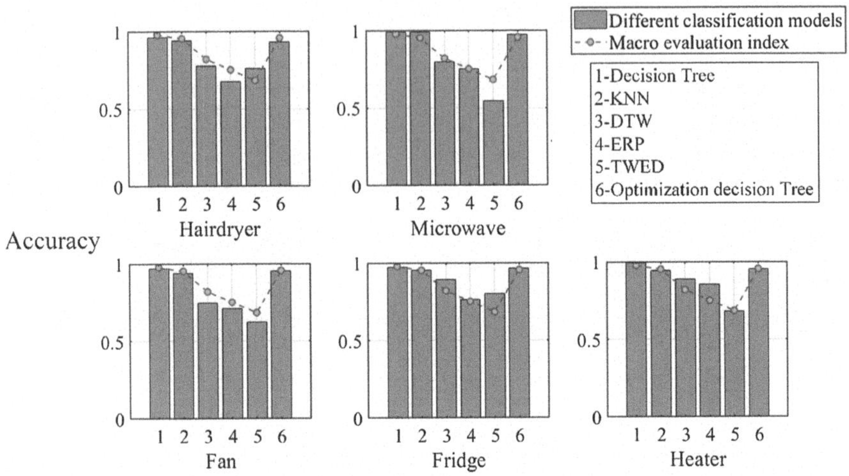

Fig. 4.15 Summary of classification models for accuracy indices

multiple different models based on the same dataset. At the same time, more issues also need to be considered when testing and evaluating the same model based on multiple datasets. The multi-classification problem also lacks the corresponding effective evaluation index. To carry out an effective evaluation, it will need macro-Precision, macro-Recall, macro-Accuracy and macro-F1-score for evaluation.

In this experiment on NILM, six algorithms were used to classify the five common types of household appliances. In the classification process, the difference in performance is compared when the same appliance is classified by different algorithm models under the same indicator. Specifically, as shown in Figs. 4.15, 4.16, 4.17 and 4.18.

The performance of the decision tree algorithm and the KNN algorithm are relatively good. However, the classification results of the three distance algorithms of DTW are not very ideal.

In general, improving the classification performance of the classifier requires optimizing every step of NILM, including event detection, feature extraction, load identification and classification. When each part is done well, an ideal classification result can be obtained.

4.5.3 Conclusion

This chapter considers several kinds of classification matching algorithms commonly used in template matching to carry out non-intrusive load monitoring experiments. The original current-voltage signal in the PLAID dataset is processed as more than 20 features such as odd harmonics of current. It is also used as template matching by processing time-series set for active power. In this experiment, the decision tree,

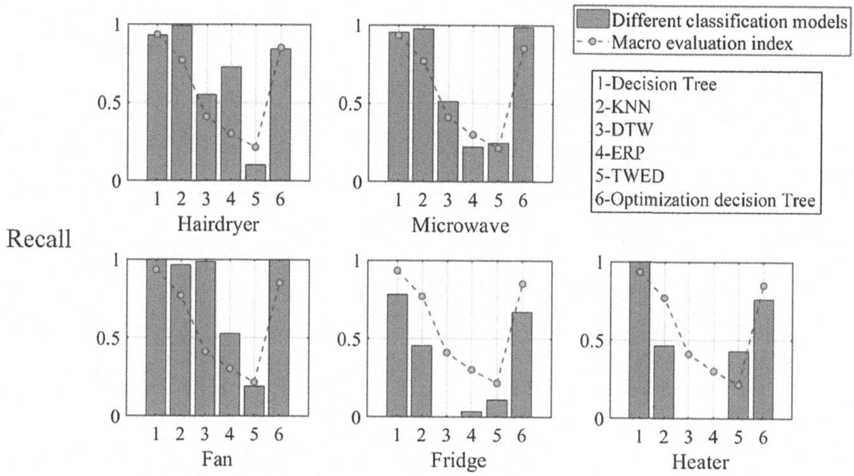

Fig. 4.16 Summary of classification models for recall indices

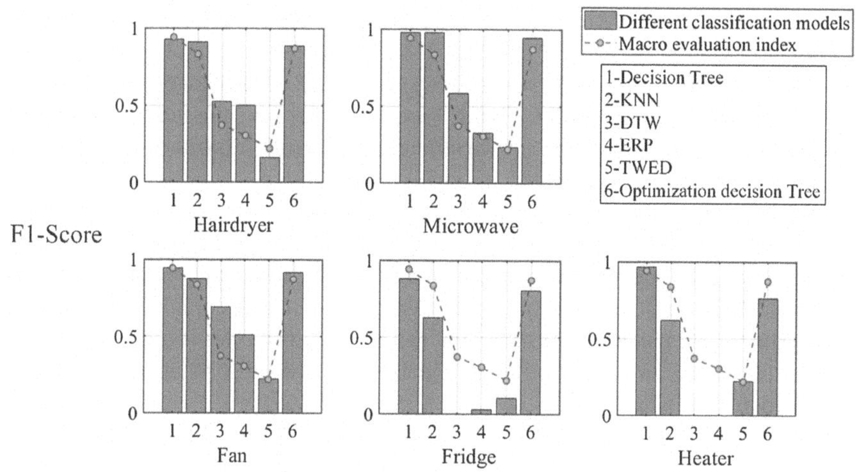

Fig. 4.17 Summary of classification models for F1-score indices

KNN and DTW algorithms were deformed to deal with the classification of NILM. The sample set was divided into a training set and a test set. The training set is 70% of the total sample size and the test set is 30%. The classification performance of six different algorithms is evaluated by certain evaluation indicators. The final conclusion is given as follows.

(a) The decision tree and its optimization algorithm have good visibility when dealing with classification problems, which can help users to understand the details of the classification process better. The decision tree classification method has

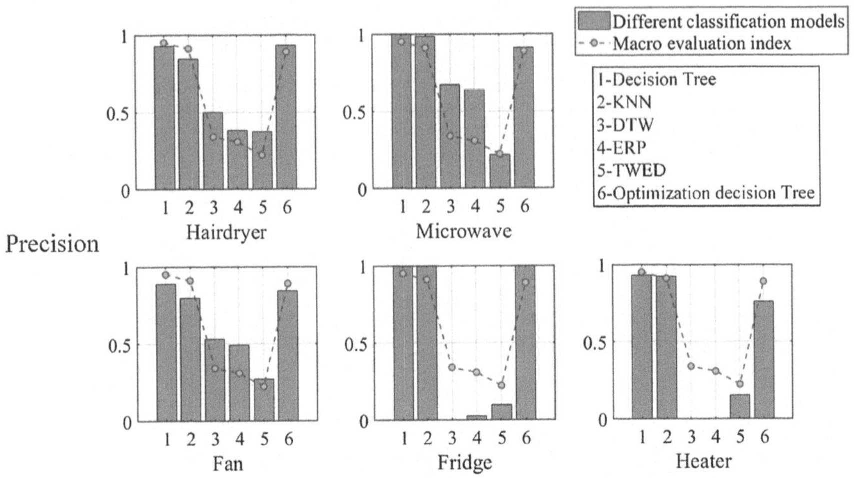

Fig. 4.18 Summary of classification models for precision indices

the advantages of fast calculation speed, high precision and low requirement for the user's prior knowledge.

(b) The KNN algorithm also has excellent performance when dealing with electrical equipment classification problems, which can be used for the processing of nonlinear problems in non-intrusive load monitoring. It has the advantages of high recognition accuracy and insensitivity to outliers.

(c) DTW, ERP and TWED algorithms are good time series distance calculation methods. However, if it is directly used for the appliance classification processing of the NILM without conversion, that would be difficult to obtain the desired effect. In general, template matching has many advantages of machine learning. At the same time, template matching is also used as a basis for machine learning classification, which lays a foundation for the subsequent experiment of Chap. 6.

References

Basu K, Debusschere V, Douzal-Chouakria A, Bacha S (2015) Time series distance-based methods for non-intrusive load monitoring in residential buildings. Energy Build 96:109–117

Bhaduri K, Qiang Z, Oza NC, Srivastava AN (2010) Fast and flexible multivariate time series subsequence search. In: IEEE international conference on data mining

Biansoongnern S, Plangklang B (2016) Nonintrusive load monitoring (NILM) using an artificial neural network in embedded system with low sampling rate. In: International conference on electrical engineering/electronics

Chen Q, Li D, Tang CK (2013a) KNN matting. IEEE Trans Pattern Anal Mach Intell 35(9):2175–2188

Chen Y, Bing H, Keogh E, Batista GEAPA (2013b) DTW-D: time series semi-supervised learning from a single example. In: Acm Sigkdd international conference on knowledge discovery & data mining

Cover T, Hart P (1967) Nearest neighbor pattern classification. IEEE Trans Inf Theory 13(1):21–27

Fu WC, Keogh E, Lau LYH, Ratanamahatana CA, Wong CW (2008) Scaling and time warping in time series querying. Vldb Journal 17(4):899–921

Han J, Kamber M (2005) Data mining: concepts and technique. In: Antimicrobial agents & chemotherapy. Morgan Kauffman Publishers

Hart GW (1992) Nonintrusive appliance load monitoring. Proc IEEE 80(12):1870–1891

Kang SJ, Ji WY (2017) Classification of home appliance by using Probabilistic KNN with sensor data. In: Signal & information processing association summit & conference

Kung DC (1994) On decision tree verification and consolidation. Inf Softw Technol 36(8):485–494

Lei C, Ng R (2004) On the marriage of Lp-norms and edit distance. In: Proceedings 2004 VLDB conference

Marteau PF (2008) Time warp edit distance with stiffness adjustment for time series matching. IEEE Trans Pattern Anal Mach Intell 31(2):306–318

Olanow CW, Koller WC (1998) An algorithm (decision tree) for the management of Parkinson's disease: treatment guidelines. American Academy of Neurology. Neurology 50(3 Suppl 3):S1

Ruggieri S (2002) Efficient C4.5 [classification algorithm]. IEEE Trans Knowl Data Eng 14(2):438–444

Singh M, Kumar S, Semwal S, Prasad RS (2015) Residential load signature analysis for their segregation using wavelet—SVM. Lecture Notes in Electrical Engineering 326:863–871

Wang R (2012) AdaBoost for feature selection, classification and its relation with svm, a review. Phys Procedia 25(22):800–807

Zhou DZ, Jiang WB (2010) Efficient clustering algorithm for multivariate time series. Comput Eng Appl 46(1):137–139

Zhou DZ, Xiao-Li WU, Yan HC (2008) An efficient similarity search for multivariate time series. J Comput Appl 28(10):2540–2541

Chapter 5
Steady-State Current Decomposition Based Appliance Identification

5.1 Introduction

With the rapid increase of electric power consumption, energy conservation is of great concern. As stated in Chap. 1, the Non-intrusive Load Monitoring (NILM) technique is a convenient and effective method, which can offer fine-grained energy-consuming information of each appliance. And the information is quite important for managing energy and improving energy efficiency. For the resident, such fine-grained information can benefit their organization of energy use. And for the appliance manufacturers, the application of the NILM will promote the appliance improvement in energy saving (Lin and Tsai 2015). Statistics show that with the message provided by the NILM system, the energy costs can be cut down by 10% to 15% (Darby 2006).

The NILM methods mainly contain source disaggregation methods and pattern recognition methods (Jian et al. 2010). In contrast to the pattern recognition methods, the source separation methods are time-based, disaggregating the total current sequences directly without detecting the switch events of appliances. And the source separation methods are more efficient and workable due to that there is no training part in these methods. To make the NILM system perform better, appropriate features and effective models are the two most momentous aspects to be concentrated on (Tabatabaei et al. 2017).

On the one hand, many load signatures have been explored in recent studies. The initial features utilized in the source separation method is the power signal, such as the power load (Fan et al. 2012) or the rated power (Tang et al. 2014). However, the accuracy of methods using these features is not satisfying due to the limited information. Thus, the current separation methods are presented with harmonic features. Bouhouras et al. employed some self-defined signatures generated from the current magnitudes (Bouhouras et al. 2012). But these load signatures neglected the phase information in the current signal, therefore the performance of NILM was

H. Liu, *Non-intrusive Load Monitoring*,
https://doi.org/10.1007/978-981-15-1860-7_5

constrained. And then, Bouhouras et al. again proposed the harmonic current vector to make use of the phase information. The impact of phase information was also explored and illustrated in the paper. But the phase information was embodied by the cosines of phase angles, losing half of the harmonic phase information. Recently, the whole harmonic current vector has been studied and regarded as the load signature (Bouhouras et al. 2019). The vector computing principle was adopted to calculate it, which can reserve all of the information in the current harmonics. At present, the weights of all harmonic vectors are identical. However, the information provided by different harmonics is various. In the chapter, a weighted current vector is designed to make full use of the information in all harmonics and different source separation features.

On the other hand, the load disaggregation methods have also attracted much attention from the researchers. Rahimpour et al. added a sum-to-k constraint to the Non-negative Matrix Factorization (NMF) and proved that the proposed model can extract meaningful sources from the aggregated current exactly (Rahimpour et al. 2017). But the method was not performed on high-frequency data and is supervised. García et al. combined the non-negative matrix factorization model with an interaction part to solve the determination of the factors (García et al. 2018). But the number of factors given manually is usually not the optimal value. The problem was not solved fundamentally. Miyasawa et al. proposed a semi-binary NMF model and took the feedback information from the users into consideration (Miyasawa et al. 2019). And the experiments showed that the method was able to achieve better results. Besides the NMF methods, the mathematical optimization models have also been intensively explored and widely applied in this domain. Bouhouras et al. put forward a geometric searching algorithm combined with the harmonic vector features (Bouhouras et al. 2019). This algorithm is very efficient. But when the selected harmonic features of the appliances are not distinguishable enough, it becomes unstable. Tang et al. applied the Sparse Switching Event Recovering (SSER) optimization model to the NILM issues and employed a Parallel Local Optimization Algorithm (PLOA) to solve the SSER (Tang et al. 2014). But the method is complicated and the single-objective optimization method is not as stable as the multi-objective methods. And some researchers have also explored genetic algorithms (Hock et al. 2018), sparse optimization (Piga et al. 2016), ant colony optimization (Gonzalez-Pardo et al. 2015), particle swarm optimization (Chang et al. 2013) and other modern objective models in NILM.

In the chapter, the load disaggregation experiments are also carried out on the WHITED dataset (Kahl et al. 2016). In this study, 22 types of appliances that are commonly-used are selected, including some high energy-consuming appliances and some appliances with the most samples. Figure 5.1 gives the power supply and area information of the appliances, which is necessary for the design of the steady-state current disaggregation experiments. And the number of samples and appliances is listed in Table 5.1.

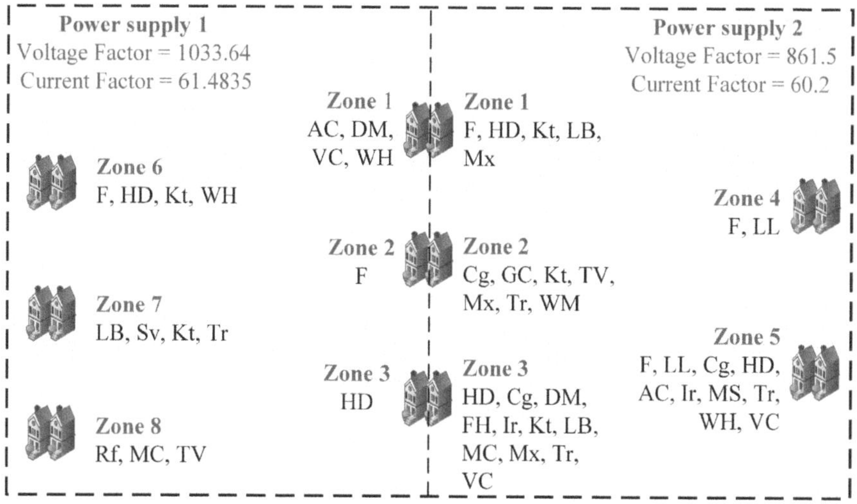

Fig. 5.1 The power supply information and area information of the appliances

5.2 Classical Steady-State Current Decomposition Models

The steady-state current decomposition is now a hotspot in the NILM and it is developed from the source separation methods based on steady-state features, such as steady-state power decomposition. These methods are time-based methods, trying to disaggregate the whole total signal into a combination of the appliance signals. Many classical features and models have been explored. In this segment, these classical features and models are introduced and compared in detail and their advantages and drawbacks are also pointed out.

5.2.1 *Model Framework*

Figure 5.2 shows the model framework for steady-state decomposition with the three classical models and two classical features to be compared. And as shown in Fig. 5.2, the steps of steady-state decomposition are clarified as follows:

(a) Normalizing the current and voltage data, extracting the features to be used in the load decomposition, and building the feature library of all appliances;
(b) Extracting the features of aggregated current or power data each second;
(c) Performing the models on the features of aggregated data and getting the disaggregation results every second;
(d) Comparing the results of the models obtained in *step (c)*. Calculating the three evaluation index values, named Precision, Recall and F1-score. Analyzing the performance of various features and models.

Table 5.1 The information of the selected appliances types in the WHITED dataset

Appliance	Abbreviation	Number of samples	Number of appliances
LED light	LL	90	9
Charger	Cg	50	5
Fan heater	FH	10	1
Light bulb	LB	60	6
Hairdryer	HD	60	6
Kettle	Kt	60	6
Toaster	Tr	40	4
Game console	GC	40	4
Washing machine	WM	10	1
TV	TV	20	2
Stove	Sv	10	1
Mixer	Mx	40	4
Fan	F	60	6
Vacuum cleaner	VC	40	4
Water heater	WH	40	4
Drilling machine	DM	20	2
Iron	Ir	30	3
Massager	MS	30	3
Microwave	MC	20	2
Refrigerator	Rf	10	1
Air conditioning	AC	10	1
Water pump	WP	10	1

5.2.2 Classical Features of Steady-State Decomposition and the Feature Extraction Method

The most classical steady-state decomposition is usually based on features extracted from the data sampled at a low frequency. Using these features, the NILM system can be easily applied to the existing meters, meanwhile the data analysis and transmission will be efficient. However, there are also common drawbacks to these features. The classical feature is usually simplex and the information contained in the feature is limited due to the low sampling rate. Even for some classical features derived from high-frequency data, the information is underutilized. But the exploration of the classical features is still of great significance.

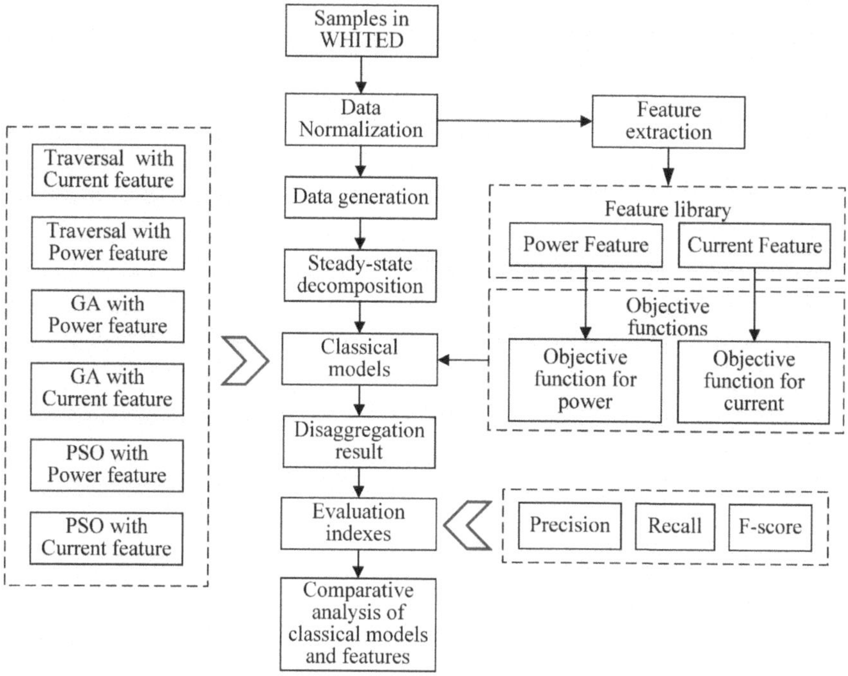

Fig. 5.2 The model framework for classical steady-state decomposition

(1) *Classical features of steady-state decomposition*

Features used in steady-state decomposition must follow two principles. One is that the feature should be superposable and the other is that the feature is stable when the appliance is under a steady operation state. Some important features are introduced as follows:

(a) Active power and reactive power

The most commonly-used classical feature in steady-state decomposition is the power sequences (Bonfigli et al. 2017), containing the active power P_a and reactive power P_r. These two features are all superposable. The power time series can reflect the information in the current and voltage of an appliance and directly provide the energy consumption information. But the disaggregation is difficult to be accomplished when there are appliances with approximate active power or reactive power, due to the singleness of the feature. And there are many cases like that. As shown in Fig. 5.3, the steady-state active power of the two appliances is very similar.

(b) Power phase angle θ_P and apparent power P_s

The apparent power combined with the power phase angle is another form of power, which can be called the power vector P_v. Unlike the active or reactive power, neither the power phase angle nor apparent power is superposable. But the power vectors can

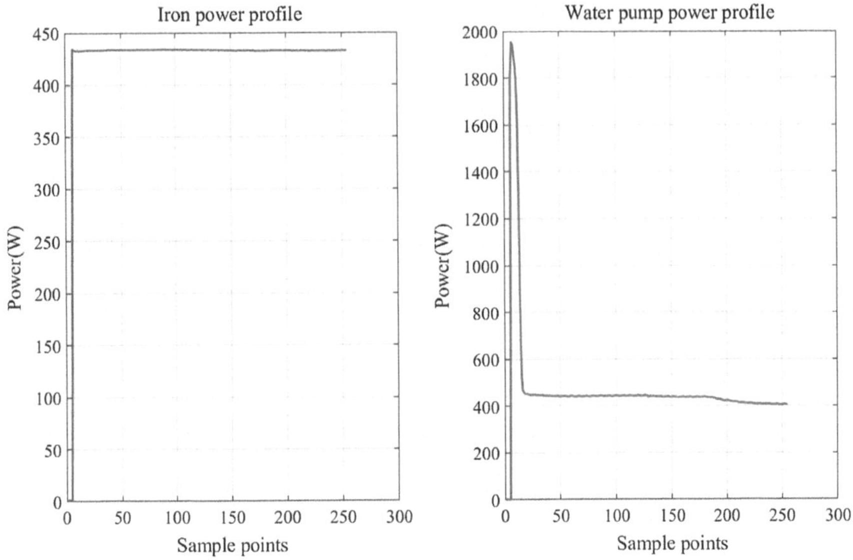

Fig. 5.3 The active power profile of two appliances

be stacked up under the principle of vector computation. And due to the complete phase information and amplitude information contained in the power vector, it is a little more differentiable than the active power or reactive power. The power vectors of some selected appliances in the WHITED dataset are displayed in Fig. 5.4 and the power vector is defined as follows:

$$P_v = P_s \angle \theta_p \tag{5.1}$$

(c) Current sequences

The current sequence sampled at a high frequency is also a classical feature in steady-state decomposition. The currents of appliances running under steady states usually have certain rules. Their characters are stable and unaltered under steady-states, like the current of iron shown in Fig. 5.5. And by applying the Fourier transform to the current sequence, the instantaneous current i_a of appliance a can be expressed as follows:

$$i_a(t) = I_{a1}\cos(\omega t + \theta_{a1}) + I_{a2}\cos(2\omega t + \theta_{a2}) + \cdots + I_{am}\cos(m\omega t + \theta_{am}) \tag{5.2}$$

where I_{a1} is the amplitude of the fundamental harmonic and θ_{a1} is the initial phase angle of the fundamental harmonic; I_{am} is the amplitude of the m-th harmonic and θ_{am} is the phase angle of the m-th harmonic. And when there are n appliances running simultaneously, the aggregated current i_z can be approximated by a linear superposition of the currents of these n kinds of appliances. It is expressed as

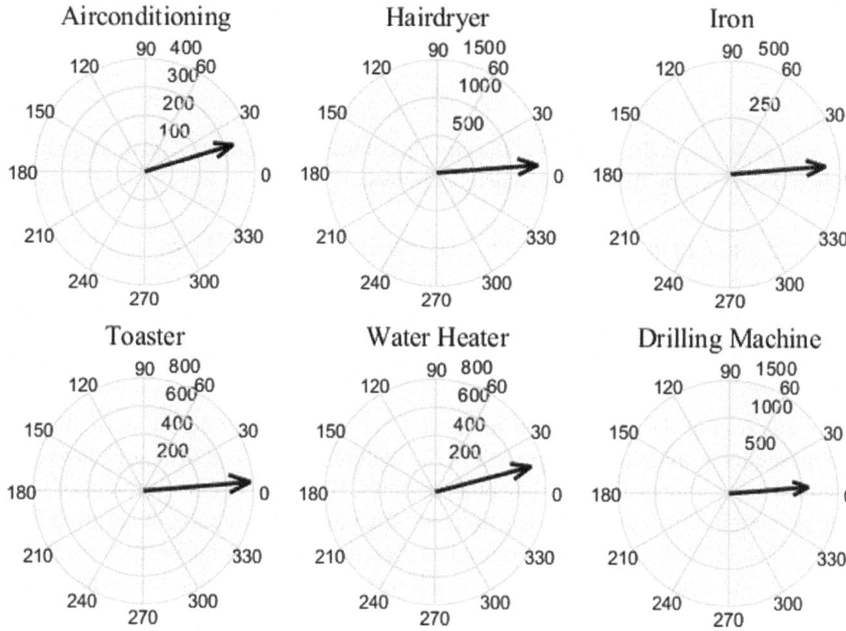

Fig. 5.4 The power vectors of some selected appliances

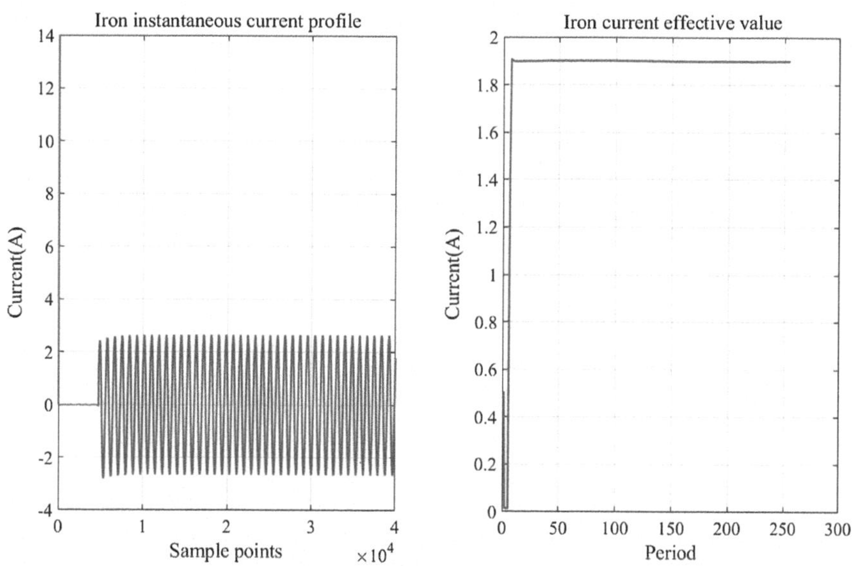

Fig. 5.5 The instantaneous current and its effective value of iron

follows:

$$i_z(t) = \alpha_1 i_1(t) + \alpha_2 i_2(t) + \cdots + \alpha_n i_n(t) \tag{5.3}$$

where $\alpha_1, \alpha_2, \ldots \alpha_n$ is the weights of the currents of the n appliances. And this equation is often described by the phasor matrix as follows:

$$\begin{bmatrix} I_{z1}\angle\theta_{z1} \\ I_{z2}\angle\theta_{z2} \\ \vdots \\ I_{zm}\angle\theta_{zm} \end{bmatrix} = \begin{bmatrix} I_{11}\angle\theta_{11} & I_{21}\angle\theta_{21} & \cdots & I_{n1}\angle\theta_{n1} \\ I_{12}\angle\theta_{12} & I_{22}\angle\theta_{22} & \cdots & I_{n2}\angle\theta_{n2} \\ \vdots & \vdots & \cdots & \vdots \\ I_{1m}\angle\theta_{1m} & I_{2m}\angle\theta_{2m} & \cdots & I_{nm}\angle\theta_{nm} \end{bmatrix} \begin{bmatrix} \alpha_1 \\ \alpha_2 \\ \vdots \\ \alpha_n \end{bmatrix} \tag{5.4}$$

By employing the information in several harmonics of appliance currents, the current sequence avoids the singleness of the feature used in load disaggregation and is proved to be more distinctive than other features. But in the classical studies of load disaggregation, this feature is underutilized. The phase information in harmonics is used by calculating its cosine value rather than the phase angle itself. And all the harmonics are added together at last, which results in the case that the information in some useful harmonics may be obscured by the first harmonic.

(d) Admittance

Admittance is the ratio of the current vector to the voltage vector. The admittance vectors of some appliances are shown Fig. 5.6. In parallel circuits, the admittance of each branch can be stacked up directly but they should be added in vector form. The feature contains the information in both current and voltage series and is more effective in the decomposition of the aggregated load. And it also desires a high sampling rate because it is an instantaneous quantity. The admittance and the current series are the features that are still under exploration and improvement. For example, some new admittance-based features have been proposed recently to improve the precision of the load disaggregation methods (Liu et al. 2018a). The basic calculation of admittance is illustrated below (Liu et al. 2018a):

$$\begin{cases} \dot{G}_t = \dot{I}_t / \dot{U}_t \\ \left|\dot{G}_t\right| = \left|\dot{I}_t\right| / \left|\dot{U}_t\right| \\ \theta_{G_t} = \theta_{I_t} - \theta_{U_t} \end{cases} \tag{5.5}$$

where $\dot{G}_t$ is the admittance vector and $\dot{I}_t$, $\dot{U}_t$ are the relevant current and voltage vectors respectively; $\left|\dot{G}_t\right|$, $\left|\dot{I}_t\right|$, $\left|\dot{U}_t\right|$ are the amplitudes of $\dot{G}_t$, $\dot{I}_t$, $\dot{U}_t$; θG_t, θI_t, θU_t are the phase angle of the admittance, current, and voltage.

(2) *Feature extraction method*

The experiments in the chapter are performed on a high-frequency dataset. The features calculated from samples resampled from the original samples do not need

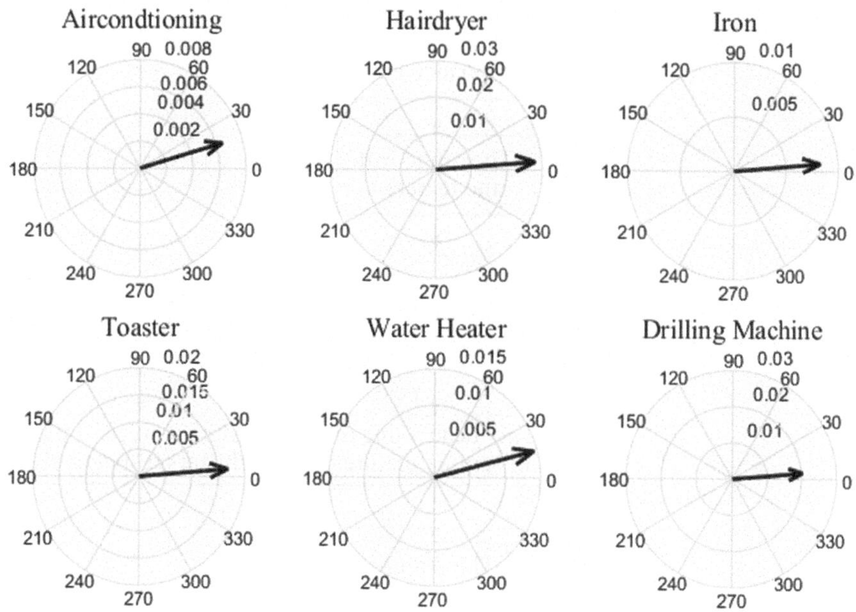

Fig. 5.6 The admittance vectors of some appliances

such a high sampling rate, like the power load. And features like the current time series and admittance are derived from the original samples.

The features like instantaneous current and admittance are calculated by frequency-domain analysis of the current and voltage signal using Fast Flourier Transform (FFT). The FFT analysis transforms the current signal into the frequency domain signal with a range of 0–22.5 kHz. But the spectrum leakage is a crucial issue to be solved in using FFT. In the work, the problem is solved by intercepting the signal into sequences which are several times of integral voltage periods. The fundamental frequency of the voltage signal in the WHITED dataset is 50 Hz. The power time series is extracted by calculating the power of each period of the voltage signal. And thus, there are 50 points in a one-second-long power time series. When building the feature database, a roughly three-second-long signal of each sample is selected. Besides, the aggregated current is analyzed each second.

Commonly, the fundamental frequency of the voltage-time series is not exactly 50 Hz, small variations always exist (0.5 Hz at most) (Liu et al. 2018b). Thus, a lobe searching method is utilized to get the real frequencies of the harmonics. The algorithm sets the frequency of 50, 150, 250, 350, 450 Hz as the search centers and then defines several search areas around the centers with a fixed width of 10 Hz. Since the signal is an integral multiple of the voltage period, there is only one lobe beside the harmonic order after it is transformed into the frequency-domain signal. By searching this lobe, the harmonic frequency is acquired. And then, the features are extracted.

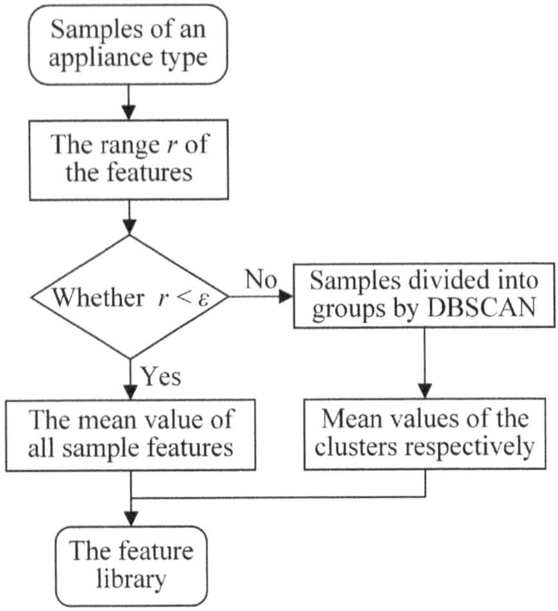

Fig. 5.7 The steps of feature library establishment

After extracting the features from each sample, how to build appropriate feature libraries becomes an important issue. For the reason that the various operation modes of an appliance and the different brands of appliances usually have obviously different harmonic features, there may be several various representative features of one kind of appliances. Thus, the feature library is built as the following two steps, which is shown in Fig. 5.7.

Step 1. Commonly, when one value is not enough to represent the character of one appliance, the features of the appliance samples will be more dispersed. The range r is selected to quantify the degree of dispersion. And a threshold ε is also created to decide whether one or more values are needed to describe an appliance. If r does not exceed ε, the mean value of the sample features is taken as the only one feature value of the appliance.

Step 2. If r exceeds ε, a Density-based Spatial Clustering of Applications with Noise (DBSCAN) clustering algorithm is adopted to cluster the sample features. DBSCAN is a state-of-the-art clustering approach, which is able to adaptively determined the cluster number and is resistant to the influence of outliers. It has been intensively studied and widely used in various domains. In NILM, the DBSCAN algorithm is usually used as an unsupervised method in the identification of the unknown appliances (De Baets et al. 2019). But in this chapter, the DBSCAN algorithm is an important part of feature extraction. By applying the DBSCAN method to the feature clustering, samples with different feature values are classified into several clusters. And the mean feature value of the cluster with enough samples is selected as the representative feature value of the appliance.

5.2.3 Classical Methods in Steady-State Current Decomposition

The most classical load disaggregation methods in steady-state decomposition are the mathematical optimization models, including modern optimal algorithms (Piga et al. 2016; Chang et al. 2013), quadratic programming and so on. In the segment, three classical models are briefly introduced and compared in order to exhibit the performance of the classical methods and the classical features in steady-state decomposition.

(1) *The theoretical basis of the models*

Traversal search. Traversal search is the simplest method in finding the optimal combination of appliances that is the most approximate to reality.

Genetic algorithm. The genetic algorithm (GA) is a highly parallel and adaptive optimization algorithm. It has been widely used in various optimization fields for its simple searching principle and fine global optimization ability. GA uses a number of genetic manipulations acting on groups to get a new generation group. In the implementation process of GA in steady-state decomposition, each individual represents a kind of combination of the appliances. And in heredity and mutation, the appliances in the combination are changeable.

Particle swarm optimization. Eberhart et al. firstly proposed the Particle swarm optimization (PSO) in 1995 when exploring social–psychological issues (Kennedy 1995). And the PSO model has been quite mature and is widely applied to various engineering optimization issues. Due to its distinctive optimal way, the convergence property and efficiency of the method are better. The algorithm searches the optimal results by changing the position in every iteration. In each iteration, the personal best and global best particles are selected to calculate the position of the particles in the next iteration. In the work, in order to enable PSO method in steady-state decomposition, an independent variable Cb is designed to reflect the sought combination of appliances. It is introduced as follows:

$$Cb \in \left(0, 2^z\right) \tag{5.6}$$

where Cb is the independent variable and it is an integer variable; z is the number of all appliances included in the feature database. In the optimization, Cb is transformed into a binary form to clarify the chosen combination of appliances. Cb is a binary number with z bits and each bit of Cb decides whether an appliance is on or off. And Cr is the binary form of Cb defined as below:

$$Cr = Cb = [b_1\ b_2 \cdots\ b_i \cdots b_z] \tag{5.7}$$

where b_i is one bit of Cr and represents the on or off state of an appliance. But actually, the PSO algorithm desires an independent variable with a continuous value. Thus, a hidden variable cb that is continuous is designed corresponding to the variable Cb.

The relationship between Cb and cb is illuminated as below:

$$Cb = [cb + 0.5], \quad cb \in [0.5,\ 2^z + 0.5) \tag{5.8}$$

where $[cb + 0.5]$ is the round-off number of cb. And by using such a hidden variable, the velocity of each particle is calculated as below (Liu et al. 2019):

$$v = u_1 \cdot cb + u_2 \cdot cb_{pbest} + u_3 \cdot cb_{gbest} \tag{5.9}$$

where the subscripts *pbest* and *gbest* stand for the personal best particle and the global best particle respectively; and the coefficient u_1, u_2, u_3, which are all within the range of 0–1, are the weights of the particle's inertia, the personal best particle's position, and the global best particle's position respectively.

(2) *The objective functions of the disaggregation scheme*

To apply the optimization methods in load decomposition, the objective functions of different features are designed as below:

(a) Power

$$f_P = \min \left| P_{at} - \sum P_{ak} \right| \tag{5.10}$$

where P_{at} is the aggregated power and P_{ak} is the power of appliance k in the combination.

(b) Current series

$$f = \sum_{j=1,3,5,7,9} \left(I_{tj} \cdot \cos\theta_{tj}\right) - \sum_{k=1}^{n} \sum_{j=1,3,5,7,9} \left(I_{kj} \cdot \cos\theta_{kj}\right) \tag{5.11}$$

where I_{tj} is the j-th harmonic amplitude of the total current and θ_{tj} is the j-th harmonic phase angle of the total current; I_{kj} is the j-th harmonic amplitude of the current of appliance k and θ_{kj} is the j-th harmonic phase angle of the current of appliance k.

5.2.4 Performance of the Various Features and Models

As shown in Fig. 5.2, the section compares six steady-state decomposition methods in order to analyze the performance of various classical models and features. Due to that the methods in this chapter are optimization approaches for load disaggregation, after the features of all appliances extracted, there is no training part of the disaggregation scheme. And the test set consists of several aggregated loads. Figure 5.8 shows the aggregated load of an instance, which is called #1. Table 5.2 lists the information of three aggregated load to be decomposed.

According to the disaggregation results which are shown in Figs. 5.9, 5.10 and 5.11, some conclusions can be summarized as follows:

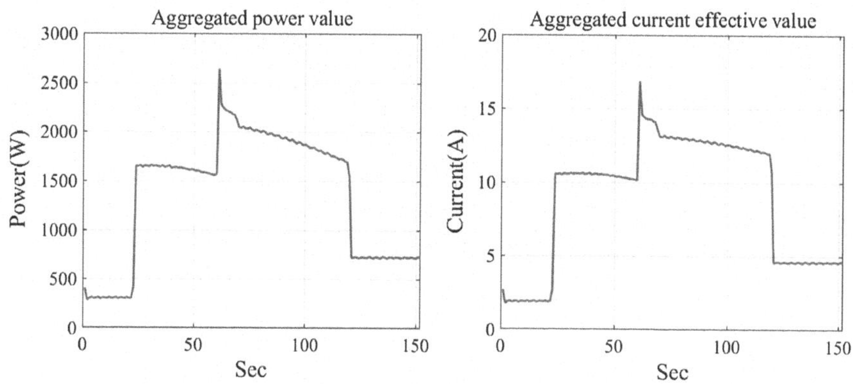

Fig. 5.8 The aggregated current and power of #1

Table 5.2 Some instances of aggregated loads

Number of aggregated load instances	Appliances in the current	Running time of each appliance	Total time
#1	AC + HD + VC	68.5 + 97.0 + 93.0	152.9
#2	AC + I + T	35.6 + 47.3 + 66.7	96.0
#3	VC + WH + WP	44.3 + 56.8 + 92.2	126.0

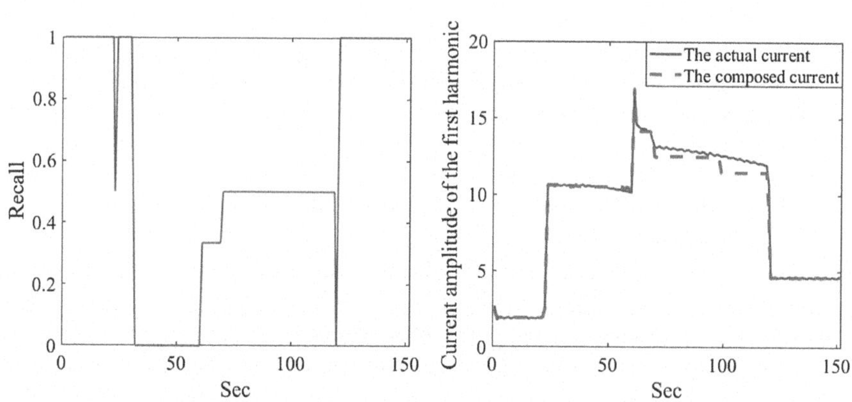

Fig. 5.9 The load #1 disaggregation results of the traversal model with current feature

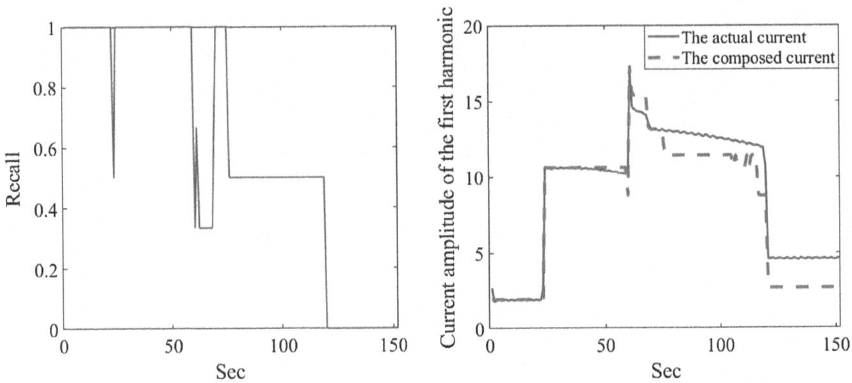

Fig. 5.10 The load #1 disaggregation results of the GA model with the current feature

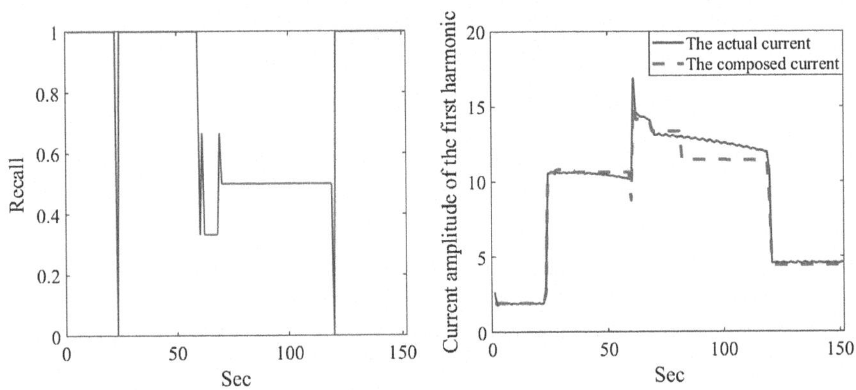

Fig. 5.11 Load #1 disaggregation results of the PSO model with the current feature

(a) The most important and difficult issue of the steady-state decomposition is that when there are similar features of appliances or the combinations of appliances in the feature database, the disaggregation becomes inaccurate. As the comparison results shown in Fig. 5.9, the values of the real current and the composed current are approximate, but the recall values are not satisfying. It is because that the incorrect appliance combination is mistaken as the best result, just the same as the case mentioned above.
(b) The current series is not very effectively utilized by this means. The results of the methods are all not accurate enough. From the recall values of the superposition parts of the currents shown in Figs. 5.9, 5.10 and 5.11, it can be concluded that the estimation of those parts is incorrect. And it is hopeful to enhance the stability and precision of these methods by improving the feature used in steady-state decomposition.

(c) The singleness of the feature may lead to the case that the appliance combination with the closest value to the real current may not be the right answer. Commonly, the traversal method which compares all the appliance combinations can reach a good result and the disadvantage is its inefficiency. But it can be seen from Fig. 5.9 that the results of the traversal method are worse than the other two methods, which may be caused by the aforementioned case.

As seen in Fig. 5.12 and Table 5.3, it can be concluded that:

(a) The power features are less effective than the current series. As shown in Table 5.3, the index values of all disaggregation experiments reveal that the overall performance of methods using active power is worse than the same methods using the current series.
(b) In Fig. 5.12, all three methods work equally poorly, which indicate that a good feature is more important in steady-state decomposition.
(c) The performance of the classical methods and features cannot meet the requirement of the NILM system. In NILM, the load decomposition accuracy is usually desired to be higher than 80% (Makonin et al. 2016). As we can see in Table 5.3, the performance of the methods is lower.

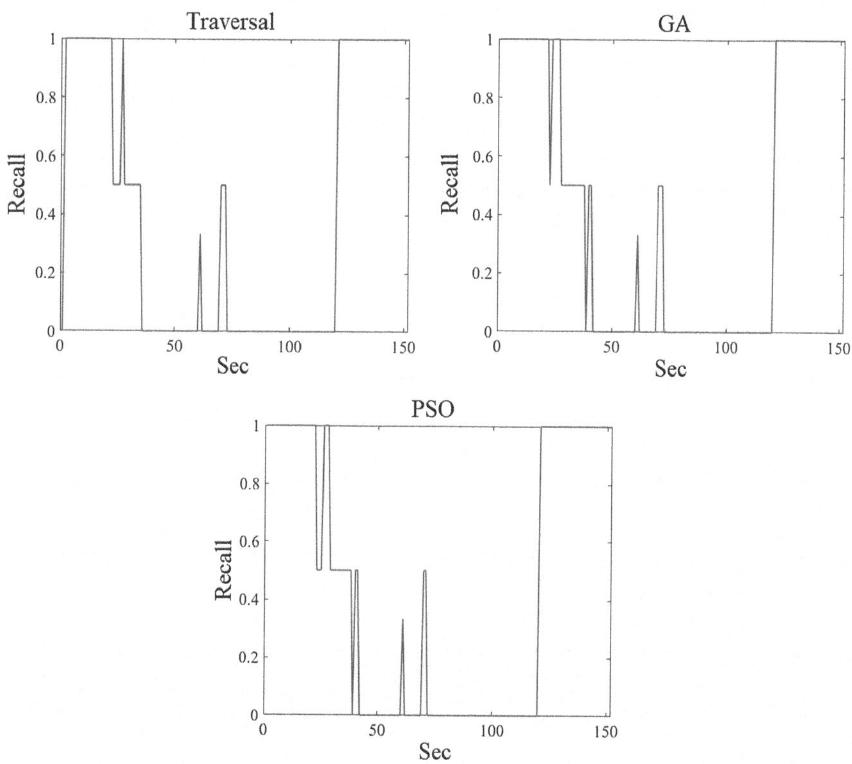

Fig. 5.12 Load #1 disaggregation recall values of the three methods using power features

Table 5.3 The disaggregation results of classical load decomposition methods

Feature		Power			Current		
Method		Traversal method	GA	PSO	Traversal method	GA	PSO
#1	Precision	1.00	1.00	1.00	1.00	1.00	0.967
	Recall	0.277	0.312	0.305	0.504	0.636	0.734
	F1-score	0.435	0.477	0.468	0.674	0.792	0.835
#2	Precision	0.795	0.869	0.960	0.977	0.950	0.960
	Recall	0.665	0.576	0.548	0.540	0.602	0.609
	F1-score	0.724	0.693	0.698	0.696	0.737	0.745
#3	Precision	0.996	0.983	0.983	0.921	0.948	0.864
	Recall	0.641	0.647	0.657	0.653	0.541	0.449
	F1-score	0.780	0.780	0.788	0.765	0.689	0.591

5.3 Current Decomposition Models Based on Harmonic Phasor

As mentioned in the above segment, the classical steady-state decomposition cannot make full use of the information in the current data. And most of the optimization models used in classical steady-state decomposition are single-objective, only caring about the magnitude of the errors between the real aggregated current and the calculated one. In fact, the approximate degree of the calculated and real currents is not only reflected by the magnitude of errors but also relevant to the dispersion degree of the errors. Thus, in the segment, the new current features and the multi-objective optimization models are applied to the steady-state decomposition

5.3.1 *Model Framework*

Figure 5.13 shows the model framework for steady-state current decomposition with novel features and multi-objective optimization models. And as shown in Fig. 5.13, the steps of the steady-state current decomposition with novel features and models are introduced as below:

(a) Normalizing the current and voltage data, extracting the current harmonic features by applying the FFT algorithm to the data, and constructing the feature library of all kinds of appliances under single-load operation scenarios;
(b) Extracting the current harmonic vectors of aggregated current each second;
(c) Performing the multi-objective models on the aggregated current harmonic vectors, determining the objective functions and breaking down the total signal in seconds;

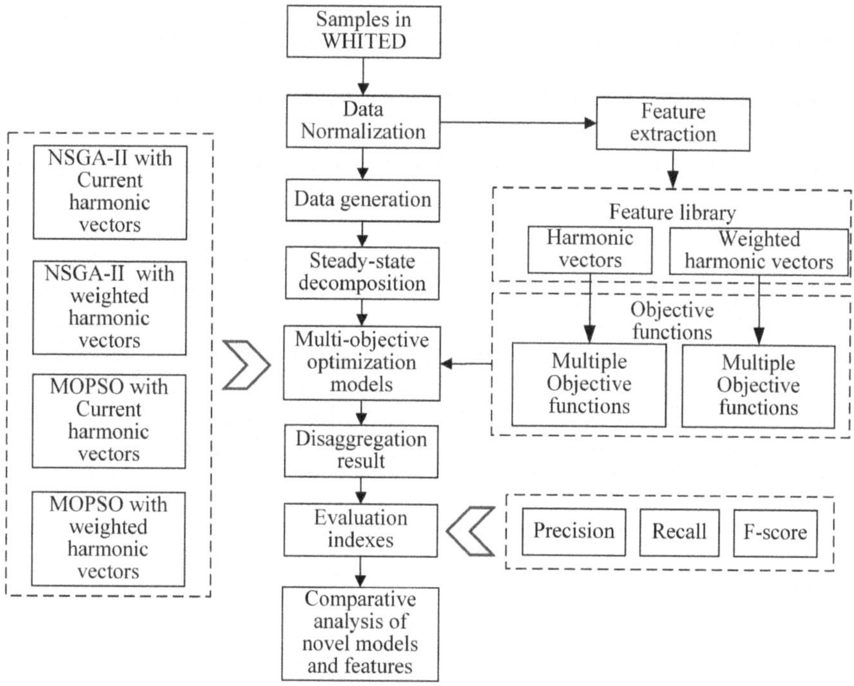

Fig. 5.13 Modeling flow chart of ELM prediction model driven by single data

(d) Comparing the results of the models obtained in *step (c)*, and calculating the three evaluation index values, named precision, recall, and F1-score.

5.3.2 Novel Features of Steady-State Current Decomposition

The traditional features in the steady-state decomposition methods of the NILM are usually derived from the low-sampling data and even for the features acquired from high-frequency data, the information is underutilized. The traditional methods which use the x-axis projections of current harmonics as the load disaggregation features leave out part of the harmonic phase information (Bouhouras et al. 2017). Thus, some novel features are introduced.

(1) *Novel features*

(a) Current harmonic vector (Bouhouras et al. 2019). Trying to make full use of the information contained in the current harmonics, the current harmonic vectors are proposed as follows:

$$LS_j = [\vec{I}_{1j}, \vec{I}_{3j}, \ldots, \vec{I}_{ij}, \ldots] \tag{5.12}$$

where LS_j is the feature of appliance j and $\vec{I}_{ij}$ is the current vector of harmonic order i appliance j, which is defined as below:

$$\vec{I}_{ij} = (A_{ij}, \varphi_{ij}) \tag{5.13}$$

where A_{ij} and φ_{ij} are the amplitude and phase angle of harmonic order i appliance j.

The current harmonic vectors are unique for its form and calculating principle. The vector form retains all the information from the harmonics and the principle of vector computation to avoid losing any information. Due to more information contained in the features, the disaggregation methods using harmonic vectors are more accurate and stable than the methods taking the current amplitudes or the x-axis projections of the harmonics. The harmonic vectors of some typical appliances are shown in Fig. 5.14.

(b) Weighted current harmonic vectors. The current harmonic vector is a novel and effective feature in the steady-state current decomposition. But in the disaggregation scheme, the information contained in various harmonic vectors is not weighted reasonably. On the one hand, the higher harmonics are quite sensitive to be affected by the fluctuation. And the fluctuation is often found in the current signals, which is resulted by many reasons, such as the errors in measurement and the instability in the power supply network. On the other hand, the information in higher harmonics is easy to be covered by the fundamental harmonic, because of the minor amplitudes of high harmonics. Thus, an appropriate weighted strategy is required to get rid of useless or incorrect information and enhance the effect of useful and accurate information.

By analyzing the standard deviation of the amplitude σ_a and that of the phase angle σ_p of all samples of an appliance type, these two indices are able to reflect the reliability of the harmonic. And then, a geometric method is utilized to calculate the standard deviation of the harmonic vectors σ of the samples from σ_a and σ_p, which is detailed as below:

$$\begin{aligned}\sigma &= \sqrt{\left[(A+\sigma_a)\cos\left(\varphi+\sigma_p\right) - A\cdot\cos\varphi\right]^2 + \left[(A+\sigma_a)\sin\left(\varphi+\sigma_p\right) - A\cdot\sin\varphi\right]^2} \\ &\approx 2\sqrt{A^2 + A\cdot\sigma_a}\sin\frac{\sigma_p}{2}\end{aligned} \tag{5.14}$$

where A, φ is the amplitude and phase angle of the harmonic. And then, the weight of each harmonic order is designed as below:

$$w_j = \gamma_j \cdot \left(1 - \frac{\sigma_j}{A_j}\right) \tag{5.15}$$

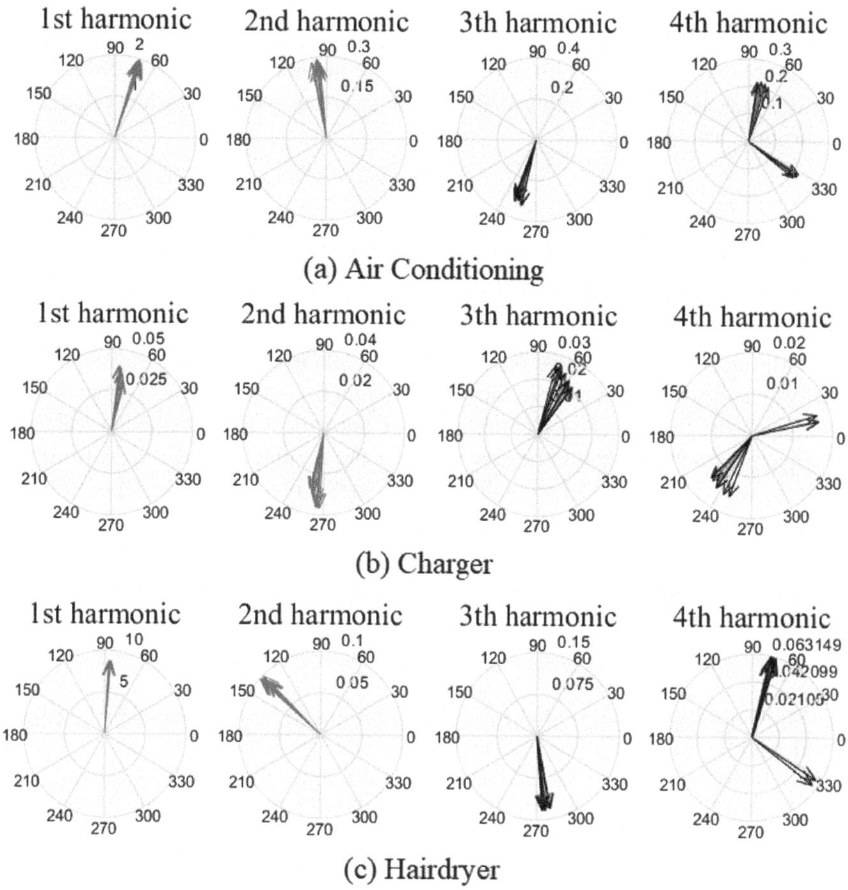

Fig. 5.14 Harmonic vectors of the appliance instances

$$w = \frac{1}{m}\sum_{j=1}^{m} w_j \tag{5.16}$$

where w_j is the weight of the harmonics of appliance j and w is the weight vector of the harmonics of the current superimposed by the appliances in an appliance combination; m is the number of appliances in an appliance combination; γ is a coefficient vector related to the harmonic amplitudes. By applying the weight vector to the harmonics, the steady-state current decomposition can be more accurate and robust.

(2) *Feature extraction*

The feature extraction methods are similar to the method introduced in Sect. 5.2.2. The difference is that the weights of all harmonics are also acquired when calculating the current harmonic vectors.

5.3.3 Multi-objective Optimization Methods in Steady-State Current Decomposition

The traditional single-objective optimization models in steady-state decomposition usually take the minimum error as the objective function, which is precise but incomplete. To ensure that the chosen combination of appliances is closest to the real case, the objective functions should be able to assess the accuracy of the results from different aspects. Thus, in order to enable more different objective functions employed in the optimization methods, the multi-objective optimization models are introduced. The section presents two multi-objective optimization methods to verify the effectiveness of the utilization of multiple objective functions. And the objective functions are also proposed.

(1) *Multi-objective optimization models*

(a) Multi-objective genetic algorithm (Non-dominated Sorting Genetic Algorithm-II, NSGA-II)

Genetic algorithm (GA) is a classical modern optimization model, which has developed quite maturely. And Deb et al. designed the NSGA-II algorithm in 2000 to improve the GA in complex and multi-objective optimization problems (Deb et al. 2000). Up till now, the NSGA-II model has also been applied widely in various aspects and it is always able to achieve good results. But in the steady-state current decomposition domain of NILM, the performance of NSGA-II model has not been verified. In the study, the NSGA-II model is carried out along with the multiple objective functions and the novel features. The basic searching principle of the NSGA-II model is similar to the GA algorithm, but several improvements have been made in the NSGA-II algorithm. Firstly, the NSGA-II model uses the fast elitist non-dominated sorting strategy, which not only improves the search efficiency but also ensures that the optimal solution already found will not be abandoned. Besides, the crowding degree is utilized in NSGA-II, which is computationally efficient and avoids using manually determined shared parameters. And the utilization of the crowding degree also guarantees the diversity of the population. By applying these improvements, the NSGA-II algorithm is more efficient and has better convergence. As to the application of the NSGA-II model in the steady-state decomposition, the input and output of the algorithm are similar to the GA algorithm introduced in Sect. 5.2.3. For that the NSGA-II model is a common algorithm in optimization issues, the details of it can be found in reference (Deb et al. 2000).

(b) Multi-objective particle swarm optimization model (MOPSO)

MOPSO model is derived from the PSO model, which is thoroughly introduced in Sect. 5.2.3. After the PSO model first proposed by Dr. Eberhart and Dr. Kennedy (Kennedy 1995). Coello et al. developed it into a multi-objective optimization method in 2002 (Coello and Lechuga 2002) and 2004 (Coello et al. 2004). The optimum value searching principle of MOPSO method is still based on the basic principle of the PSO method.

As the MOPSO method has multiple objective functions, the selection of the best particle becomes the main issue in the MOPSO model. And many approaches have been proposed to settle down the problem. In the model, the Pareto-optimal algorithm is utilized to find the global best article (Yang et al. 2009). And a probabilistic and domination based rule is employed to determine the local best particle. In the application of MOPSO, the population size and the maximum iteration should be preset and the steps of the method are clarified as below:

Step 1. Randomly choosing some combinations of appliances as the initial positions, calculate the adaptive values, and then the grids are created to find the dominant particles.
Step 2. Building the group, calculate the values of objective functions of all selected combinations and find the local best particle.
Step 3. Calculating the density of the non-dominated particles group and for each particle choose one corresponding global best particle;
Step 4. Recalculating the velocities and the positions of all particles, and then updating the group. Judge whether the function value is stable or the maximum iteration is reached, if not, go back to Step 3.

(2) *Objective functions*

The objective functions are designed for the current harmonic vectors and the weighted current harmonic vectors respectively, which are universal to both of the NSGA-II model and MOPSO model.

(a) Objective functions for current harmonic vectors

As the current harmonic vectors are calculated based on vector computing principles, the real current vectors are subtracted by the composite vectors under the vector principle when calculating the errors of estimation. Thus, the first objective function is proposed as follows:

$$\min \ f_1 = \sum_{j=1,3,5,7} \left| \dot{I}_{aj} - \sum_{i=1}^{n} \dot{I}_{ij} \right| = \sum_{j=1,3,5,7} \delta_j \tag{5.17}$$

where $\dot{I}_{aj}$ is the j-th harmonic vector of the aggregated current and $\sum_{i=1}^{n} \dot{I}_{ij}$ is the estimation of the aggregated current by the j-th harmonic vectors of n appliances in an appliance combination; δ_j is the error between the calculated value and the real value of harmonic j.

Additionally, it is insufficient to seek out the most accurate combination of appliances just by using the objective function of minimum errors. Since the final error is the summation of the errors of all harmonic orders, it is incapable to reflect the estimation accuracy of all harmonic orders. Thus, an index is needed to embody the global approximation between the calculated values and the real ones of all harmonics. The second function f_2 adopts the standard deviation of the errors of all harmonics, which is defined as below:

$$\min \quad f_2 = \sigma_e = \sqrt{\sum_{j=1,3,5,7} \frac{\left(\delta - \overline{\delta}\right)^2}{\left|\dot{I}_{aj}\right|}} \tag{5.18}$$

where σ_e is the standard deviation of the errors, $\overline{\delta}$ is the mean value of the errors, which is defined as follows:

$$\overline{\delta} = \frac{1}{4} \sum_{j=1,3,5,7} \left| \dot{I}_{aj} - \sum_{i=1}^{n} \dot{I}_{ij} \right| \tag{5.19}$$

(b) Objective functions for weighted current harmonic vectors

The difference between the objective functions of the weighted harmonic vectors and the harmonic vectors are the weights applied in the minimum error objective function, which is proposed as follows:

$$\min \quad f_1 = \sum_{j=1,3,5,7} w_j \cdot \left| \dot{I}_{aj} - \sum_{i=1}^{n} \dot{I}_{ij} \right| \tag{5.20}$$

And the other objective function is the same as the one which is mentioned above.

5.3.4 *Performance of the Novel Features and Multi-objective Optimization Models*

As shown in Fig. 5.13, the performance of the two proposed multi-objective methods and two special features are evaluated in the experimental part. The instances of the aggregated currents are introduced in Table 5.2. Analyzing the results of the steady-state current decomposition in Figs. 5.15, 5.16, 5.17 and 5.18 and Table 5.4, the advantages of multi-objective optimization and weighted harmonic vectors are verified.

(a) The utilization of multiple objective functions and multi-objective optimization models is proved to be able to enhance the precision of steady-state current

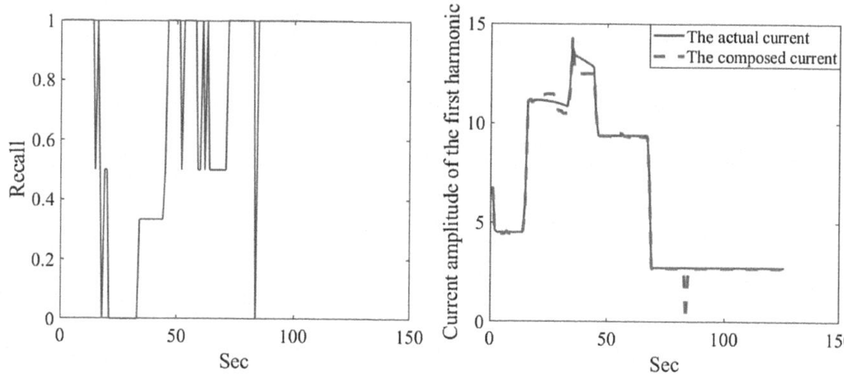

Fig. 5.15 Load #3 disaggregation results of the NSGA-II model with current harmonic vectors

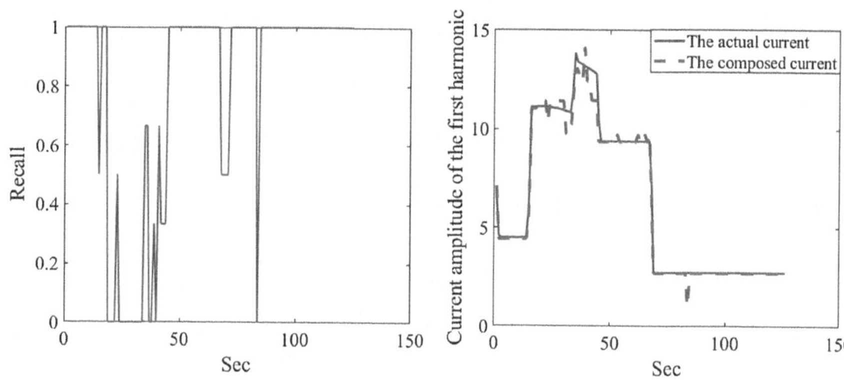

Fig. 5.16 Load #3 disaggregation results of the MOPSO model with current harmonic vectors

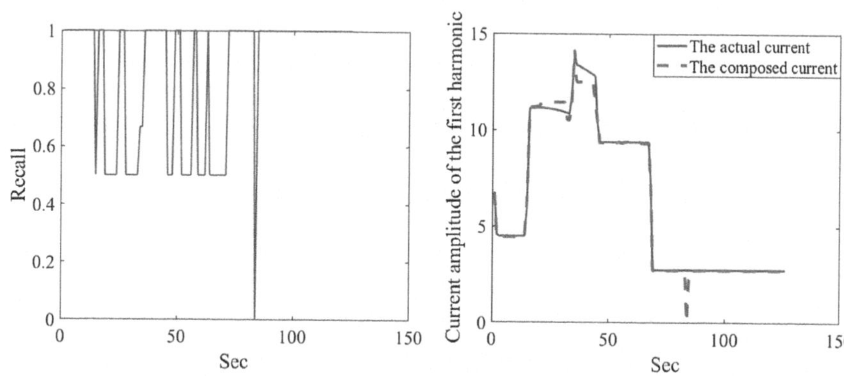

Fig. 5.17 Load #3 disaggregation results of the NSGA-II model with weighted harmonic vectors

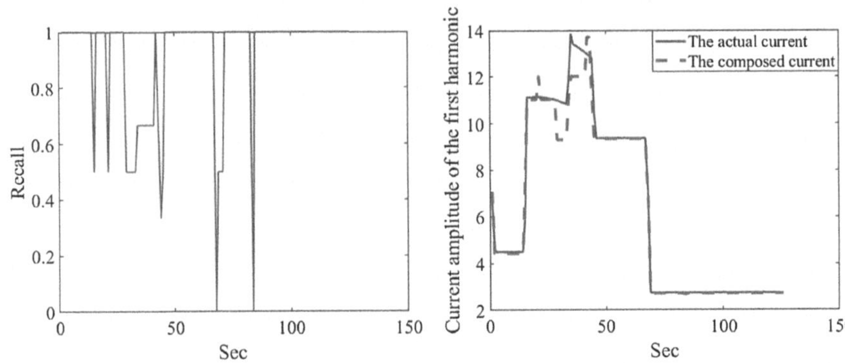

Fig. 5.18 Load #3 disaggregation results of the MOPSO model with weighted harmonic vectors

Table 5.4 The disaggregation results of multi-objective optimization models

Feature		Current harmonic vectors		Weighted harmonic vectors	
Method		NSGA-II	MOPSO	NSGA-II	MOPSO
#1	Precision	1.00	1.00	0.936	0.943
	Recall	0.688	0.664	0.859	0.871
	F1-score	0.819	0.802	0.896	0.906
#2	Precision	0.949	0.930	0.929	0.921
	Recall	0.685	0.712	0.802	0.795
	F1-score	0.796	0.85	0.861	0.853
#3	Precision	0.950	0.999	0.931	0.988
	Recall	0.657	0.710	0.825	0.885
	F1-score	0.777	0.830	0.875	0.933

decomposition. Most of the F1-score values of the experiments are higher than 0.80, which are satisfying. And compared to the methods exhibited in Sect. 5.2, these methods are more effective and accurate.

(b) The weighted current harmonic vectors are more differentiable than the common harmonic vectors. Analyzing the results in Table 5.4 and comparing the performances shown in Figs. 5.15 and 5.17, it can be discovered that methods using the weighted harmonic vectors as the load disaggregation features can achieve better results.

(c) When the features are distinctive, NSGA-II and MOPSO are both capable to find the best combination of appliances. Comparing Fig. 5.15 with Figs. 5.16 and 5.17 with Fig. 5.18, we find the performances of NSGA-II and MOPSO are similar. And comparing Figs. 5.15, 5.16 and Figs. 5.17, 5.18, the performance improvements of the two methods are both evident.

In summary, the weighted harmonic vectors and the multi-objective optimization methods are effective in steady-state current decomposition.

5.4 Current Decomposition Models Based on Non-negative Matrix Factor

The non-negative matrix factorization (NMF) was first proposed by Lee on Nature (Lee and Seung 1999). And after it was published, a great number of researchers were attracted to further explore it. The initial purpose of the NMF algorithm was to learn and extract the component parts of an image. While up to now, NMF has been widely applicated in many aspects, such as pattern recognition, signal analysis and so on. But few studies have verified the performance of NMF in the load disaggregation domain. NMF is a novel method in NILM with a bright perspective because the non-negative constraints of the method improve the interpretability of the load disaggregation results. In the section, the NMF technique and some variations of this method used in load disaggregation are discussed.

5.4.1 *Model Framework*

As shown in Fig. 5.19, the model framework for current decomposition using non-negative matrix factorization and reconstructed current data is exhibited. And the steps are introduced as follows:

(a) Reconstructing the current series into matrix form;
(b) Designing the objective function of the non-negative matrix factorization;
(c) Performing the non-negative matrix factorization models on the aggregated current signals, apply the optimization algorithms to the NMF models and break down the total current signal;
(d) Comparing the results obtained in *step (c)*, and calculate the evaluation indices. Then analyze the performance of non-negative matrix factorization models with different data reconstruction approaches.

5.4.2 *Reconstruction of the Data*

In order to apply NMF methods to the current decomposition, the current series should be reconstructed into two-dimension matrixes. And in the segment, two reconstruction methods are introduced as follows:

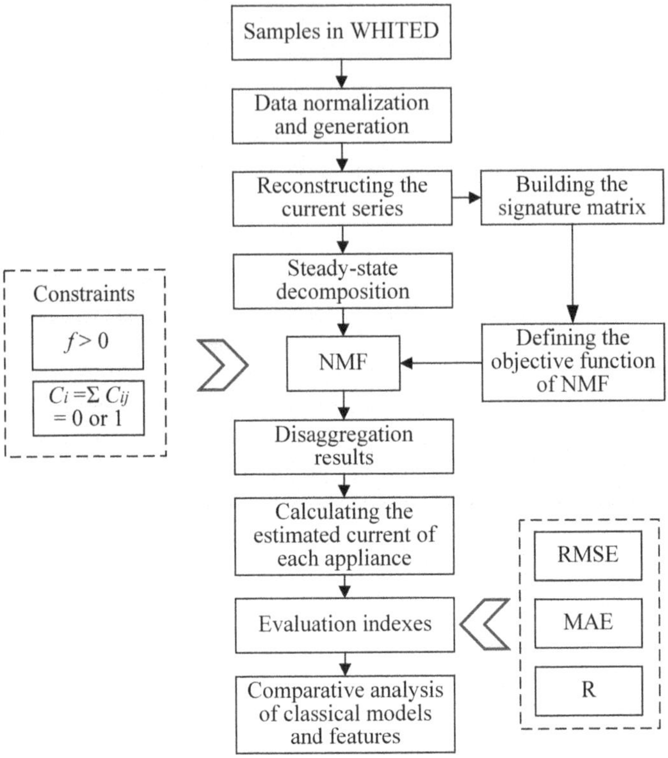

Fig. 5.19 The model framework of the NMF load decomposition method

(a) The phase space reconstruction method. If we consider the current as a time series $i(t)$, we can restructure the sequence by breaking it up into N contiguous windows of length M:

$$I = \left\{ \begin{array}{l} i(0), i(1), \ldots, i(M-1) \\ i(M), i(M+1), \ldots, i(2M-1) \\ \quad\vdots \\ i((N-1)M), i((N-1)M+1), \ldots, i(NM-1) \end{array} \right\} \tag{5.21}$$

(b) Reconstruction by decomposing the current series into several harmonic series. Since by FFT analysis the current series can be broken down into time series of various harmonics, the two-dimension form of the current is constructed as follows:

$$I = \begin{Bmatrix} i_1(t) \\ i_2(t) \\ \vdots \\ i_k(t) \end{Bmatrix} \tag{5.22}$$

where $i_k(t)$ is the time series of the current harmonic k.

5.4.3 Non-negative Matrix Factorization Method of the Current Decomposition

(1) *The basic theory of NMF*

NMF is a special matrix factorization that approximately decomposes the matrix with a non-negative constraint (García et al. 2018), which can be defined as follows:

$$V \approx WH \tag{5.23}$$

where W, H are the approximate factorization of V.

The elements in V, W, H are all non-negative, and all the vectors in V are weighted sums of the vectors in W. Analyzing the meaning of elements in W by treating them as the components of V. Besides, the elements of the vectors in H are clarified as that of the corresponding vectors in W. Thus, when NMF applied in the current decomposition, the components in W are relevant to the currents of the running appliances. But there are usually unfixed solutions of NMF decomposition since the estimation of V is alterable. This problem may lead to decomposition results that are not corresponding to the correct appliances.

(2) *Application in current decomposition*

As for the NMF applicated in current decomposition, Eq. 5.20 can be defined as follows:

$$I \approx SC \tag{5.24}$$

where I is the aggregated current signal; S is the signature matrix, which can be treated as the basic components of the signal; and C is the coefficient matrix. There are two important issues to be solved. One is the selection of the signature matrix, which is the basis of the non-negative matrix factorization. The signature matrix of this method is different from the features aforementioned. The signature matrix contains different current signals of appliances and is used to estimate the aggregated current signal. To establish a thorough signature matrix, most kinds of appliance samples should be contained in the signature matrix and the samples of the same appliance should also be various. The other issue is the determination of the objective function and the solution to this optimization problem. The objective function is defined as below:

$$f = \min_{f>0} \left\| I - [S_1, S_2, \ldots, S_i, \ldots S_n] \begin{bmatrix} C_1 \\ C_2 \\ \vdots \\ C_i \\ \vdots \\ C_n \end{bmatrix} \right\|^2 \tag{5.25}$$

where S_i is the signature matrix of appliance i and C_i is the corresponding coefficient matrix. In each signature matrix of an appliance, there are several different current samples, and the coefficient matrix is the same. The matrixes are shown as below:

$$\begin{cases} S_i = \left[S_{i1}, S_{i2}, \ldots, S_{ij}, \cdots, S_{ik}\right] \\ C_i = \left[C_{i1}, C_{i2}, \ldots, C_{ij}, \ldots, C_{ik}\right] \end{cases} \tag{5.26}$$

where S_{ij} is the current sample j of the appliance i in the signature matrix; same thing with the coefficient matrix. But as the unfixed solutions of NMF, another constraint should be added to the optimization. That is the sum of coefficients of an appliance should be equal to one (Rahimpour et al. 2017), which is embodied in the new objective function as below:

$$\begin{cases} f = \min_{f>0} \left\| I - [S_1, S_2, \ldots, S_i, \ldots S_n] \begin{bmatrix} C_1 \\ C_2 \\ \vdots \\ C_i \\ \vdots \\ C_n \end{bmatrix} \right\|^2 \\ C_i = \sum C_{ij} = 0 \; or \; 1 \end{cases} \tag{5.27}$$

And by a simple alternative non-negative least square method the optimal coefficient matrix is able to be reached. After the coefficient matrix is calculated, the estimated current signal can be acquired by the following equation (Rahimpour et al. 2017):

$$I_i = S_i C_i \tag{5.28}$$

5.4.4 Evaluation of the NMF Method in Current Decomposition

(1) *Evaluation indexes*

The evaluation indexes used in this segment is different from the aforementioned indexes. To make the analysis more clearly, the MAE, RMSE, and R-square are selected to evaluate the performance of NMF.

(a) MAE (Mean Absolute Error). The MAE index is able to reflect the precision of the models, which is formulated as follows (Willmott and Matsuura 2005):

$$MAE = \frac{1}{N}\sum_{i=1}^{N}|y_i - \hat{y}| \tag{5.29}$$

(b) RMSE (Root Mean Squared Error). The RMSE index can reflect the errors of the model more authentically. It is given below (Willmott and Matsuura 2005):

$$RMSE = \sqrt{\frac{1}{N}\sum_{i=1}^{N}(y_i - \hat{y}_i)^2} \tag{5.30}$$

(c) R-square. The index reflects the correlation between the two series. And it is calculated as below (Jawlik 2016):

$$R^2 = 1 - \frac{\sum_{i=1}^{N}(y_i - \hat{y}_i)^2}{\sum_{i=1}^{N}(y_i - \tilde{y}_i)^2} \tag{5.31}$$

(2) *Experiments*

In the experimental part, the first current reconstruction method is utilized. And due to the difficulty of the whole signal separation, the types of appliances contained in the signature matrix are cut down to three. The estimation results of the NMF method are exhibited as below:

As shown in Fig. 5.20 the estimated current is not very approximate to the real current. It is because the NMF utilized here is a little different from the original NMF method. The base matrix of the NMF models consists of several signals of the appliances in order to make the disaggregation more realistic but, as a result, the optimization direction is limited. Table 5.5 displays the estimation of the currents contained in the aggregated current. It can be seen from Fig. 5.20, the results are not very satisfying. The current values are not very precise.

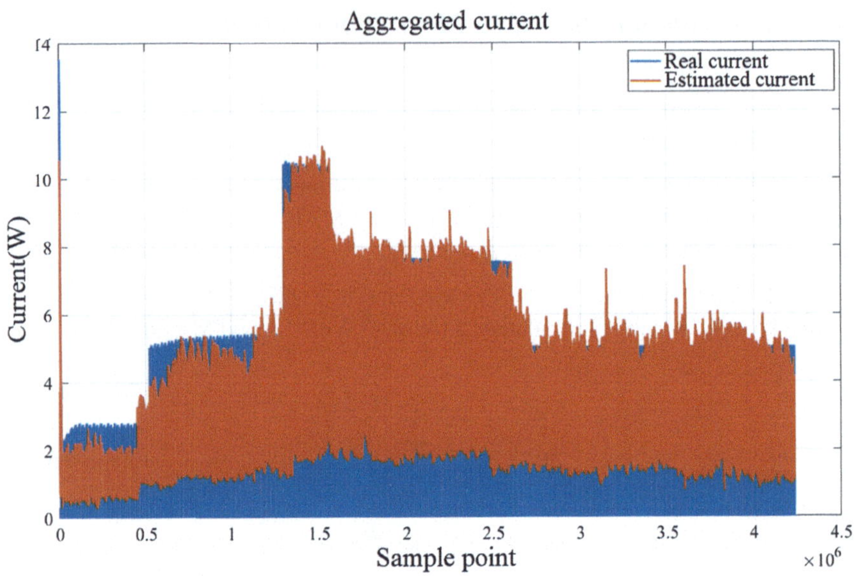

Fig. 5.20 The estimation of the aggregated current

Table 5.5 The disaggregation results of NMF

Feature		Aggregated current	Air conditioning	Iron	Toaster
#2	MAE	0.454	0.175	0.223	0.234
	RMSE	0.590	0.245	0.305	0.421
	R	0.966	0.958	0.944	0.979
#2 with current effective value	MAE	0.141	0.177	0.212	0.094
	RMSE	0.253	0.319	0.391	0.149
	R	0.985	0.975	0.962	0.991

Additionally, a decomposition experiment is added, using the current effective value. The results in Fig. 5.21 show that the error of the method is minor. This reflects the fact that it is easier to estimate when the curve is flatter. In summary, the NMF model is a method that still needs further research in order to be applicated in NILM. But as a blind source separation method, NMF has a great perspective on the load decomposition domain.

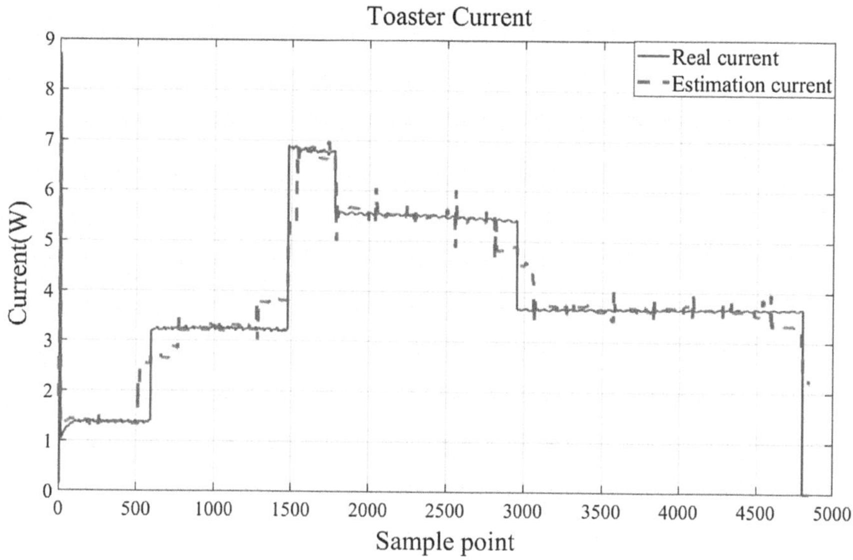

Fig. 5.21 The estimation of aggregated current effective value

5.5 Experiment Analysis

5.5.1 Data Generation

As stated in Sect. 5.1, the WHITED dataset is not enough for the steady-state decomposition study. In steady-state decomposition, the aggregated currents are important, which makes up the test set. However, the WHITED dataset lacks it. In the section, a man-made aggregated current generation method is adopted.

Because the samples in WHITED were measured in different areas, the voltage data of different appliances have various differences in the amplitude and period, which causes great trouble to the current superimposing method. Thus, a data normalization method is proposed. The data with larger voltage periods are processed by cutting down the points in the voltage signal at regular intervals. The interval N is defined as below:

$$N = \frac{s}{s - s_{\min}} \tag{5.32}$$

where s is the length of the voltage period; $s_{\min}$ is the length of the shortest period of all samples. As for the samples of all appliances in WHITED, the data only lasts for 5 s. This is also unbefitting for the current superimposing. To expand the original data, the data of the steady running state of an appliance is repeated and connected behind the start state of the appliance. By finding the identical zero-crossing points in the original series and the repeated series, the new data series is derived.

$$\begin{cases} \text{Find } u_j \in U_0, u_j \cdot u_{j+1} \leq 0 \\ \text{Find } u_i \in U_1, u_i \cdot u_{i+1} \leq 0 \end{cases} \tag{5.33}$$

where U_0 and U_1 are the original and repeated voltage series; u_j and u_i are the voltage points in U_0, U_1.

Finally, as for data generation, the currents superimposing method should obey Kirchhoff's Law in the AC circuit. Saleh et al. proposed a current composing method, which has already been verified by lab measurements (Saleh et al. 2018). The method superimposed the currents of different appliances by period, and then connected all the compounded one-period-long current series. The aggregated currents are available by this means.

5.5.2 Comparison Analysis of the Features Used in the Steady-State Decomposition

In the segment, all the disaggregation experiments are performed using the PSO method and various features, in order to compare the influence of different load signatures in the steady-state decomposition. As shown in Fig. 5.22 and Table 5.6, the results indicate that the current harmonic vector and the weighted harmonic vector are the most differentiable features. The disaggregation schemes using the current series and power features are not very stable and the precision of them is also lower.

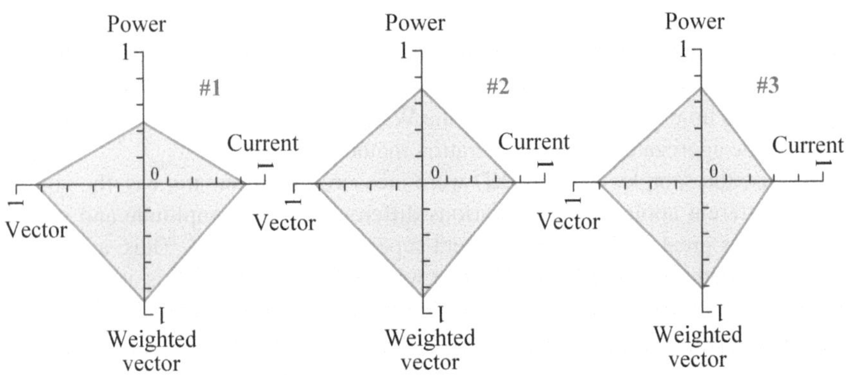

Fig. 5.22 The F1-scores of PSO method with various features

Table 5.6 The disaggregation results of PSO method with various features

Feature		Power	Current	Current harmonic vector	Weighted harmonic vector
#1	Precision	1.00	0.961	0.996	0.925
	Recall	0.315	0.744	0.729	0.872
	F1-score	0.479	0.838	0.841	0.898
#2	Precision	0.969	0.937	1.00	0.928
	Recall	0.560	0.628	0.685	0.786
	F1-score	0.709	0.751	0.813	0.851
#3	Precision	0.917	0.839	0.966	0.921
	Recall	0.603	0.461	0.693	0.727
	F1-score	0.728	0.595	0.807	0.812

5.5.3 *Comparison Analysis of the Models Used in the Steady-State Decomposition*

In order to analyze the performance of various models in steady-state decomposition, the experiments all employ the weighted harmonic vectors as their features. As shown in Fig. 5.23 and Table 5.7, the methods all have good performance using the weighted harmonic vectors, and two multi-objective optimization methods can achieve higher F1-score values. The results prove that the multi-objective methods and the weighted harmonic vectors contribute to a satisfying load disaggregation.

Fig. 5.23 The F1-score of different models

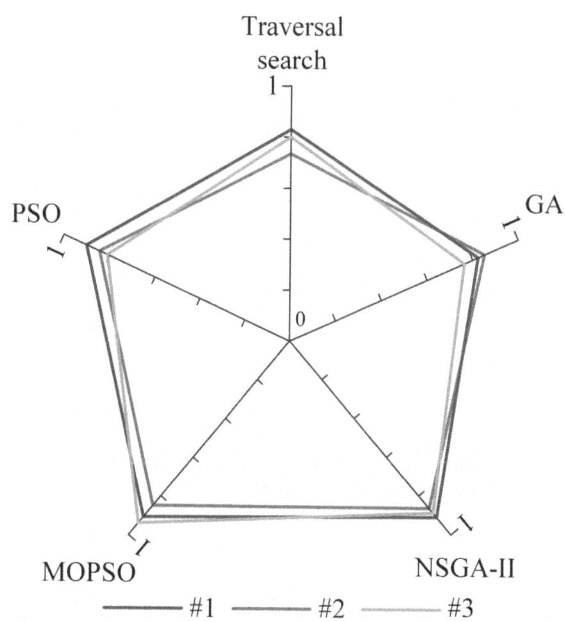

Table 5.7 The disaggregation results of different methods

Feature		Traversal search	GA	PSO	MOPSO	NSGA-II
#1	Precision	1.00	0.916	0.925	0.943	0.936
	Recall	0.746	0.741	0.872	0.871	0.859
	F1-Score	0.854	0.820	0.898	0.906	0.896
#2	Precision	0.819	0.977	0.928	0.921	0.929
	Recall	0.667	0.770	0.786	0.795	0.802
	F1-Score	0.735	0.861	0.851	0.853	0.861
#3	Precision	0.905	0.914	0.921	0.988	0.931
	Recall	0.714	0.653	0.727	0.885	0.825
	F1-Score	0.798	0.762	0.812	0.933	0.875

5.5.4 *Conclusion*

(a) The harmonic vectors contain more information about the appliances and are more distinctive in load disaggregation. The weighted harmonic vectors can further improve the precision of the current decomposition models.
(b) The multi-objective optimization methods usually have better performance than the single-objective optimization methods.
(c) The NMF method, as a blind source signal separation method, has been widely applicated in many aspects. But its application in NILM is not yet mature. How to improve the performance of NMF in NILM is still a hot spot in the future.

References

Bonfigli R, Principi E, Fagiani M, Severini M, Squartini S, Piazza F (2017) Non-intrusive load monitoring by using active and reactive power in additive factorial hidden Markov models. Appl Energy 208:1590–1607. https://doi.org/10.1016/j.apenergy.2017.08.203

Bouhouras AS, Chatzisavvas KC, Panagiotou E, Poulakis N, Parisses C, Christoforidis GC (2017) Load signatures development via harmonic current vectors. In: International universities power engineering conference

Bouhouras AS, Gkaidatzis PA, Panagiotou E, Poulakis N, Christoforidis GC (2019) A NILM algorithm with enhanced disaggregation scheme under harmonic current vectors. Energy Build 183:392–407. https://doi.org/10.1016/j.enbuild.2018.11.013

Bouhouras AS, Milioudis AN, Andreou GT, Labridis DP (2012) Load signatures improvement through the determination of a spectral distribution coefficient for load identification. In: 2012 9th international conference on the European energy market, 10–12 May 2012, pp 1–6. https://doi.org/10.1109/eem.2012.6254662

Chang HH, Lin LS, Chen N, Lee WJJIToIA (2013) Particle-swarm-optimization-based nonintrusive demand monitoring and load identification in smart meters. IEEE Trans Ind Appl 49(5):2229–2236

Coello CAC, Lechuga MS (2002) MOPSO: a proposal for multiple objective particle swarm optimization. In: Proceedings of the 2002 congress on evolutionary computation CEC '02 (Cat.

No.02TH8600), vol 1052, 12–17 May 2002, pp 1051–1056. https://doi.org/10.1109/cec.2002.1004388

Coello CAC, Pulido GT, Lechuga MS (2004) Handling multiple objectives with particle swarm optimization. IEEE Trans Evol Comput 8(3):256–279

Darby S (2006) The effectiveness of feedback on energy consumption. A Rev DEFRA Lit Metering, Billing Direct Disp 486(2006):26

De Baets L, Develder C, Dhaene T, Deschrijver D (2019) Detection of unidentified appliances in non-intrusive load monitoring using siamese neural networks. Int J Electr Power Energy Syst 104:645–653. https://doi.org/10.1016/j.ijepes.2018.07.026

Deb K, Agrawal S, Pratap A, Meyarivan T (2000) A fast elitist non-dominated sorting genetic algorithm for multi-objective optimization: NSGA-II. In: International conference on parallel problem solving from nature. Springer, pp 849–858

Fan YC, Liu X, Lee WC, Chen ALP (2012) Efficient time series disaggregation for non-intrusive appliance load monitoring. In: International conference on ubiquitous intelligence and computing and international conference on autonomic and trusted computing, pp 248–255

García D, Díaz I, Pérez D, Cuadrado AA, Domínguez M, Morán A (2018) Interactive visualization for NILM in large buildings using non-negative matrix factorization. Energy Build 176:95–108. https://doi.org/10.1016/j.enbuild.2018.06.058

Gonzalez-Pardo A, Ser J, Camacho D (2015) On the applicability of ant colony optimization to non-intrusive load monitoring in smart grids. In: Conference of the Spanish association for artificial intelligence

Hock D, Kappes M, Ghita B (2018) Non-intrusive appliance load monitoring using genetic algorithms. In: Conference series materials science and engineering, vol 366(1), p 012003

Jawlik AA (2016) r, Multiple R, r2, R2, R square, R2 adjusted. In: Statistics from A to Z. Wiley

Jian L, Ng SKK, Kendall G, Cheng JWMJIToPD (2010) Load signature study—part II: disaggregation framework, simulation, and applications. IEEE Trans Power Deliv 25(2):561–569

Kahl M, Haq AU, Kriechbaumer T, Jacobsen HA (2016) WHITED—a worldwide household and industry transient energy data set. In: International workshop on non-intrusive load monitoring, pp 1–5

Kennedy J (1995) Particle swarm optimization. In: Sammut C, Webb GI (eds) Encyclopedia of machine learning. Springer US, Boston, MA, pp 760–766. https://doi.org/10.1007/978-0-387-30164-8_630

Lee DD, Seung HSJ (1999) Learning the parts of objects by non-negative matrix factorization. Nature 401(6755):788

Lin Y-H, Tsai M-S (2015) The integration of a genetic programming-based feature optimizer with fisher criterion and pattern recognition techniques to non-intrusive load monitoring for load identification. Int J Green Energy 12(3):279–290

Liu H, Yu C, Wu H, Chen C, Wang Z (2019) An improved non-intrusive load disaggregation algorithm and its application. Sustain Cities Soc: 101918. https://doi.org/10.1016/j.scs.2019.101918

Liu Y, Wang X, Zhao L, Liu YJE, Buildings (2018a) Admittance-based load signature construction for non-intrusive appliance load monitoring. Energy Build 171:209–219

Liu Y, Xue W, Lin Z, Liu YJE, Buildings (2018b) Admittance-based load signature construction for non-intrusive appliance load monitoring. Energy Buil 171:209–219

Makonin S, Popowich F, Bajić IV, Gill B, Bartram LJIToSG (2016) Exploiting HMM sparsity to perform online real-time nonintrusive load monitoring (NILM). IEEE Trans Smart Grid 7(6):2575–2585

Miyasawa A, Fujimoto Y, Hayashi Y (2019) Energy disaggregation based on smart metering data via semi-binary nonnegative matrix factorization. Energy Build 183:547–558. https://doi.org/10.1016/j.enbuild.2018.10.030

Piga D, Cominola A, Giuliani M, Castelletti A, Rizzoli AEJIToCST (2016) Sparse optimization for automated energy end use disaggregation. IEEE Trans Control Syst Technol 24(3):1044–1051

Rahimpour A, Qi H, Fugate D, Kuruganti TJIToPS (2017) Non-intrusive energy disaggregation using non-negative matrix factorization with sum-to-k constraint. IEEE Trans Power Syst (99):1
Saleh A, Held P, Benyoucef D, Abdeslam DO (2018) A novel procedure for virtual measurements generation suitable for training and testing in the context of non intrusive load monitoring. In: Signal 2018 editors, p 36
Tabatabaei SM, Dick S, Xu WJIToSG (2017) Towards non-intrusive load monitoring via multi-label classification. IEEE Trans Smart Grid (99):1
Tang G, Wu K, Lei J, Tang J (2014) A simple model-driven approach to energy disaggregation. In: IEEE international conference on smart grid communications, pp 566–571
Willmott CJ, Matsuura KJ (2005) Advantages of the mean absolute error (MAE) over the root mean square error (RMSE) in assessing average model performance. Climate Res 30(1):79–82
Yang J, Zhou J, Liu L, Li YJC, Mw Applications (2009) A novel strategy of pareto-optimal solution searching in multi-objective particle swarm optimization (MOPSO). Comput Math Appl 57(11):1995–2000

Chapter 6
Machine Learning Based Appliance Identification

6.1 Introduction

The smart grid is a trend for future global grid development. As an important techniques of the smart grid, the Non-Intrusive Load Monitoring (NILM) was proposed by Professor Hart in the 1980s (Hart 1992). The target of NILM is load decomposition. The total load information of the user is obtained through the entrance detecting device, and then the information corresponding to each electrical device in the household is decomposed. Load characteristics identification and matching is an important part of NILM. It compares the extracted electrical characteristics with the load characteristic quantities in the established template feature library. If the similarity of the two characteristics meets a certain threshold range, it can be considered as an input or cut-out event of the electrical equipment. The NILM technology can mine electricity information from a deeper level. So users can optimize their power usage behavior and save electricity under the guidance of a new generation of electricity list (Froehlich et al. 2010).

Machine learning is widely used in NILM. Kamat adopted the fuzzy pattern recognition theory to calculate the running time of the electrical equipment according to the switching time information of the equipment (Kamat 2004). The load decomposition is finally completed. In 2005, Srinivasan et al. used artificial neural networks to identify different devices and the performance of Multi-Layer Perception (MLP), Support Vector Machine (SVM) and Radial Basis Function network (RBF) are compared (Srinivasan et al. 2005).

The chapter mainly uses machine learning methods to study the non-intrusive load identification of five common household appliances.

It can be found that the process of non-intrusive load identification can be divided into several parts: data preprocessing, load event detection, feature extraction, and load classification model establishment.

H. Liu, *Non-intrusive Load Monitoring*,
https://doi.org/10.1007/978-981-15-1860-7_6

The chapter analyzes the performance of different machine learning classifiers. After feature extraction and preprocessing of the PLAID dataset, the model of non-intrusive load identification and corresponding evaluation index system are established. Three machine learning classification models are trained based on the five types of selected equipment. In the trained model, the feature vector is the input and the device classification result is the output. Finally, the performance of different algorithms in NILM is analyzed in the comparison experiment.

6.2 Appliance Identification Based on Extreme Learning Machine

6.2.1 The Theoretical Basis of ELM

The Extreme Learning Machine (ELM) proposed by Huang belongs to Single Hidden Layer Feedforward Neural Network (SLFN) (Huang et al. 2004). The input and output layer of the ELM are connected only through one layer of neural network. The thresholds and weights of the ELM hidden layer are randomly generated. The above advantages make the learning speed of the ELM much higher than the traditional artificial neural network (Yoan et al. 2010).

The N input neurons, m output neurons and L hidden layer neurons determined the basic structure of the ELM. The nonlinear mapping functions $h = g(wx+b)$ and input samples $(x_i, y_i) \in R_n \times R_m$, $(i = 1, 2, 3, \ldots, N)$ are defined. The neuron output expression of the output layer is $h(x_i)\beta = y_i^T$. Where, the input weight is represented as w, the output weight is represented as β, and b is an offset corresponding to the hidden layer node. Assuming that the output of the hidden layer is represented as a matrix H, and the target matrix of the training data is represented as Y, then the neuron output of the output layer is expressed as follows:

$$H^*\beta = Y \tag{6.1}$$

$$H^* = \begin{bmatrix} g(w_1, b_1, x_1) & g(w_2, b_2, x_1) & \ldots & g(w_L, b_L, x_1) \\ \vdots & \vdots & \vdots & \vdots \\ g(w_1, b_1, x_N) & g(w_2, b_2, x_N) & \ldots & g(w_L, b_L, x_N) \end{bmatrix}_{N \times L} \tag{6.2}$$

where,

$$\beta = \begin{bmatrix} \beta_1^T \\ \beta_2^T \\ \vdots \\ \beta_L^T \end{bmatrix}_{L \times m} \tag{6.3}$$

$$Y = \begin{bmatrix} y_1^T \\ y_2^T \\ \vdots \\ y_N^T \end{bmatrix}_{N \times m} \tag{6.4}$$

6.2.2 *Steps of Modeling*

(a) Pre-processing the PLAID common dataset of load identification research. More than 20 feature quantities such as active power, reactive power, harmonics, etc., were extracted from the original dataset;
(b) Dividing the processed sample set into a training set and testing set. The input data of the training set and testing set are normalized and equalized;
(c) Training the ELM after initialization of parameters and get the trained ELM model;
(d) Inputting the testing set into the trained the ELM model for electrical equipment classification, and the prediction results are obtained;
(e) Calculating the corresponding evaluation indices for the output of *step (d)*. Accuracy, Recall, F1-score, and Precision are utilized to evaluate the performance of the classifier.

6.2.3 *Classification Results*

The classification results of extreme learning machines with different kernel functions are shown in Fig. 6.1 and Tables 6.1, 6.2 and 6.3.

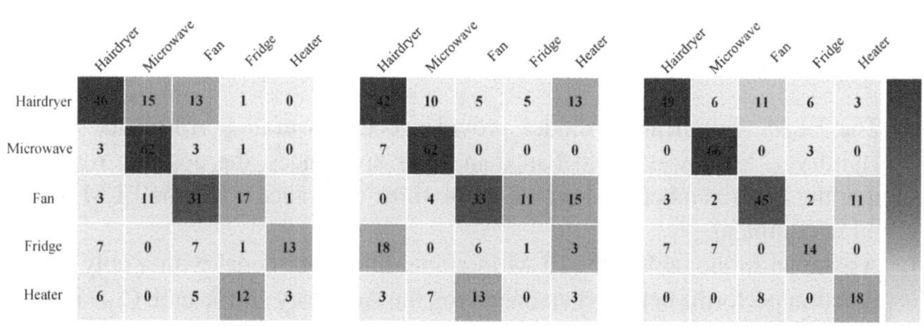

Fig. 6.1 Confusion matrix of the ELM-RBF/ELM-lin/ELM-poly model load identification results

Table 6.1 The classification performance indices of ELM-RBF model

Model	Type of appliance	Accuracy	Recall	F1-score	Precision
ELM-RBF	Hairdryer	0.8385	0.6133	0.6866	0.7797
	Microwave	0.9691	0.8986	0.9394	0.9841
	Fan	0.8649	0.4921	0.6392	0.9118
	Fridge	0.8962	0.0357	0.0690	1.0000
	Heater	0.9115	0.1154	0.2069	1.0000
	Total	0.8960	0.4310	0.5082	0.9351

Table 6.2 The classification performance indices of ELM-lin model

Model	Type of appliance	Accuracy	Recall	F1-score	Precision
ELM-lin	Hairdryer	0.8077	0.5600	0.6269	0.7119
	Microwave	0.9537	0.8986	0.9118	0.9254
	Fan	0.8456	0.5238	0.6226	0.7674
	Fridge	0.8962	0.0357	0.0690	1.0000
	Heater	0.9077	0.1154	0.2000	0.7500
	Total	0.8822	0.4267	0.4861	0.8309

Table 6.3 The classification performance indices of ELM-poly model

Model	Type of appliance	Accuracy	Recall	F1-score	Precision
ELM-poly	Hairdryer	0.7115	0.6533	0.5665	0.5000
	Microwave	0.9421	0.9565	0.8980	0.8462
	Fan	0.7104	0.7143	0.5455	0.4412
	Fridge	0.8231	0.5000	0.3784	0.3043
	Heater	0.7231	0.6923	0.3333	0.2195
	Total	0.7820	0.7033	0.5443	0.4622

According to the results in Fig. 6.1 and Tables 6.1, 6.2 and 6.3, the following conclusions can be drawn:

(a) The ELM classification model provides better overall performance on the Hairdryer, Microwave, and Fan than other appliances, the possible reason is that the large number of samples of the three appliance make the ELM model more complete.
(b) As shown in the tables, the ELM load identification model with a poly kernel function performs better when balancing the Accuracy and Recall. Considering the four indices, the ELM load recognition model with the RBF kernel function is more stable.
(c) The core of ELM is to transform complex iterative processes into hidden layer parameters. In addition, the choice of the weight of the ELM hidden layer

node and the choice of the kernel function greatly affects the final classification results. Prior knowledge can solve the performance degradation caused by the wrong choice. In general, the ELM can be used in the field of non-intrusive load identification.

6.3 Appliance Identification Based on Support Vector Machine

6.3.1 *The Theoretical Basis of SVM*

Support Vector Machine (SVM) theory is actually a convex quadratic programming problem. It has some advantages over many traditional methods. For nonlinear classification problems, SVM has better classification performance than linear classifiers, and it does not cause "dimensionality disaster". The support vector machine can handle the problem of small sample data. In addition, it does not fall into the local optimum and the final result does not deviates from the expected result (Jia-Yun et al. 2014).

The pioneering of support vector machine theory was first inspired by the idea of optimal classification hyperplane. The optimization problem of solving the optimal classification hyperplane in the feature space can be expressed as follows:

$$\begin{cases} \max W(\alpha) = \sum_{i=1}^{n} \alpha_i - \frac{1}{2} \sum_{i,j=1}^{n} y_i y_j \alpha_i \alpha_j K(x_i \cdot x_j) \\ s.t. \sum_{i=1}^{n} \alpha_i y_i = 0, 0 \le \alpha_i \le C, i = 1, \ldots, n \end{cases} \tag{6.5}$$

where $\alpha \geq 0$, $K(x_i \cdot x_j)$ represents a kernel function. So the corresponding decision function is expressed as follows:

$$f(x) = \mathrm{sgn}[\sum_{i=1}^{n} \alpha_i^* y_i K(x_i \cdot x) + b^*] \tag{6.6}$$

The parameter selection of the kernel function has a significant impact on the support vector machine load classification results. In the early days, the problem that the vector set of the low-dimensional space is hard to be divided is usually solved by mapping the low-dimensional space to the high-dimensional space. But such a move would dramatically increase the computational complexity. This problem can be effectively solved by choosing the appropriate kernel function.

6.3.2 *Steps of Modeling*

(a) Pre-processing the PLAID dataset of load identification research. As described in Sect. 6.2.2, more than 20 characteristic quantities are extracted from the original sequence;
(b) Dividing the processed sample set into training set and testing set. The input data of training set and testing set are normalized and equalized;
(c) Training the SVM model after initializing kernel function value, penalty function type and other parameters of the model. In this way, the trained SVM model is obtained;
(d) Inputting the divided testing set into the trained SVM model. Then, the prediction sample sequence relative to the real classification sequence is obtained;
(e) Calculating the corresponding evaluation indices which including Accuracy, Recall, F1-Score, and Precision for the output sequence of *step (d)*.

6.3.3 *Classification Results*

The classification results of support vector machine with different kernel functions are shown in Fig. 6.2 and Tables 6.4, 6.5 and 6.6.

According to the results in Fig. 6.2 and Tables 6.4, 6.5 and 6.6, the following conclusions can be drawn:

(a) It can be seen from the figures that all the three types of SVM models perform well for the identification of Microwave. The SVM load identification model with tan-h kernel function has a lower recognition accuracy rate for the real positive case when identifying the other four types of appliances. The effect is not good enough.

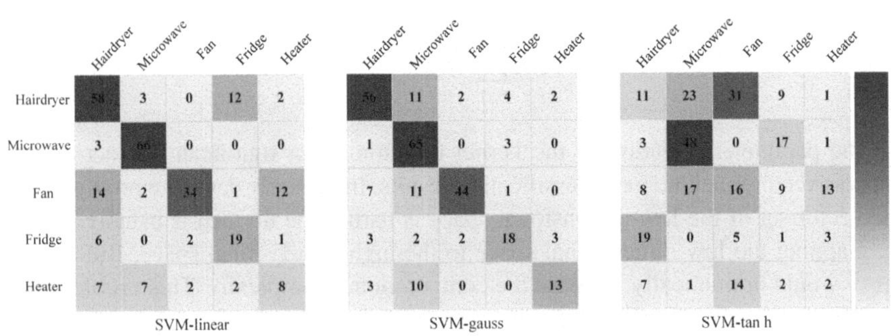

Fig. 6.2 Confusion matrix of the SVM-linear/SVM-gauss/SVM-tan h model load identification results

Table 6.4 The classification performance indices of SVM-linear model

Model	Type of appliance	Accuracy	Recall	F1-score	Precision
SVM-linear	Hairdryer	0.8423	0.7733	0.7389	0.7073
	Microwave	0.9768	0.9565	0.9565	0.9565
	Fan	0.8571	0.5397	0.6476	0.8095
	Fridge	0.9423	0.6786	0.7170	0.7600
	Heater	0.8808	0.3077	0.3404	0.3810
	Total	0.8999	0.6512	0.6801	0.7229

Table 6.5 The classification performance indices of SVM-gauss model

Model	Type of appliance	Accuracy	Recall	F1-score	Precision
SVM-gauss	Hairdryer	0.8500	0.7467	0.7417	0.7368
	Microwave	0.9846	0.9420	0.9701	1.0000
	Fan	0.8687	0.6984	0.7213	0.7458
	Fridge	0.9231	0.6429	0.6429	0.6429
	Heater	0.9192	0.5000	0.5532	0.6190
	Total	0.9091	0.7060	0.7258	0.7489

Table 6.6 The classification performance indices of SVM-tanh model

Model	Type of appliance	Accuracy	Recall	F1-score	Precision
SVM-tan h	Hairdryer	0.6115	0.1467	0.1789	0.2292
	Microwave	0.7992	0.6957	0.6486	0.6076
	Fan	0.7066	0.2540	0.2963	0.3556
	Fridge	0.7269	0.0357	0.0274	0.0222
	Heater	0.7115	0.0769	0.0506	0.0377
	Total	0.7111	0.2418	0.2404	0.2505

(b) The SVM load identification model with the tan-h kernel function is difficult to balance the Accuracy and Recall. If the two indices have the same weight, the F1-Score value is low. The SVM load identification model with a linear kernel function performs slightly better in this respect. It can be seen from the accuracy, recall, F1-Score, and precision that the gauss kernel function has the best over all performance.

(c) The above results may be caused by the imbalance of samples. The number of samples of the Fridge and Heater is significantly less than that of the other three electrical appliances. It is also because SVM classifiers require higher prior knowledge of users. The selection of kernel function and its initial parameters have a great impact on the performance of the final classifier.

In order to get better performance of the SVM classification model, the application of the optimization algorithm is a potential way to optimize each parameter. Another method is to add the pre-processing in the feature extraction part to improve the adaptability of the features. Besides, the feature selection method can sort the priority of the proposed features to avoid data redundancy. Generally speaking, the SVM model can be used in the field of non-intrusive load monitoring and recognition.

6.4 Appliance Identification Based on Random Forest

6.4.1 The Theoretical Basis of Random Forest

The Random Forest (RF) proposed by Leo Breiman is a statistical mathematical theory which is based on the decision tree. It combines bagging integrated learning theory (Kwok and Carter 1990) and random subspace method (Ho 1998).

The random forest adopts a bootstrap heavy sampling method to extract multiple samples from original samples and conduct decision tree modeling for each bootstrap sample. Then the prediction of multiple decision trees is combined and the final prediction result is obtained.

Supposing that the given data set is $D = \{X_i, Y_i\}$, $X_i \in R^K$, $Y_i \in \{1, 2, \ldots, C\}$, the random forest is made up of M decision trees and $\{g(D, \theta_m), m = 1, 2, \ldots, M\}$ is the base classifier. Then a combinatorial classifier is obtained after integration learning. When the samples to be classified are input, the classification results of random forest output are determined by the majority vote of the classification results of each decision tree. In the construction of the RF, there are some important definitions which are given as follows.

Sample Bagging: From the original sample set D, through the bootstrap, the M training samples D_m are randomly selected. Then a corresponding decision tree is built accordingly.

The random subspace of features: When splitting each node of the decision tree, a feature subset (usually Log_2K) is uniformly randomly extracted from all K features. Then choose an optimal split feature from this subset to build the tree.

The process of randomly selecting the training sample set and feature subset is independent. Therefore, $\{\theta_m, m = 1, 2, \ldots, M\}$ is an independent and identically distributed sequence of random variables.

Since the training of each decision tree is independent from each other, the training of random forests can be realized by parallel processing, which effectively guarantees the efficiency and scalability of the random forest algorithm.

6.4.2 Steps of Modeling

(a) Pre-processing the PLAID dataset of load identification research. As described in Sect. 6.2.2, 22 features are extracted from the original sequence;
(b) Dividing the processed sample set into the training set and testing set. The input data of the training set and testing set are normalized and equalized;
(c) Training the RF after initializing the number of decision trees and other parameters in the model. In this way the RF model is obtained;
(d) Inputting the segmented testing set samples into the trained RF model. Then, the prediction sample sequence relative to the real classification sequence is obtained;
(e) Calculating the corresponding evaluation indices which including Accuracy, Recall, F1-Score, and Precision for the output sequence of *step (d)*.

6.4.3 Classification Results

The results of the random forest classification model are shown as follows (Fig. 6.3, Table 6.7).

Machine learning is widely used in NILM. As the integrated algorithm is adopted, the accuracy of the random forest algorithm is better than most single algorithms.

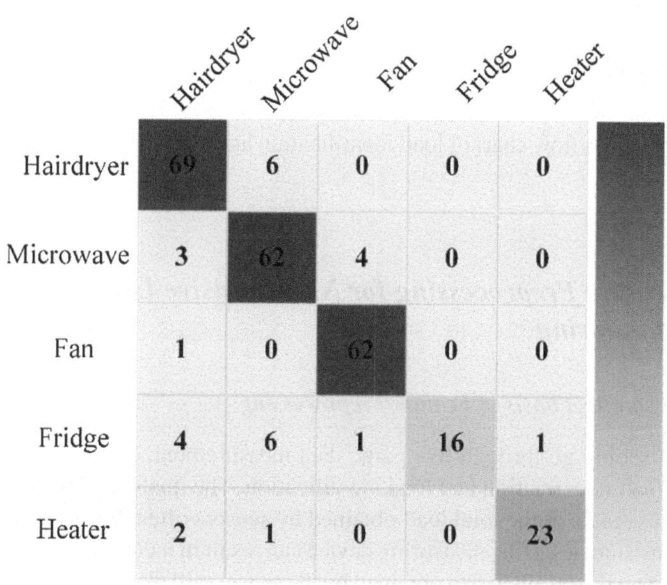

Fig. 6.3 Confusion matrix of the RF model load identification results

Table 6.7 The classification performance indices of the RF model

Model	Type of appliance	Accuracy	Recall	F1-score	Precision
RF	Hairdryer	0.9654	0.9200	0.9388	0.9583
	Microwave	0.9730	0.8986	0.9466	1.0000
	Fan	0.9807	0.9841	0.9612	0.9394
	Fridge	0.9538	0.5714	0.7273	1.0000
	Heater	0.9692	0.8846	0.8519	0.8214
	Total	0.9684	0.8517	0.8852	0.9438

Due to the introduction of sampling randomness and feature selection randomness of bootstrap, the random forest can avoid overfitting. Random forests can also be used to process high dimensional data without feature selection. In the case of the sample unbalance and random sample in the experiment, the classification performance loss is small. Due to the combination of trees, random forest can process nonlinear data. The random forest algorithm has strong anti-interference ability, which can reduce the dependence of the model on the feature data.

In general, random forest algorithms can be applied to non-intrusive load monitoring. In addition, the classification performance of random forests is superior than most single algorithms.

6.5 Experiment Analysis

6.5.1 Model Framework

The whole process flow chart of load identification and classification model is shown in Fig. 6.4.

6.5.2 Feature Preprocessing for Non-intrusive Load Monitoring

(1) *The theoretical basis of Feature Preprocessing*

The NILM should go through five parts: data measurement, data processing, event detection, feature extraction and load identification. The quality of the transient and steady-state signals of the total load obtained by sensors often fails to meet expectations. Inconsistencies in the measuring device can result in inconsistent measurement standards. And the original data obtained by the sensor will also cause data loss in the process of compression and transmission. These unfavorable factors make data preprocessing important. The essence of the NILM is data mining, so the pre-processing

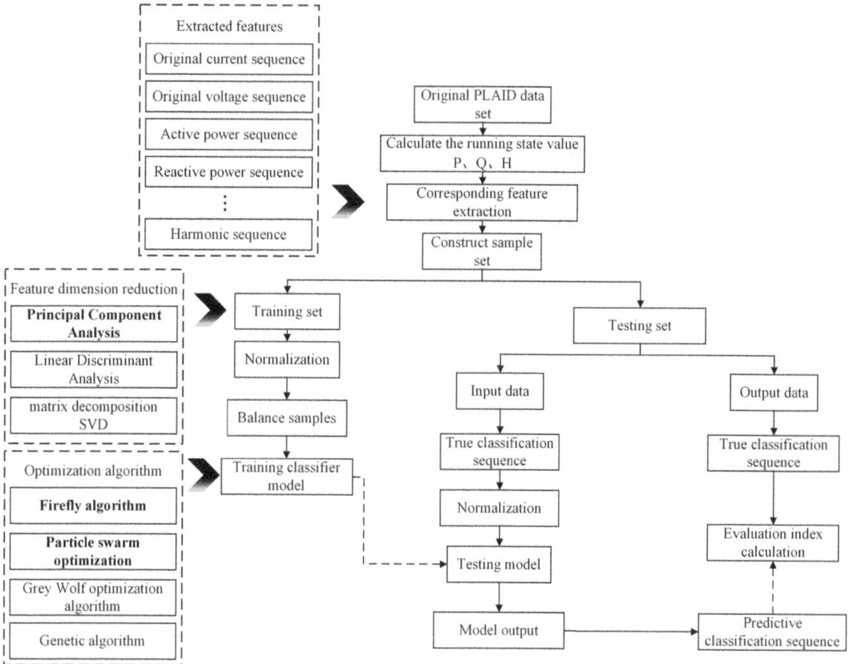

Fig. 6.4 Load identification and classification model

of data is not only the processing of original data but also the feature extraction before the final load identification operation. Some problems may occur when the extracted features are not processed.

In the practical case of the NILM, there might be too many features in the dataset. If a large number of these features are not screened before training the classification model, problems such as poor prediction effect, over-fitting or too long training time may occur. In this section, experiments are carried out on the dimensionality reduction in feature selection.

The data dimension reduction method is mainly divided into the linear method and the nonlinear method. The former mainly includes Principal Component Analysis (PCA), Linear Discriminant Analysis (LDA), Uniform Manifold Approximation and etc. The PCA and LDA have many similarities. Their essence is to map the original sample to a lower-dimensional sample space. However, the mapping goals of the PCA and LDA are different. The PCA is to maximize the divergence of the mapped samples while the LDA is to make the mapped samples have the best classification performance. The LDA is a supervised dimensionality reduction method.

This section uses the PCA as an example to perform dimensionality reduction experiments on 22 features extracted from Sect. 4.1.2.

The PCA is a non-marking method that can be used to reduce data dimensions. The goal of PCA is to map high-dimensional data to low-dimensional spatial representations by linear projection. But at the same time, it can make the subspace in the projection space satisfy the variance of the data in each dimension to the maximum. This allows the raw data to be represented in fewer dimensions while still retaining the distribution characteristics of the original data. Due to the retention of variance information, the PCA is also the most complete linear dimension reduction method for data retention in data reduction technology (Jolliffe and Cadima 2016).

(2) *Steps of modeling*

The main steps of the PCA can be divided into five steps.

(a) Calculating the sample covariance matrix;
(b) Calculating the eigenvalues and eigenvectors of the sample covariance matrix;
(c) Descend order according to the size of the eigenvalues;
(d) Selecting the number of principal components to be extracted;
(e) Calculating the corresponding m principal components.

(3) *The classification results after Feature Preprocessing Optimization*

The results of the classification model after PCA dimensionality reduction are shown as follows (Fig. 6.5, Tables 6.8, 6.9 and 6.10).

In this section, the experiments perform the PCA dimensionality reduction on 22 types of features. The features are extracted from Sect. 4.1.2. It is used to test whether the linear dimensionality reduction of features in this example can improve the classification performance of the classifier. It can be seen from the classification results of the classifier after dimension reduction processing. Compared with the recognition accuracy of the three classifiers that did not perform the PCA dimensionality reduction, the classification accuracy after dimensionality reduction decreased by about 5%. Other classification evaluation indicators have also declined to varying degrees. The specific reasons are analyzed as follows.

The effect of the PCA is to compress and save the most important information. In this case, the dimension of data is low, which leads to feature loss and incomplete

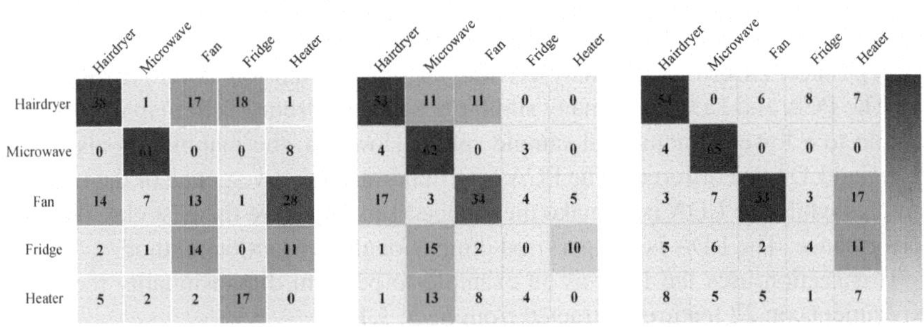

Fig. 6.5 Confusion matrix of the PCA-ELM-RBF/PCA-SVM-linear/PCA-RF model load identification results

Table 6.8 The classification performance indices of PCA-ELM-RBF model

Model	Type of appliance	Accuracy	Recall	F1-score	Precision
PCA-ELM-RBF	Hairdryer	0.8308	0.5067	0.6333	0.8444
	Microwave	0.9498	0.8841	0.9037	0.9242
	Fan	0.7876	0.2063	0.3210	0.7222
	Fridge	0.8923	0.0000	NaN	NaN
	Heater	0.9000	0.0000	NaN	NaN
	Total	0.8721	0.3194	0.6193	0.8303

Table 6.9 The classification performance indices of PCA-SVM-linear model

Model	Type of appliance	Accuracy	Recall	F1-score	Precision
PCA-SVM-linear	Hairdryer	0.7846	0.7067	0.6543	0.6092
	Microwave	0.9498	0.8986	0.9051	0.9118
	Fan	0.7915	0.5397	0.5574	0.5763
	Fridge	0.8923	0.0000	NaN	NaN
	Heater	0.9000	0.0000	NaN	NaN
	Total	0.8636	0.4290	0.7056	0.6991

Table 6.10 The classification performance indices of PCA-RF model

Model	Type of appliance	Accuracy	Recall	F1-score	Precision
PCA-RF	Hairdryer	0.8692	0.7200	0.7606	0.8060
	Microwave	0.9691	0.9420	0.9420	0.9420
	Fan	0.8378	0.5238	0.6111	0.7333
	Fridge	0.9038	0.1429	0.2424	0.8000
	Heater	0.9269	0.2692	0.4242	1.0000
	Total	0.9014	0.5196	0.5961	0.8563

training model after dimension reduction. In addition, the low linearity of the model can lead to poor results because the PCA can only solve the problem of linearly well-correlated models.

Figure 6.6 shows the variation of the ROC curve under different load identification models.

The ROC curve has a very good property: the ROC curve can remain unchanged when the distribution of positive and negative samples changes. Area Under Curve (AUC) is defined as the area under the ROC curve. The AUC value is used as the evaluation standard because in some cases the ROC curve cannot clearly explain which classifier has a better effect.

From Fig. 6.6 and Table 6.11, the advantages and disadvantages of the above-mentioned load identification models can be analyzed. In addition to the SVM-tan-h,

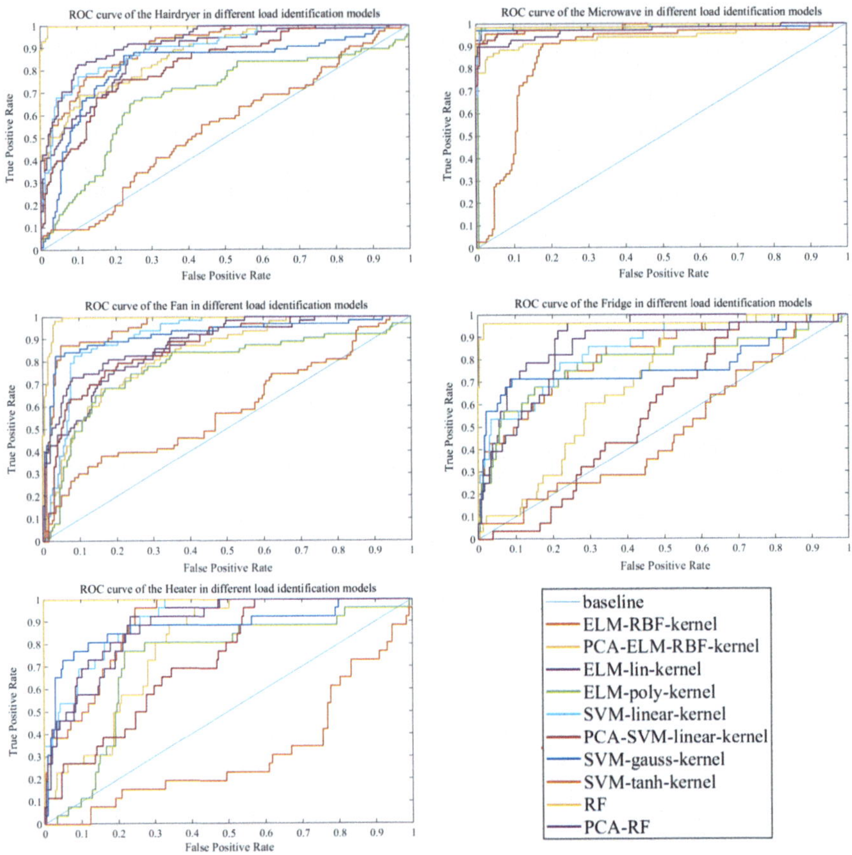

Fig. 6.6 ROC curve of the Hairdryer/Microwave/Fan/Fridge/Heater in different load identification models

PCA-SVM-linear and ELM-poly-kernel, the AUC values calculated by other models are all above 0.8, which proves the usability of the models. In this experiment, the SVM-linear, ELM-RBF, and RF models are trained using the features of the PCA dimensionality reduction process, which ultimately leads to different degrees of degradation of the classifier performance.

In general, the PCA has limited conditions and needs to be determined based on the circumstances of the extraction.

Table 6.11 AUC value comparison table for each classifier

Type of appliance		Hairdryer	Microwave	Fan	Fridge	Heater	Total
AUC	RF	0.9985	0.9934	0.9913	0.9721	1.0000	0.9911
	SVM-linear	0.9005	0.9906	0.9135	0.8522	0.9096	0.9133
	ELM-RBF	0.9089	0.9867	0.9535	0.8188	0.8905	0.9117
	SVM-gauss	0.8229	0.9846	0.9220	0.7749	0.8785	0.8766
	SVM-tan-h	0.5453	0.8673	0.5840	0.4880	0.3105	0.5590
	PCA-SVM-linear	0.8275	0.9782	0.8640	0.5771	0.7295	0.7953
	PCA-RF	0.9305	0.9696	0.8963	0.8439	0.8925	0.9066
	PCA-ELM-RBF-kernel	0.8739	0.9400	0.8303	0.6994	0.7950	0.8277
	ELM-poly-kernel	0.6859	0.9846	0.7760	0.7897	0.7234	0.7919
	ELM-lin-kernel	0.8713	0.9861	0.8565	0.9141	0.8861	0.9028
	PSO-SVM-linear	0.9369	0.9768	0.9321	0.9654	0.9247	0.9472
	FA-SVM-linear	0.9431	0.9988	0.9647	0.9721	0.9389	0.9635

6.5.3 *Classifier Model Optimization Algorithm for Non-intrusive Load Monitoring*

(1) *The theoretical basis of the optimization algorithm*

Sections 6.2 and 6.3 show that the kernel function of ELM and SVM has a great impact on the identification performance of the model. In the chapter, firefly optimization algorithm and particle swarm optimization algorithm are used to optimize the parameters of the kernel function.

Firefly algorithm (FA) proposed by Yang in 2008 is a meta-heuristic algorithm (Yang 2008). Fireflies move toward fireflies that are brighter and closer to their own. In this way, population evolution is carried out to achieve optimization.

FA is mainly based on the characteristics of fireflies illuminating for random optimization. Use the firefly individual to simulate the feasible solution. The target function value represents the brightness of fireflies. Brighter fireflies will attract other individuals to move into the direction. The attraction between them is inversely proportional to the distance. If there are no brighter individuals around a firefly, it chooses not to move or to randomly change its position.

The Particle Swarm Optimization (PSO) proposed by Kennedy and Eberhart in 1995 are inspired by group feeding behaviors such as flocks or fish schools (Eberhart and Kennedy 1995). The algorithm is a random search evolutionary algorithm for group collaboration. The model has the advantages of simple operation, easy realization of results and good parallelism. It has been used in many engineering fields (Rauf and Aleisa 2015; Khatami et al. 2017; Kumar and Inbarani 2017). The PSO is a random search algorithm based on group collaboration developed by simulating the foraging behavior of birds. The basic idea is to find the optimal solution through cooperation and information sharing among individuals in a group.

The PSO treats a bird as a particle, and the entire flock constitutes a particle swarm. Similar to other evolutionary algorithms, there are "groups" and "individuals". In the process of foraging, each bird seeks food information and provides it to the flock to share. The flock focuses on the direction of food and finally comes to a point, which is the optimal solution to seek. Through bionics to imitate the foraging process of birds, individuals in the particle group collaborate and compete to achieve optimal search for complex solutions in complex spaces.

(2) *Steps of modeling*

The optimization algorithm in the section is mainly for the Lagrangian multiplier upper limit c (penalty function) of the SVM and the kernel radius g of the kernel function. The FA and the PSO are used for multi-objective optimization. The specific optimization process is given as follows:

(a) Pre-processing the PLAID public dataset for load identification studies. 22 features are extracted from the original sequence;
(b) The processed sample set is divided into a training set and a testing set, and the input data of the training set and the testing set are normalized and equalized;
(c) Training the SVM after initializing the kernel function value, penalty function, and other parameters of the model. Here, the PSO and the FA are respectively introduced to train the two sets of optimal parameters;
(d) Initializing the parameters of the PSO, and the specific parameter settings are shown in Tables 6.12 and 6.13;
(e) The iterative calculation of the optimization algorithm yields the values of the penalty function c and the kernel radius g when the best recognition accuracy is obtained;
(f) Inputting the segmented testing set samples into the trained SVM model, and substitute the optimal parameters c and g obtained after the iteration. So a

Table 6.12 Initial parameter setting of particle swarm optimization algorithm

Model	PSO-SVM	
c1	1.5	PSO parameter local search ability
c2	1.7	PSO parameter global search capability
Max-gen	200	Maximum evolutionary quantity
Size-pop	20	Maximum number of population
k	0.6	Rate versus x (V = kX)
w-V	1	Elastic coefficient in the rate update formula
w-P	1	Elastic coefficient in the population update formula
v	3	Parameters for SVM cross validation
Pop-c-max	100	Maximum variation of SVM parameter c
Pop-c-min	0.1	The minimum value of the variation of SVM parameter c
Pop-g-max	1000	Maximum variation of SVM parameter g
Pop-g-min	0.01	The minimum value of the variation of SVM parameter c

Table 6.13 Initial setting of parameters for the firefly optimization algorithm

Model	FA-SVM	
n	20	number of fireflies
Max-generation	100	number of pseudo time steps
alpha	0.25	Randomness 0–1 (highly random)
Beta-m-n	0.2	minimum value of beta
gamma	1	Absorption coefficient

predicted sample sequence relative to the actual classification sequence is obtained;

(g) Calculating the evaluation indices corresponding to the output sequence of *step (f)*, including Accuracy, Recall, F1-Score, Precision, and AUC.

(3) *The classification results of the classifier model optimization algorithm*

The results of the classification model after PSO and FA parameter optimization are shown in Fig. 6.7 and Tables 6.14 and 6.15.

In the experiment, the punishment function c and kernel radius g of the SVM is optimized by the FA and PSO. From the above results, it can be seen that the classification performance of FA-SVM and PSO-SVM is improved when compared with the SVM-linear load identification classification model. The results of the confusion matrix show that the number of appliances identified by the optimized model is more concentrated on the main diagonal of the matrix. It can be seen from Tables 6.14 and 6.15 that the PSO algorithm improves the recognition accuracy of the SVM model by nearly 3%, and the Firefly algorithm improves the recognition accuracy of the SVM model by more than 4%. The value of several other evaluation indicators has

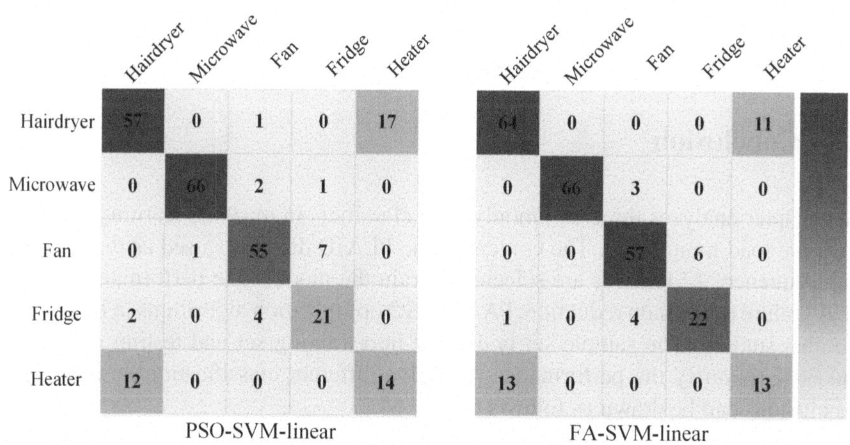

Fig. 6.7 Confusion matrix of the PSO-SVM-linear/FA-SVM-linear model load identification results

Table 6.14 The classification performance indices of PSO-SVM-linear model

Model	Type of appliance	Accuracy	Recall	F1-score	Precision
PSO-SVM-linear	Hairdryer	0.8769	0.7600	0.7808	0.8028
	Microwave	0.9846	0.9565	0.9706	0.9851
	Fan	0.9423	0.8730	0.8800	0.8871
	Fridge	0.9462	0.7778	0.7500	0.7241
	Heater	0.8885	0.5385	0.4912	0.4516
	Total	0.9277	0.7812	0.7745	0.7701

Table 6.15 The classification performance indices of FA-SVM-linear model

Model	Type of appliance	Accuracy	Recall	F1-score	Precision
FA-SVM-linear	Hairdryer	0.9077	0.8667	0.8442	0.8228
	Microwave	0.9885	0.9565	0.9778	1.0000
	Fan	0.9500	0.9048	0.8976	0.8906
	Fridge	0.9500	0.8148	0.8000	0.7857
	Heater	0.9115	0.5000	0.5306	0.5652
	Total	0.9415	0.8086	0.8100	0.8129

also been improved. This shows the superiority of the two optimization algorithms. The FA is better than PSO in the experiment.

Table 6.11 summarizes the comparison of AUC values for all models in the chapter. In general, better classification performance can be obtained by using the optimization algorithm. Figures 6.8, 6.9, 6.10, 6.11 and 6.12 show the results of classifier classification performance with different classification models, in which the performance of all the model are compared.

6.6 Conclusion

The chapter analyzes three commonly used classifiers in machine learning for non-intrusive load monitoring. The content of the PLAID dataset is used as the original data sequence. 22 features are selected to train the model. The performance of the PCA feature dimension reduction, FA and PSO optimization algorithms in the model are also studied. The sample set is divided into training set and testing set by 70 and 30% to verify the performance of twelve different classification models. The conclusions can be drawn as follows.

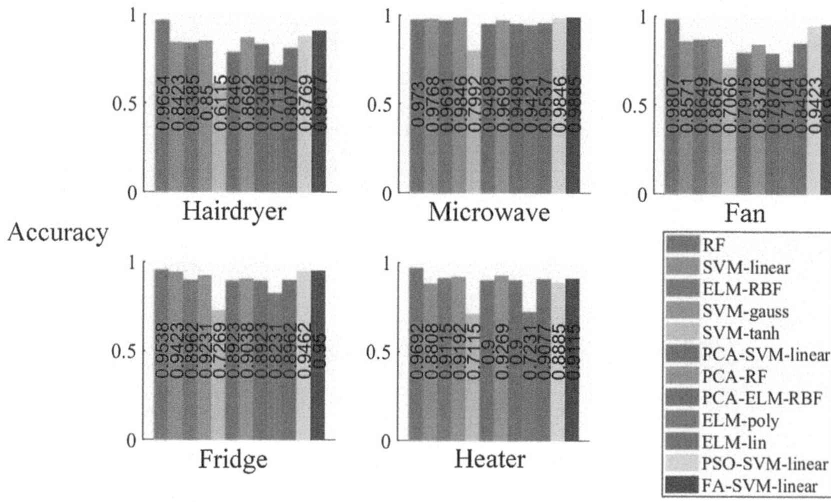

Fig. 6.8 Summary of classification models for accuracy indicators

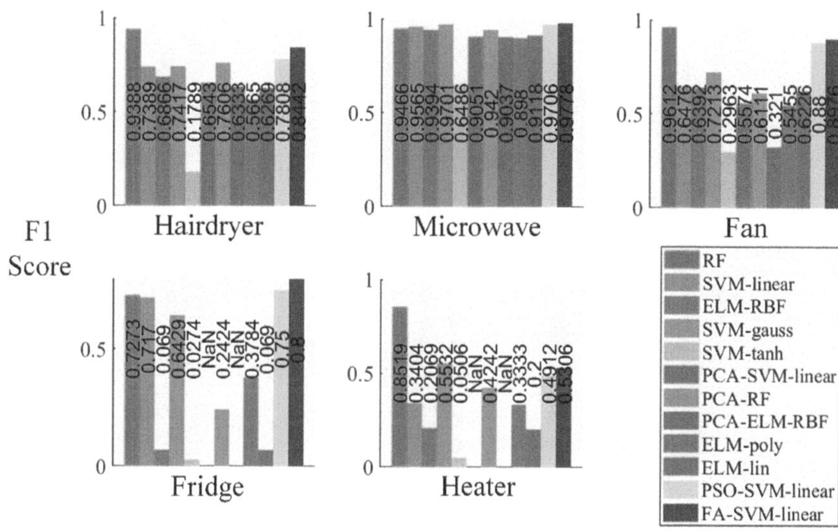

Fig. 6.9 Summary of classification models for F1-score indicators

(a) The ELM, SVM and random forest can be applied to the NILM with good classification results. Among them, the classification of the RF is the best.

(b) In order to reduce the adverse effects of data redundancy, the PCA dimensionality reduction is proposed to preprocess the feature data. The results show that the dimensionality reduction can improve the performance under certain conditions.

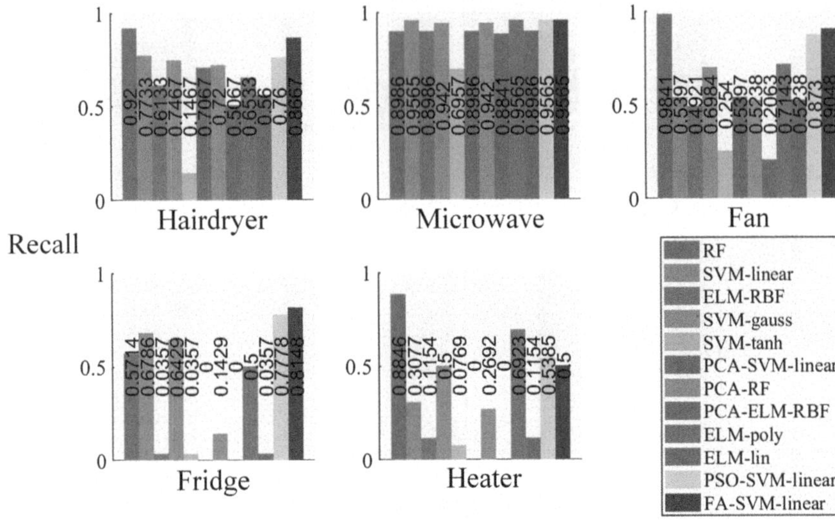

Fig. 6.10 Summary of classification models for recall indicators

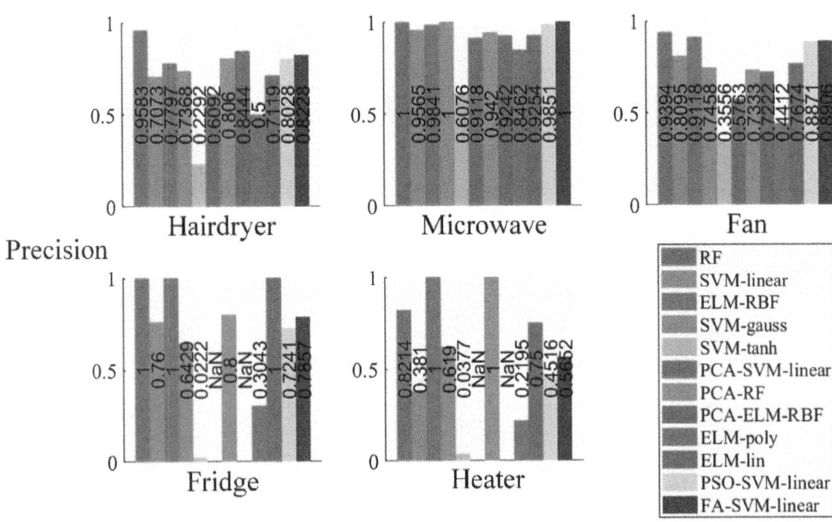

Fig. 6.11 Summary of classification models for precision indicators

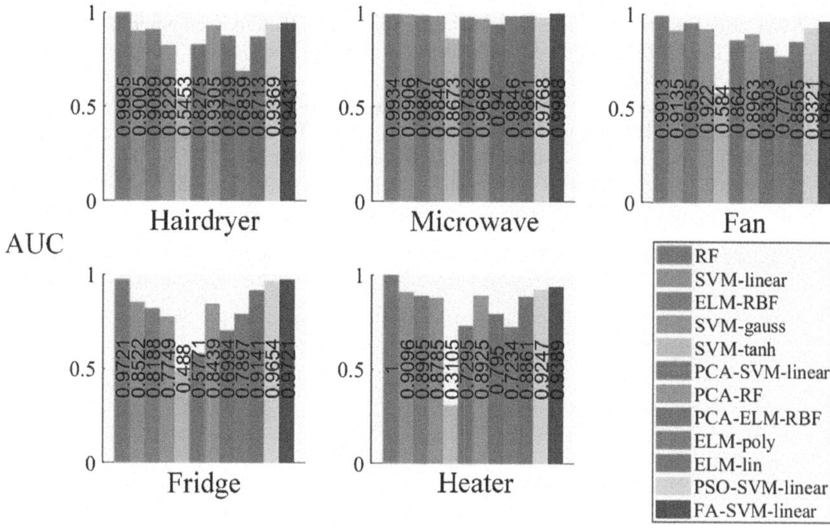

Fig. 6.12 Summary of classification models for AUC indicators

(c) In order to reduce the dependence on parameter selection, the optimization algorithm is used to optimize the penalty function and kernel radius of the SVM. The results show that the SVM models optimized by the PSO and FA have better performance.

References

Eberhart RC, Kennedy J (1995) A new optimizer using particle swarm theory. In: MHS'95 sixth international symposium on micro machine & human science, pp 39–43

Froehlich J, Larson E, Gupta S, Cohn G, Reynolds MS, Patel SN (2010) Disaggregated end-use energy sensing for the smart grid. IEEE Pervasive Comput 10(1):28–39

Hart GW (1992) Nonintrusive appliance load monitoring. Proc IEEE 80(12):1870–1891

Ho TK (1998) The random subspace method for constructing decision forests. IEEE Trans Pattern Anal Mach Intell 20(8):832–844

Huang GB, Zhu QY, Siew CK (2004) Extreme learning machine: a new learning scheme of feedforward neural networks. In: IEEE international joint conference on neural networks

Jia-Yun GU, Liu JF, Chen M (2014) A modified regression prediction algorithm of large sample data based on SVM. Comput Eng

Jolliffe IT, Cadima J (2016) Principal component analysis: a review and recent developments. Philos Trans A Math Phys Eng Sci 374(2065):20150202

Kamat SP (2004) Fuzzy logic based pattern recognition technique for non-intrusive load monitoring. In: Tencon IEEE region 10 conference

Khatami A, Mirghasemi S, Khosravi A, Lim CP, Nahavandi S (2017) A new PSO-based approach to fire flame detection using K-medoids clustering. Expert Syst Appl 68(C):69–80

Kumar SU, Inbarani HH (2017) PSO-based feature selection and neighborhood rough set-based classification for BCI multiclass motor imagery task. Neural Comput Appl 28(11):3239–3258

Kwok SW, Carter C (1990) Multiple decision trees. Mach Intell Pattern Recognit 9:327–335
Rauf A, Aleisa EA (2015) PSO based automated test coverage analysis of event driven systems. Intell Autom Soft Comput 21(4):491–502
Srinivasan D, Ng WS, Liew AC (2005) Neural-network-based signature recognition for harmonic source identification. IEEE Trans Power Deliv 21(1):398–405
Yang XS (2008) Nature-inspired metaheuristic algorithms. Luniver Press
Yoan M, Antti S, Patrick B, Olli S, Christian J, Amaury L (2010) OP-ELM: optimally pruned extreme learning machine. IEEE Trans Neural Netw 21(1):158–162

Chapter 7
Hidden Markov Models Based Appliance

7.1 Introduction

The application of unsupervised learning methods in NILM is developing more and more rapidly. Kong et al. proposed a smart meter data filter solver based on bootstrap to solve NILM problem (Kong et al. 2016a). Aiad and Lee proposed a clustering split method and tested it on the REDD common dataset (Aiad and Lee 2018b). Kong et al. presented a quick method in which an accurate optimal solution can be obtained (Kong et al. 2016b). Zhong et al. proposed a staggering factor inhomogeneous hidden Markov model (IFNHMM) by introducing two constraints (Zhong et al. 2014). Chen et al. proposed a method of NILM based on FHMM (Chen et al. 2017). Aiad and Lee proposed a quantized continuous state hidden Markov model based on a new transition matrix estimation method (Aiad and Lee 2018a). Choi and Yang proposed an improved FHMM method based on electrical correlation (Choi and Yang 2016). Song et al. proposed a non-invasive load monitoring and decomposition method based on factor hidden Markov model (Song et al. 2018). Wang et al. proposed a factor hidden Markov model (FHMMs) as an alternative to the generalized additive model (GAMs) method (Wang et al. 2013).

In the chapter, the active power is adopted as the input features. The HMM, FHMM, and HSMM are utilized to identify the running state of household electrical equipment, so as to model the power consumption of each appliance. The non-intrusive load decomposition experiment is carried out based on the AMPds data set. The experiment is conducted on the MATLAB 2018a running on a notebook with a 64-bit operating system, Intel i7 6700-2.8 GHz CPU, 8 GB RAM. The active power of three loads, including the basement (BME), dishwasher (DWE) and fridge (FGE) was selected as the research object. According to the appliance running state, the load is divided into ON/OFF load, Finite State Machines (FSM) load, and Continuously Variable Devices (CVD) load (Zoha et al. 2012). The three selected loads correspond to different types of loads. The BME belongs to ON/OFF load, the DWE belongs to FSM load, and the FGE belongs to the CVD load. Among the three types of loads,

H. Liu, *Non-intrusive Load Monitoring*,
https://doi.org/10.1007/978-981-15-1860-7_7

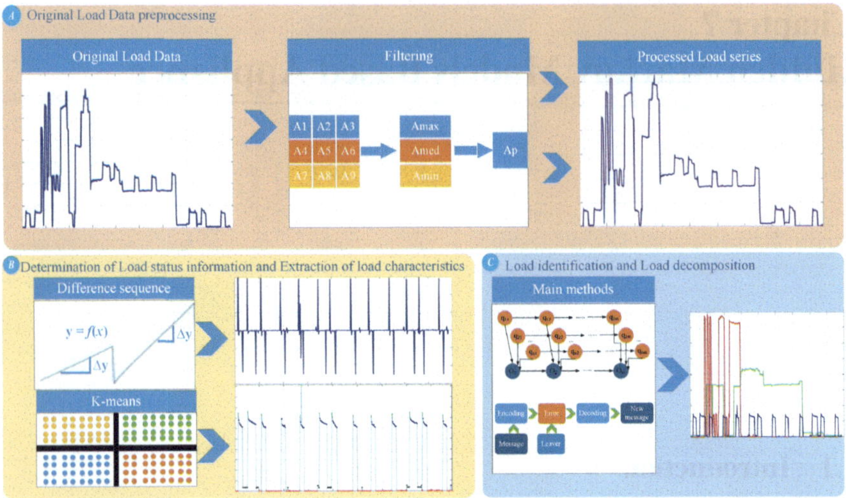

Fig. 7.1 The framework of the hidden Markov models based NILM model

there are not only simple ON/OFF load, but also complex CVD load and FSM load, which makes the research more real and comprehensive.

As shown in Fig. 7.1, the Hidden Markov Models based model is constructed as follows. Firstly, median filtering was used to preprocess the original active power data and remove noise data. Then, the k-means clustering algorithm was used to calculate the running state number and the corresponding power level of each load. The active power difference value was used for event detection, so as to determine the actual running state of each load. The training set and testing set were selected from the AMPds dataset and the experimental comparative analysis was conducted based on the FHMM and HSMM respectively.

7.2 Appliance Identification Based on Hidden Markov Models

7.2.1 Basic Problems Solved by HMM

The HMM is mainly used to solve evaluation, decoding and learning problems in reality (Rabiner and Juang 1986). The commonly used algorithms to solve the above problems are the forward-backward algorithm (Yu and Kobayashi 2003), the Viterbi algorithm (Hagenauer and Hoeher 1989), and the Expectation-Maximization (EM) algorithm (Moon 1996).

(a) Evaluation problem

Given the observation sequence $S(s_1, s_2, \ldots, s_r)$ and the model parameters $\lambda = (\pi, A, B)$, the probability of generating the observation sequence S under a given condition λ is calculated, which is equivalent to the evaluation of the model matching degree and a given observation sequence. When calculating the probability of the observation sequence, the forward-backward algorithms are generally adopted. The basic idea is to define the forward variable, according to the recursion of every moment and every state in order. In this way, the relationship between the forward variable at the tth time and the forward variable at the $t + 1$th time is set up and the value of forwarding variable in every moment is calculated. According to the definition of the forward variable, the calculation results of $P(S|\lambda)$ are obtained. Similarly, by defining the backward variable, the calculation results of the value of the backward variable in every moment can also be obtained according to its definition and recursive iteration formula (Manmatha and Feng 2006).

(b) Decoding problem

Given the observation sequence $S(s_1, s_2, \ldots, s_r)$ and model parameters $\lambda = (\pi, A, B)$, the optimal implicit state sequence $Q = q_1, q_2, \ldots, q_r$ for generating this observation sequence under given conditions is calculated. The Viterbi algorithm is usually used to solve this problem. The Viterbi algorithm can find the most likely implied sequence (Lleida and Rose 2000).

(c) Learning problem

Given the observation sequence $S(s_1, s_2, \ldots, s_r)$, the optimal estimation of HMM parameters $\lambda = (\pi, A, B)$ is calculated to acquire the maximum value of $P(S|\lambda)$. The most common method to solve such problems is the Expectation-Maximization algorithm (EM) (Bengio and Frasconi 1995).

7.2.2 *Data Preprocessing*

The AMPds dataset is utilized in this chapter. The active power of a single load and total load is shown in Fig. 7.2 and Fig. 7.3, respectively. The reliability and accuracy of data have a great impact on the accuracy of identification results. AMPds dataset measured from the real house is an improved version of the AMPds dataset. Thus, the AMPds dataset has strong reliability. However, it was found that 5–10% of abnormal data would be generated during the sampling process of data sets, causing deviation or error in the measured value relative to the true value, which is normal and inevitable. Therefore, the sampled data should be pre-processed to eliminate noise and interference factors in the data before state identification.

The commonly used data denoising methods include the mean filtering, the median filtering and so on (Yagou et al. 2002). In the chapter, the median filtering is adopted to denoise the active power data of load. The median filtering is usually used to

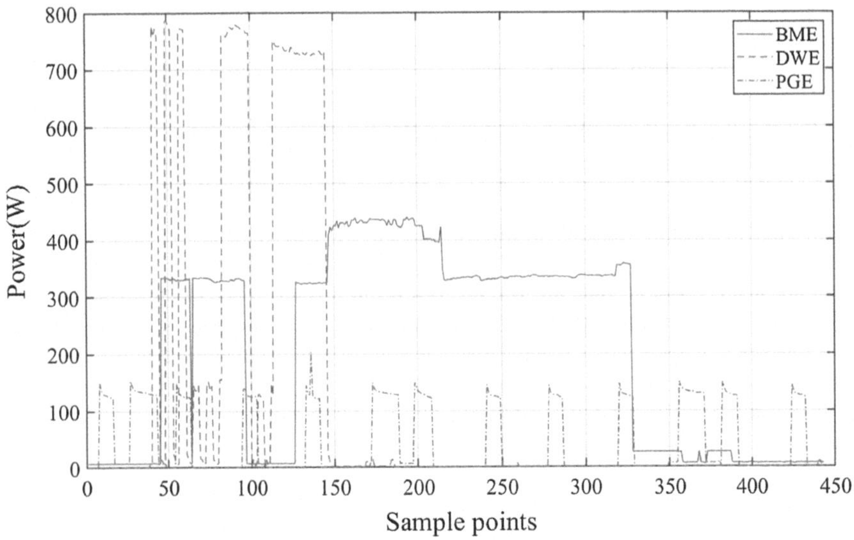

Fig. 7.2 The active power of single load

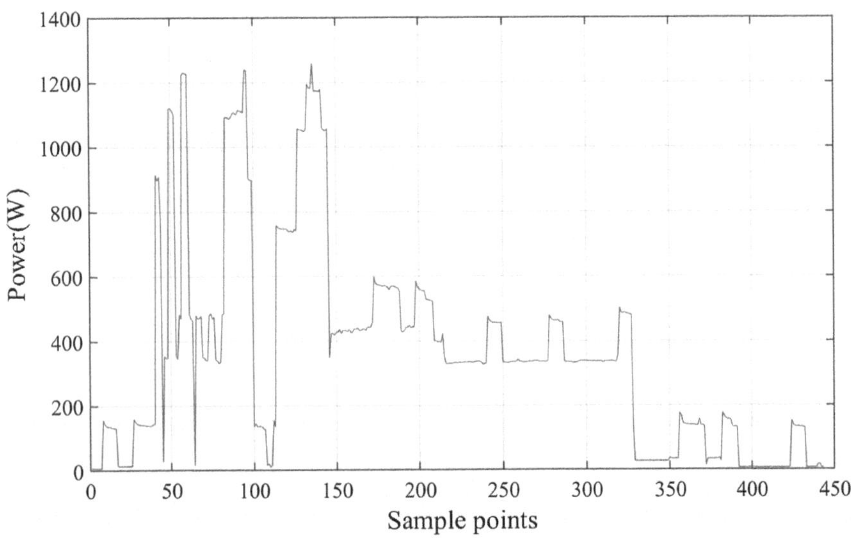

Fig. 7.3 The active power of total load

deal with nonlinear signal, whose theoretical basis is sorting statistics theory. It can effectively suppress the noise signals contained in the data. Median filtering has the advantages of simple theory, fast calculation speed and good performance in dealing with superimposed white noise and long-tail superimposed noise. Moreover, the self-adaptive ability of median filtering is very strong, which can further improve

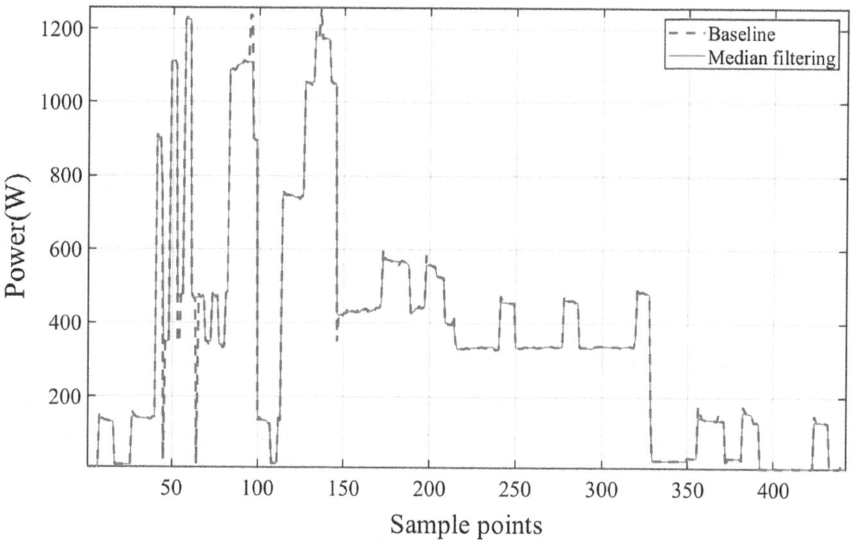

Fig. 7.4 The comparison of total load active power before and after the median filtering

the denoising performance. The calculation method of the median value in the data is shown as follows:

A set of number sequence: $x_1, x_2, \ldots, x_n$, the number is sorted by the size order of the value of n data, and then the median is (Lee and Kassam 1985):

$$y = med\{x_1, x_2, \ldots, x_n\} = \begin{cases} x_{i\left(\frac{n+1}{2}\right)} & n \text{ is singular} \\ \frac{1}{2}\left[x_{i\left(\frac{n}{2}\right)} + x_{i\left(\frac{n+1}{2}\right)}\right] & n \text{ is even number} \end{cases} \tag{7.1}$$

The load active power before and after the median filtering is shown in Fig. 7.4.

Compared to the original observation sequence, the pre-processed observation sequence has the following advantages:

(a) It retains the global variation advantage of the original observation sequence;
(b) It filters out the noise part of the original observation sequence;
(c) It enhances the information content of each data point in the transformed observation sequence.

7.2.3 Determination of Load Status Information

In the power load decomposition model based on FHMM, the hidden sequence of each HMM is the running state of the corresponding load, and the observation sequence is the active power of the corresponding load. When using FHMM for decomposition, it is necessary to know the hidden state set and observed state set in advance. The total operating state set and active power set of the load are the unions of the sets corresponding to a single load. It is necessary to figure out the number of running states and the corresponding active power level of each load in a typical working scenario. In addition, in order to evaluate the accuracy of load state identification, event detection in the process of the load operation is desired to know the real state sequence of the load. For instance, when a single load turns on, the operational state information of the load needs to be extracted from the total power signal. The AMPds dataset provides the real-time active power data of a single load, through which the load state information can be easily determined.

(1) Determination of the number of running states and active power level

The active power level estimation associated with the active power time series, but the active power time series in the same state presents continuous random changes within a certain range. In order to find the average value of the corresponding active power in the same running state, the K-means clustering algorithm is adopted for analysis. Due to its simple and efficient performance, it has been widely used in all clustering problems. In addition, the K-means clustering algorithm doesn't need to use the distribution information of samples in the clustering, as it chooses the center of mass that satisfies all the data convergence rules. In fact, a single sample may be generated between the clusters while the K-means clustering algorithm can distinguish the clusters that may be confused.

The basic steps of the K-means clustering algorithm are given as follows (Hartigan and Wong 1979):

Step 1: Selecting k cluster center points: $u_1, u_2, \ldots, u_k \in R^n$ respectively.
Step 2: Calculating the category c_i of each sample i, which is the category with the minimum Euclidean distance to the sample.
Step 3: Updating the center u_i of each category.
Step 4: Repeating step 2 and 3, until the quadratic sum of Euclidean distances converges.

Since the load power in the shutdown state is around zero and not involved in the K-means clustering algorithm, there is one more category of the real running state of the load beside the results of K-means clustering algorithm. In this article, the active power data of DWE, FGE, and BME were clustered by the K-means clustering algorithm respectively. The power data of one week was selected as clustering samples. The distribution of clustering results is shown in Fig. 7.5. The number of running states and active power levels of these three loads are shown in Table 7.1.

(2) Event detection

In order to evaluate the identification results of load state, we need to get the real state sequence of loads. It is necessary to determine the time when the running state changes, by detecting the event of the load (Leeb et al. 1995). The active power at time t is set as P_t, and the active power at load time $t-1$ is set as P_{t-1}. The difference value of the active power at time t is ΔP_t. And the difference value ΔP_t

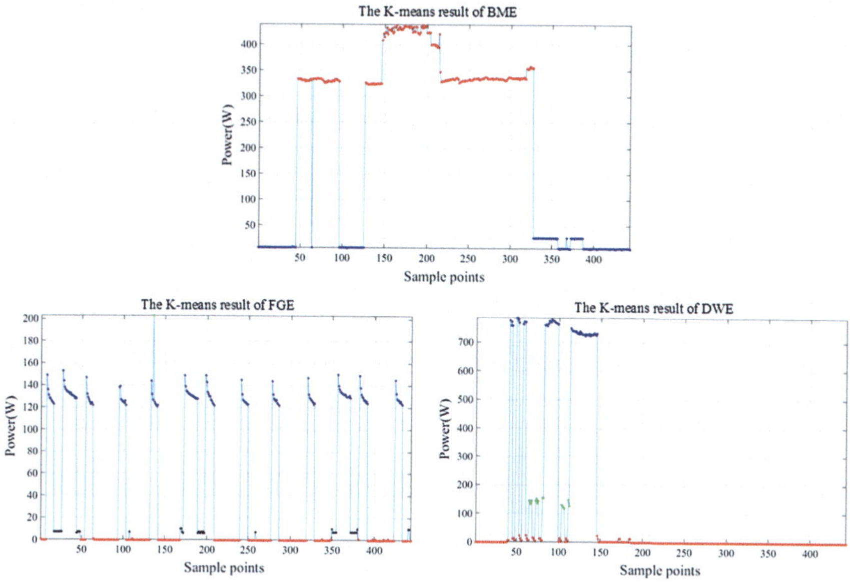

Fig. 7.5 Distribution of clustering results

Table 7.1 Running states and active power levels of each load

The serial number	Load	Number of states	Active power level (W)		
			State 0	State 1	State 2
1	BME	2	7	365	
2	DWE	3	2	144	770
3	FGE	3	0	7	130

of the active power is taken as the basis for event detection. The larger mutation value of ΔP_t is generally corresponding to the occurrence of the event. There are positive and negative mutation values of ΔP_t, and the corresponding events can be judged as rising edge events or falling edge events. If the result is positive, it is the rising edge event, indicating that the active power increases and the running operation or state of the load have changed. On the contrary, it is judged as the falling edge event, indicating that the active power decreases. Generally, it means that the load exits the running state or the running state changes to the low power. The state between different events is the stable operation interval of the load, corresponding to steady operation states of the load. The active power differential waveform of the total loads is shown in Fig. 7.6.

The state number and corresponding active power level of each load are obtained by the K-means clustering algorithm, and the occurrence point of load events is determined by the difference value of active power. Through the analysis of these, the state sequence of the load in the process of operation can be determined.

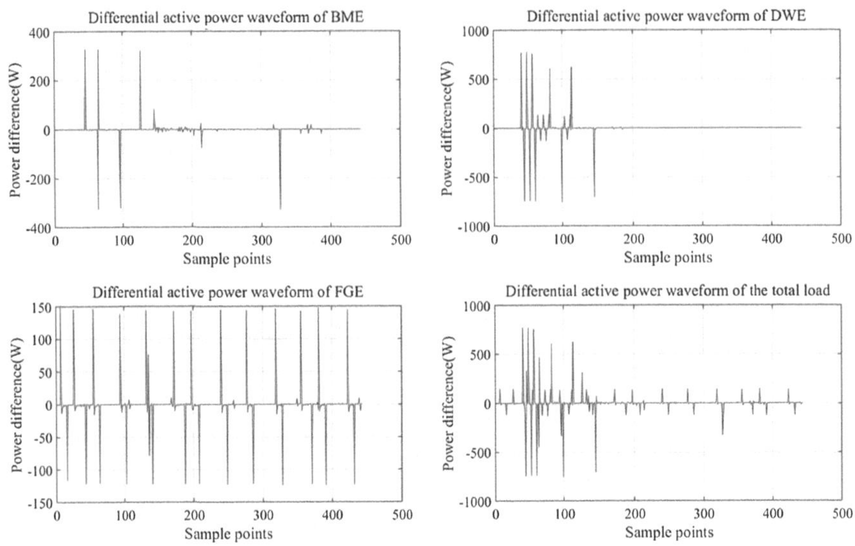

Fig. 7.6 Differential active power waveform of the total load

7.3 Appliance Identification Based on Factorial Hidden Markov Models

7.3.1 The Theoretical Basis of the FHMM

The power load model based on the FHMM can be used to model and analyze the fluctuating load by using the dual stochastic process. At the same time, the FHMM belongs to the unsupervised pattern recognition algorithm, which does not need the tag data of power load. Thus, reducing manual intervention and enhancing the practicability of the algorithm are important. The solution process does not need to carry on the complicated state combination. In addition, the combination of active power features and FHMM not only conforms to the structural characteristics of the FHMM algorithm itself but also conforms to the operation characteristics of load under actual conditions.

For load decomposition of N power equipment, its model can be described by FHMM. Each load can build its model based on an HMM, for load $i \in N$, the working state sequence can be represented by a Markov chain $Q^i\{q_1^i, q_2^i, \ldots, q_T^i\}$, and the power sequence of load i can be represented by $S^i\{s_1^i, s_2^i, \ldots, s_T^i\}$. Different from the single power load model, the active power value of a single load in this model is not observable, and the observed active power data $S^i\{s_1^i, s_2^i, \ldots, s_T^i\}$ are the total load at the power inlet. Load model parameters of N power equipment based on the FHMM can also be defined as $\lambda = (\pi, A, B)$, where

(a) π is the initial state probability, namely: $p\{q_1^1, q_1^2, \ldots, q_1^N\}$
(b) A is the probability matrix of state transition

$$P(q_t|q_{t-1}) = P\left(q_t^1, q_t^2, \ldots q_t^N|q_{t-1}^1, q_{t-1}^2, \ldots q_{t-1}^N\right) \tag{7.2}$$

(c) B is the matrix of output probability $p\{s_t|q_1^1, q_1^2, \ldots, q_1^N\}$

7.3.2 Load Decomposition Steps Based on FHMM

(1) Model parameter estimation

The parameter estimation problem of the power load model based on FHMM is similar to the learning problem of HMM. It is necessary to use the observed load power data as training samples to calculate the model parameters. The decomposition model adopted in this section is an unsupervised algorithm, and the estimation of parameters only requires the total active power data of the load, rather than the active power data of the single load.

In the power load model described by FHMM, each HMM is independent of each other. Therefore, the initial state probability and state transition probability $P(q_t|q_{t-1})$ can be expressed as:

$$\pi = \left\{q_1^1, q_1^2, \ldots q_1^N\right\} = \prod_{i=1}^{N} P\left(q_1^i\right) = \prod_{i=1}^{N} \pi^i \tag{7.3}$$

$$P(q_t|q_{t-1}) = \prod_{i=1}^{N} P\left(q_t^i|q_{t-1}^i\right) \tag{7.4}$$

In the power load model, the total active power sequence S is the linear superposition of each HMM output, it can be explained as below:

$$S = \sum_{i=1}^{N} S^i \tag{7.5}$$

At this point, the active power sequence of the state satisfies:

$$S|\left(q^1, q^2, \ldots q^n\right) \cdot N(u, C) \tag{7.6}$$

(2) Load state estimation

The state estimation problem of the load is to find the optimal state sequence of load in the model when the total active power data of load $S\{s_1, s_2, \ldots, s_T\}$ and model parameters λ are known. As $S\{s_1, s_2, \ldots, s_T\}$ is known, the optimization objective function equal to the maximum joint probability of state sequence and observation sequence which can be explained as follows:

$$\max_{Q} g(Q) = P(q_1, q_2, \ldots q_T, s_1, s_2, \ldots s_T) \tag{7.7}$$

This kind of problem is similar to the decoding problem in the standard HMM problem, which is generally solved by the Viterbi algorithm. This article uses extended Viterbi algorithm to solve the load state estimation problem.

7.3.3 Load Power Estimation

After obtaining the working status sequence of the load, it is possible to estimate the output active power sequence of each load. That is, given the model parameters λ, the total active power data of the load $S\{s_1, s_2, \ldots, s_T\}$ and the operation status of each load $Q^i\{q_1^i, q_2^i, \ldots, q_T^i\}$, the active power of each load $S^i\{s_1^i, s_2^i, \ldots, s_T^i\}$ should be solved. Considering the probability relationship between single load state and observed power values, as well as the linear addition relationship between total load and single load, the optimization model can be concluded as follows (Kolter and Johnson 2011):

Objective function:

$$\max \prod_{t=1}^{T} \prod_{i=1}^{N} P\left(s_t^i | q_t^i\right) \tag{7.8}$$

Constraints:

$$s_t = \prod_{i=1}^{N} s_t^i \tag{7.9}$$

$$s_t^i \geq 0 \tag{7.10}$$

The output probability in the objective function is expressed as:

$$P\left(s_t^i | q_t^i\right) = \sigma_i^{-1} (2\pi)^{-\frac{1}{2}} \exp\left(-\frac{1}{2\sigma_i^2}\left(s_t^i - u_{qt}^i\right)^2\right) \tag{7.11}$$

The flow chart of FHMM is shown as below (Fig. 7.7):

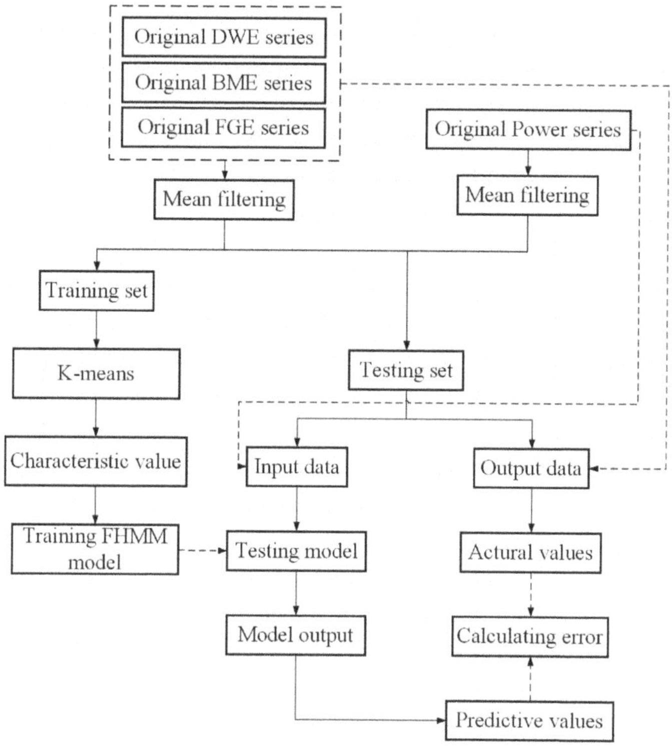

Fig. 7.7 The flow chart of the FHMM model

7.3.4 Decomposition Experiment Based on FHMM

In this section, the FHMM algorithm and the HSMM algorithm are used to model the simultaneous operation of multiple loads. The active power data of two weeks in AMPds are selected as the training set and the data of the following one week as the test set. Since the result of load decomposition depends on the type of load considered and the number of loads involved, the decomposition experiment was conducted under a different number of loads to evaluate the decomposition performance under different circumstances. For this purpose, two test groups were created for the three kinds of loads selected. The number of loads in the test group is N ($N = 2, 3$), and each test group contains test sets composed of different loads. The combination of loads is shown in Table 7.2.

(1) Combination and decomposition results of BME and DWE

The decomposition results of BME and DWE are given in Figs. 7.8 and 7.9. It can be found that since BME belongs to the ON/OFF load, there are two operating states of total copolymerization when K-mean clustering is carried out, and the power of the open state fluctuates from 310 to 440 W. DWE belongs to the FSM load, which

Table 7.2 The combination of 3 kinds of load

N	Combination	Number of combinations
2	(BME, DWE) (BME, FGE) (DWE, FGE)	3
3	(BME, DWE, FGE)	1

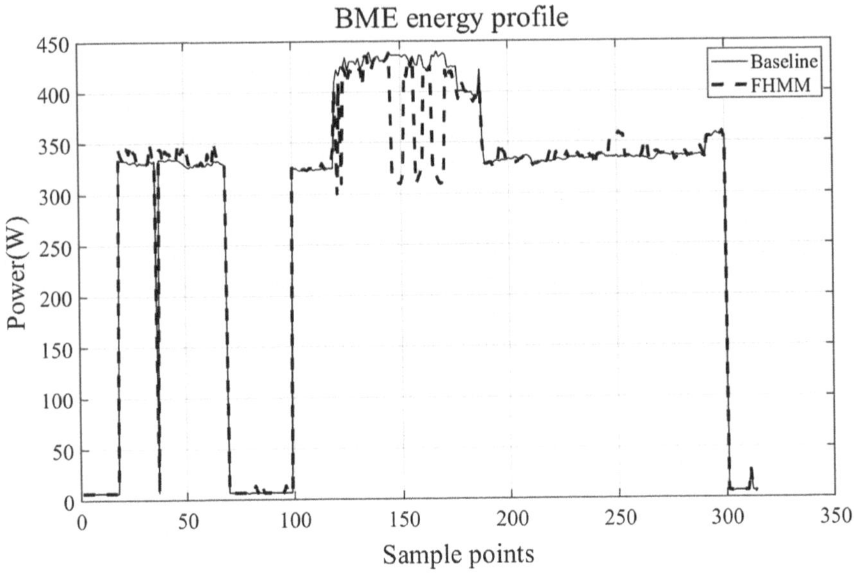

Fig. 7.8 Comparison of BME decomposition and actual active power

has three operating states, and the second state power fluctuates between 110 and 140 W.

Due to a certain over-fitting phenomenon in the modeling of HMM in the FHMM algorithm, when the total transient state power of the load is around 436 W, the load state of the BME is decomposed into the second state. During that state the power of the BME is about 310 W. Similarly, the load state of the BME is decomposed into the second state where the power is about 135 W.

(2) Combination decomposition results of BME and FGE

The decomposition results of BME and FGE are given in Figs. 7.10 and 7.11, respectively. It can be found that BME belongs to the FSM load and FGE belongs to the CVD load. FGE has three states according to the K-mean clustering results, in which the average active power of the second state is 7 W. The second state is difficult to be found that results in low accuracy.

(3) Combination decomposition result diagram of electric appliance DWE and FGE

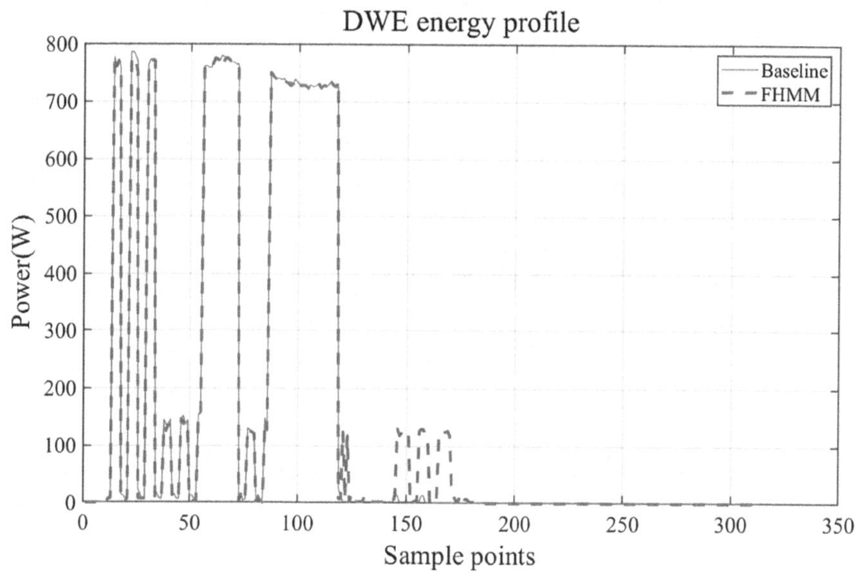

Fig. 7.9 Comparison of DWE decomposition and actual results

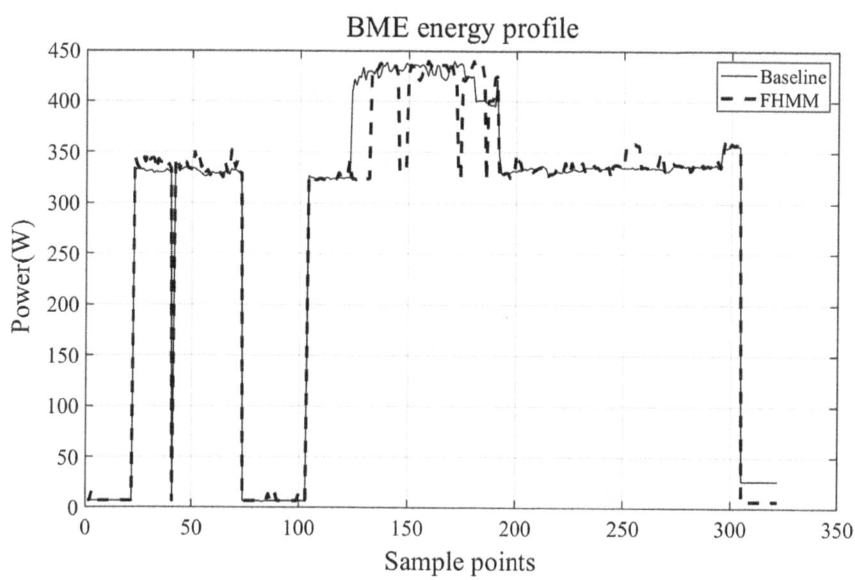

Fig. 7.10 Comparison of BME decomposition and actual active power

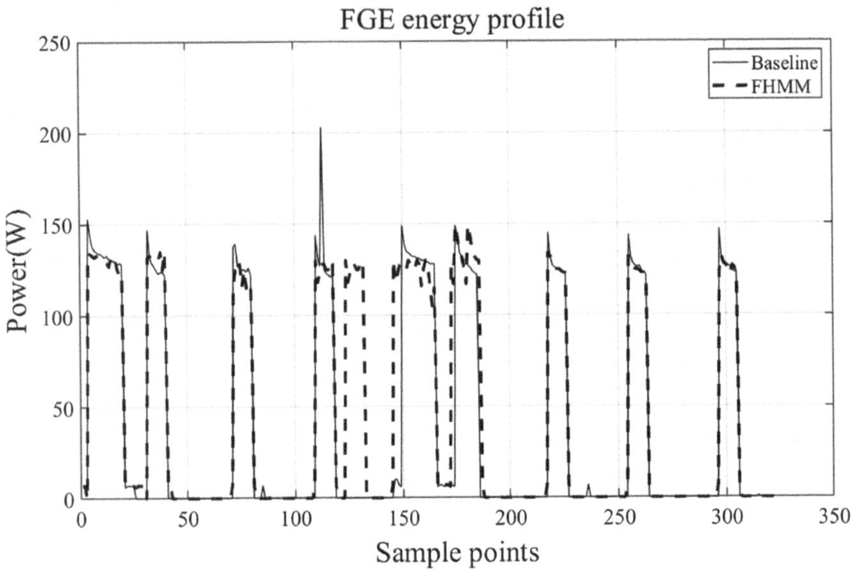

Fig. 7.11 Comparison of FGE decomposition and actual active power

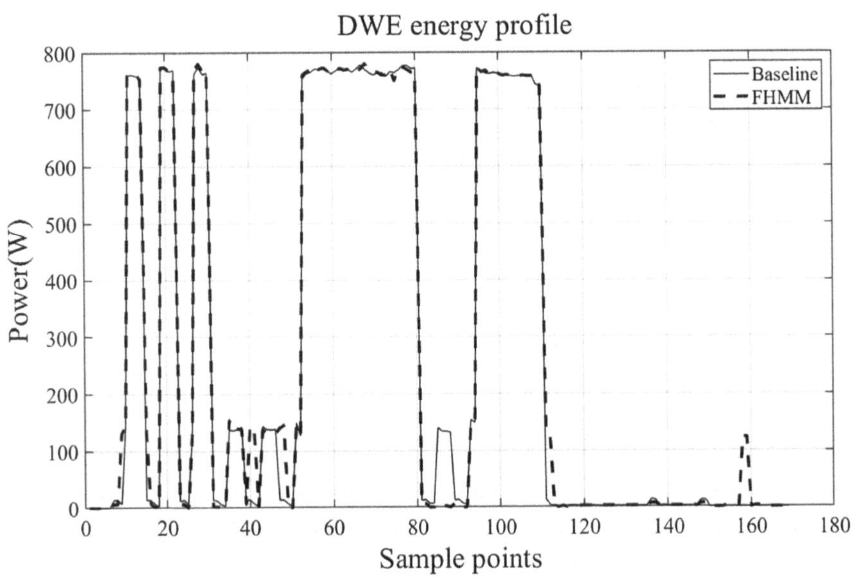

Fig. 7.12 Comparison of DWE decomposition and actual active power

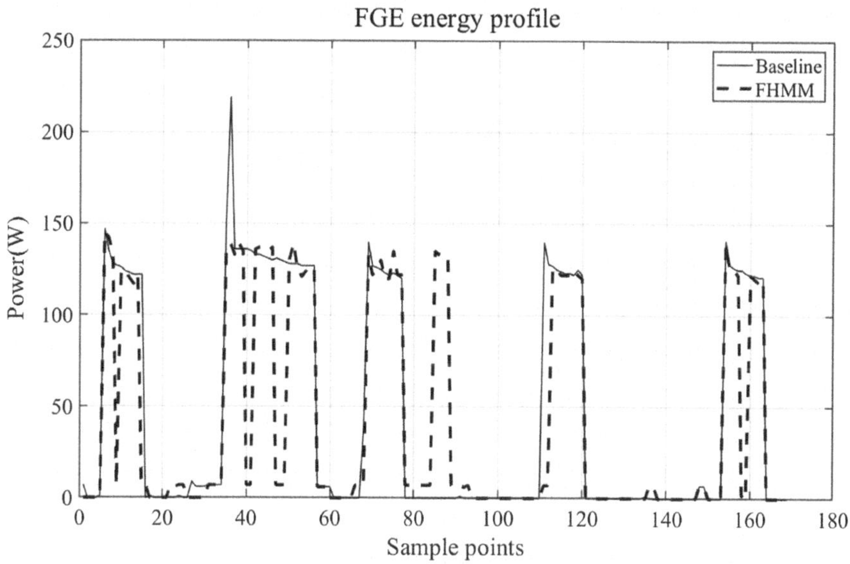

Fig. 7.13 Comparison of FGE decomposition and actual active power

The decomposition results of DWE and FGE are given in Figs. 7.12 and 7.13, respectively. In the decomposition using DWE and FGE, there are also over-fitting phenomena of HMM. The second state of FGE is difficult to define.

(4) Combination decomposition results of three electric appliances BME, DWE and FGE

The decomposition results of BME, DWE and FGE are given in Figs. 7.14, 7.15 and 7.16. When the total load of three electrical appliances is decomposed at the same time, the result is poorer than the binary classification result. Due to the increase of load quantity, it becomes more difficult to identify the second state of DWG and the third state of FGE. It is even more difficult to identify the low-power second state of FGE.

7.3.5 Evaluation Criteria and Result Analysis

(1) Evaluation criteria

After realizing household electrical load decomposition, it is necessary to evaluate the precision of load decomposition with scientific evaluation standards. In the case of load decomposition, the decomposition performance is mainly evaluated in two aspects. The two aspects are the precision of load decomposition and the proportion of energy consumed of each load in the total energy consumption. The identification

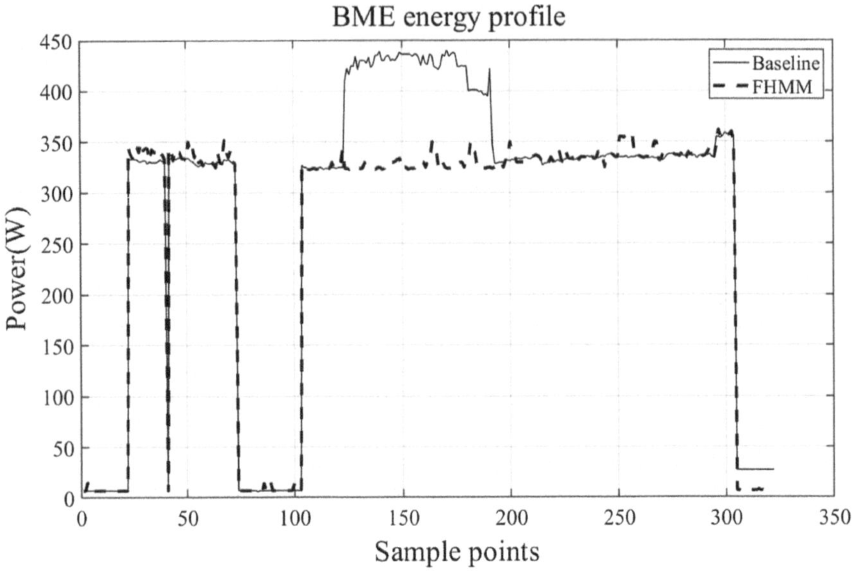

Fig. 7.14 Comparison of BME decomposition and actual active power

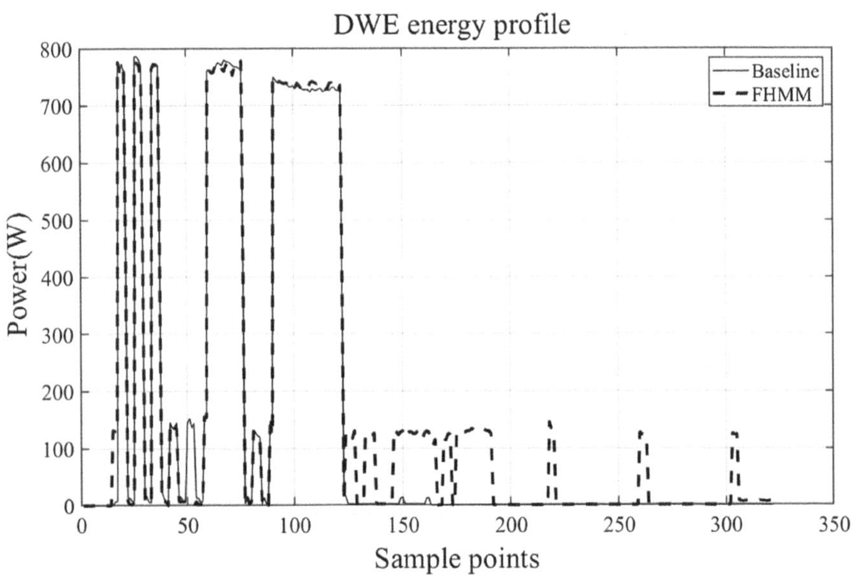

Fig. 7.15 Comparison of DWE decomposition and actual active power

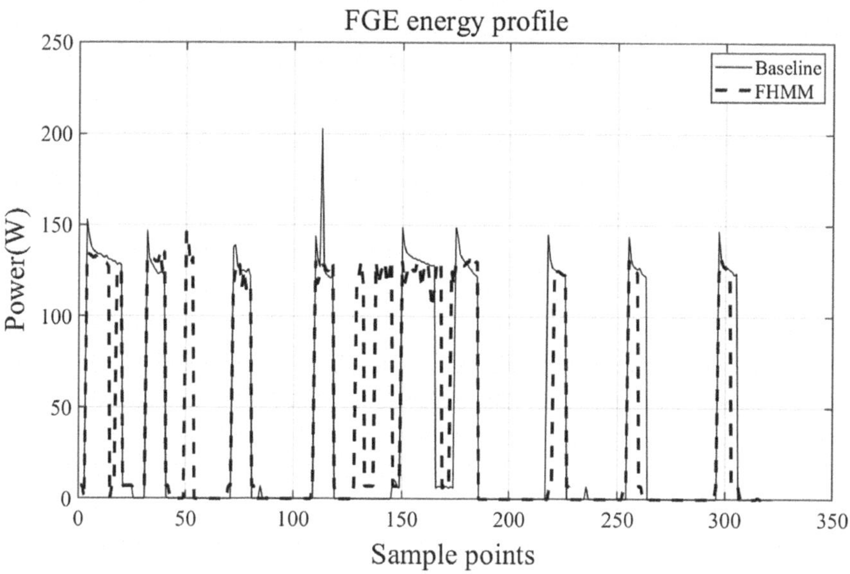

Fig. 7.16 Comparison of FGE decomposition and actual active power

of the load running state is evaluated by the accuracy of state estimation, which is defined as follows (Hart 1992):

$$Acc = \frac{n_T}{n} \tag{7.12}$$

where n_T is the number of correctly identified states, and n is the total number of predicted samples. In the study of the multi-load operation, the average value of A_{cc} is used to represent the corresponding accuracy.

Kolter and Jaakkola defined two evaluation standards of accuracy *Pre* and recall rate *Rec*, and the calculation formulas of the two evaluation standards are shown as follows (Kolter and Jaakkola 2012):

$$Pre^i = \frac{\sum_{t=1}^{n} \min(\widehat{y}_t^i, y_t^i)}{\sum_{t=1}^{n} \widehat{y}_t^i} \tag{7.13}$$

$$Rec^i = \frac{\sum_{t=1}^{n} \min(\widehat{y}_t^i, y_t^i)}{\sum_{t=1}^{n} y_t^i} \tag{7.14}$$

Among them, $\widehat{y}_t^i$ represents the predicted power value of load i at the moment t obtained by the decomposition algorithm, y_t^i represents the real power value of the load i at the moment t, and n refers to the total number of predicted samples.

In order to analyze the overall performance of the decomposition algorithm, it is necessary to calculate the *Pre* and *Rec* of all loads in the prediction set. Taking the average value of the accuracy and recall of each load as follows:

$$Pre = \frac{1}{N}\sum_{i=1}^{N} Pre^{i} \tag{7.15}$$

$$Rec = \frac{1}{N}\sum_{i=1}^{N} Rec^{i} \tag{7.16}$$

where N is the total load in the prediction set. The F_1-Measure which is the geometric average values of *Pre* and *Rec* is obtained as follows:

$$F_1 = 2\frac{Pre \times Rec}{Pre + Rec} \tag{7.17}$$

In addition to the A_{cc} and F_1-measure mentioned above, the total decomposition error T_{err} is also used in the estimation of decomposition accuracy, which is defined as follows:

$$T_{err} = \sqrt{\frac{\sum_{t,i}(y_t^i - \widehat{y}_t^{\,i})^2}{\sum_{t,i}(\widehat{y}_t^{\,i})^2}} \tag{7.18}$$

Besides, the absolute error of the energy consumed by each load is also used to evaluate the electric energy consumed by each load. The calculation formula is shown as follows (Cole and Albicki 1998):

$$E_i = |D_i - T_i| \tag{7.19}$$

where D_i is the ratio of energy consumed by load i. T is the ratio of energy consumed by the load i in a real situation. E_i is the absolute error of these two proportions.

(2) Comparison of results

Based on the above evaluation indices, the results of four groups of decomposition experiments are shown in Table 7.3. According to the real energy consumption ratio of each load and the energy consumption ratio obtained by decomposition, the absolute errors of these two are shown in Table 7.4.

It can be found from Tables 7.3 and 7.4 that the energy consumption error of the FHMM algorithm of two kinds of load classification is small. In the case of three kinds of appliances classification, the error significantly increases. The error of the FGE is the smallest, while the errors of the DWE and FGE are larger. Based on the evaluation results, the following conclusions can be drawn:

Table 7.3 Load decomposition evaluation results of the four groups

	BME and DWE		BME and FGE		DWE and FGE		Three loads		
	BME	DWE	BME	FGE	DWE	FGE	BME	DWE	FGE
A_{cc} (%)	99.04	93.63	96.89	89.13	91.71	78.70	92.27	83.22	75.58
F_1 (%)	98.26	97.26	98.21	90.07	97.62	84.53	95.55	92.27	78.09
T_{err}	0.0920	0.0889	0.0763	0.3709	0.0826	0.5292	0.1557	0.1742	0.5640

Table 7.4 Absolute errors of load energy consumption ratio

	Loads	True value T_i (%)	Predicted value D_i (%)	Absolute error E_i (%)
BME and DWE	BME	65.13	63.70	1.43
	DWE	34.87	36.30	1.43
	MAE			1.43
BME and FGE	BME	87.49	85.75	1.74
	FGE	12.51	14.25	1.74
	MAE			1.74
DWE and FGE	DWE	85.04	86.56	1.52
	FGE	14.96	13.44	1.52
	MAE			1.52
Three kinds of load	BME	59.66	54.84	4.82
	DWE	31.81	36.21	4.40
	FGE	8.53	8.95	0.42
	MAE			3.21

(a) According to the A_{cc} and F_1-Measure, the state recognition accuracy of the BME is the highest, followed by DWE, and the state recognition accuracy of the FGE is the worst. This is because BME only has two switching states, i.e. the ON and OFF states. DWE and FGE have three states, i.e. the OFF, low-power ON and high-power ON states and the low-power and high-power states of DWE are easier to identify than that of the FGE. It is difficult to identify the low-power state of FGE, and the high-power state of FGE is similar to the low-power state of DWE.

(b) The T_{err} shows that the performance of the DWE is slightly higher than the BME, and the effect of FGE is still the worst. This is because the power of ON state of DWE fluctuates greatly, resulting in a large prediction error. In addition, compared with DWE, BME has more OFF states, resulting in small errors.

(c) In the case of three-load decomposition, the identification accuracy decreases a lot compared to the two-load decomposition. This is due to the increase in the number of combinations of ON and OFF states.

7.4 Appliance Identification Based on Hidden Semi-Markov Models

7.4.1 *Hidden Semi-Markov Model*

The HSMM is an extended model of HMM (Yu and Kobayashi 2003) which was proposed by combining discrete HMM and continuous HMM (Huang and Jack 1989a). When the current state transforms to the next state, it is related not only to the current state but also to the dwell time of the current state. When solving practical problems, HSMM overcomes the limitations of HMM modeling caused by the Hidden Markov Chain hypothesis, and HSMM provides better modeling and analyzing capabilities (Huang and Jack 1989b). Different from the conventional HMM, where a state corresponds is to only one observed value, a state in HSMM corresponds is to a series of observed values.

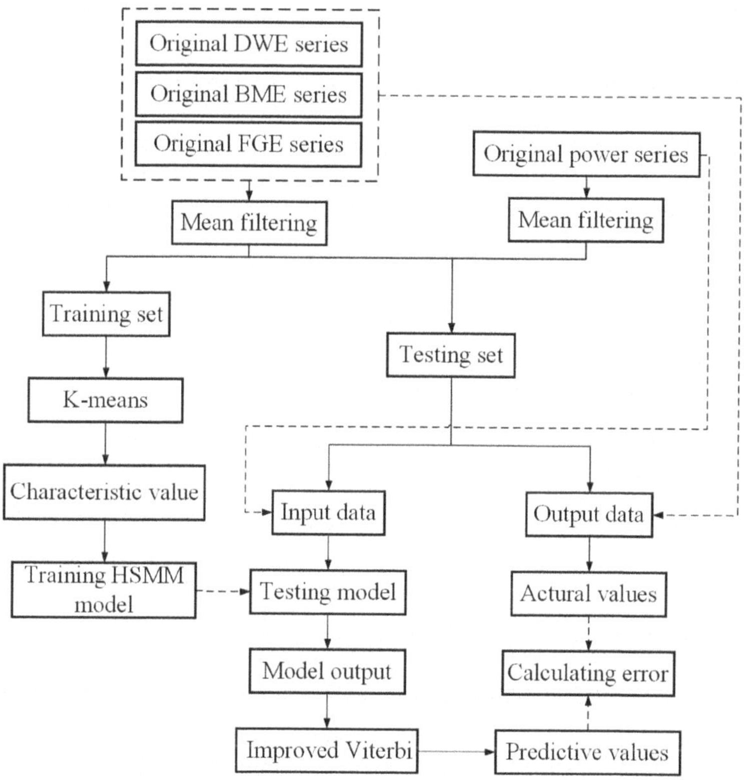

Fig. 7.17 The flow chart of HSMM

7.4.2 Improved Viterbi Algorithm

According to the given observation value sequence, namely the differential active power sequence, the improved Viterbi (Liu and Han 2010) algorithm can be used to obtain the optimal state transfer sequence of each type of appliance, which is the state sequence determined at the maximum value of $P(Q, S|\lambda)$. This section utilized an improved Viterbi algorithm based on HSMM to identify the state of household appliance and to obtain the power consumption of each appliance (Datta et al. 2008). The flow chart of HSMM model is shown in Fig. 7.17.

7.4.3 Evaluation Criteria and Result Analysis

The results of four groups of decomposition experiments are analyzed and evaluated as shown in Table 7.5. According to the real energy consumption ratio of each load and the energy consumption ratio obtained by decomposition, the absolute errors of these two results are calculated, as shown in Table 7.6.

As shown in Tables 7.5 and 7.6, the energy consumption error of the HSMM algorithm of two kinds of load classification is relatively small. In the case of three kinds of load classification, the error significantly increases. The error generated by FGE is the smallest, while the errors generated by DWE and FGE are larger. This is because the total energy consumption generated by FGE is smaller and the absolute error generated is smaller in the test set.

7.5 Experiment Analysis

In order to compare the results of the decomposition, Figs. 7.18, 7.19, 7.20 and 7.21 show the decomposition results under four combinations. Comparing with the decomposition results of FGE load, the performance of the FHMM algorithm is obviously better than that of the HSMM algorithm. However, other load decomposition results show that the HSMM algorithm performs better than the FHMM algorithm.

As shown in Table 7.7, the decomposition results of the HSMM based model and the FHMM based model are analyzed and evaluated,. The absolute errors of load energy consumption ratio are shown in Table 7.8.

Based on the above experimental results, the conclusions can be drawn as follows:

(a) A_{cc} and F_1-*Measure* evaluation indices show that the BME has the best decomposition effect and the FGE has the worst decomposition effect. This is due to the low power of FGE, which makes it difficult to identify the state.
(b) The T_{err} evaluation index shows that the identification error of the DWE is smaller than that of the BME under the FHMM algorithm. The error of the

Table 7.5 Load decomposition evaluation results of the four groups

	BME and DWE		BME and FGE		DWE and FGE		Three loads		
	BME	DWE	BME	FGE	DWE	FGE	BME	DWE	FGE
A_{cc} (%)	99.14	94.23	97.16	89.93	92.09	77.58	92.47	83.52	75.08
F_1 (%)	98.67	97.37	98.64	90.39	98.56	84.31	95.51	91.11	77.75
T_{err}	0.0668	0.0847	0.0574	0.3695	0.0586	0.4400	0.1562	0.1545	0.6086

Table 7.6 Absolute errors of load energy consumption ratio

	Loads	True value T_i (%)	Predicted value D_i (%)	Absolute error E_i (%)
BME and DWE	BME	65.13	64.01	1.12
	DWE	34.87	35.99	1.12
	MAE			1.43
BME and FGE	BME	87.49	86.20	1.29
	FGE	12.51	13.80	1.29
	MAE			1.29
DWE and FGE	DWE	85.04	87.16	2.12
	FGE	14.96	12.84	2.12
	MAE			2.12
Three kinds of load	BME	59.66	54.67	4.99
	DWE	31.81	35.64	3.83
	FGE	8.53	9.69	1.16
	MAE			3.32

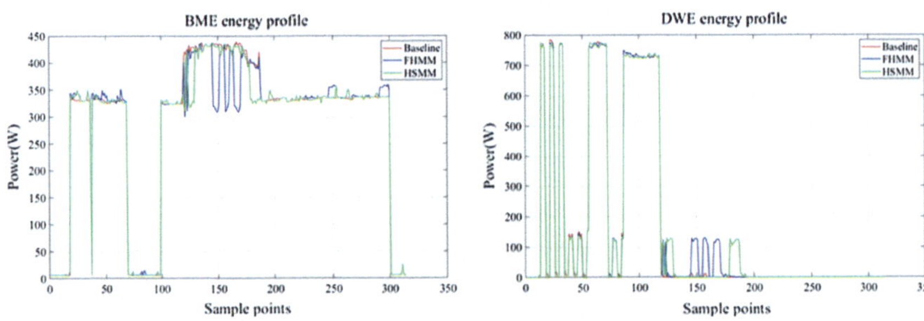

Fig. 7.18 Comparison of BME and DWE decomposition and actual active power

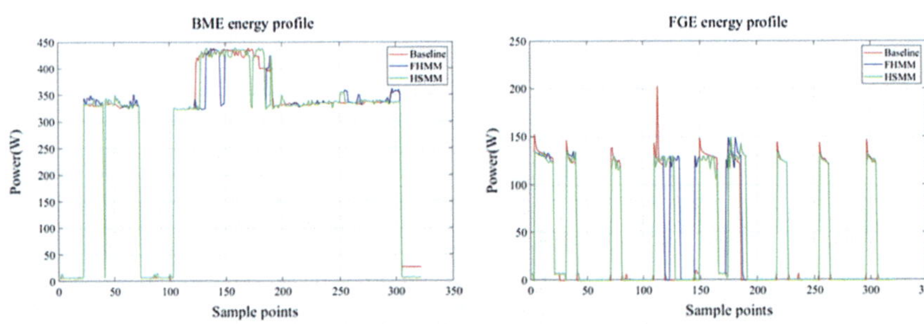

Fig. 7.19 Comparison of BME and FGE decomposition and actual active power

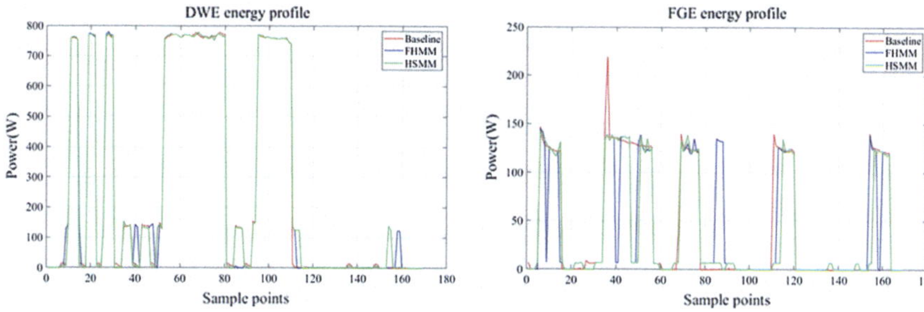

Fig. 7.20 Comparison of DWE and FGE decomposition and actual active power

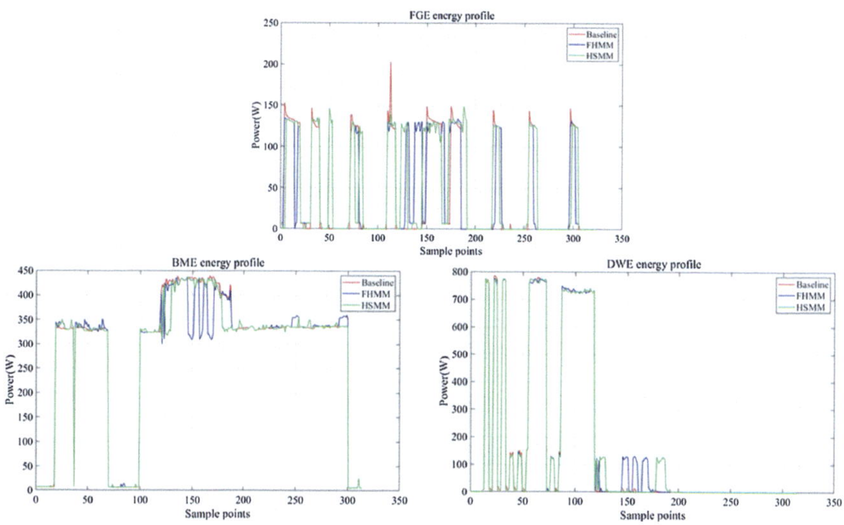

Fig. 7.21 Comparison of BME, DWE and FGE decomposition and actual active power

BME is less than that of the DWE with the HSMM algorithm. This is because HSMM considers the time factor in state identification.

(c) Comparing the results of two-load decomposition with three-load decomposition, the decomposition effect of two loads is better than that of three loads. Therefore, it is concluded that the more loads are, the more difficult it is to identify them.

(d) The evaluation indices show that the decomposition performance of the HSMM algorithm is worse than the FHMM algorithm. The main reason is that the HSMM algorithm is not effective in identifying the energy consumption state of the FGE load, which causes the error to be large.

(e) Generally speaking, the FHMM and HSMM have their own advantages in pattern recognition and load decomposition. The FHMM is slightly better than the

Table 7.7 Load decomposition evaluation results of the two methods

		BME and DWE		BME and FGE		DWE and FGE		Three loads		
		BME	DWE	BME	FGE	DWE	FGE	BME	DWE	FGE
HSMM	A_{cc}	99.14	94.23	97.16	89.93	92.09	77.58	92.47	83.52	75.08
	F_1	98.67	97.37	98.64	90.39	98.56	84.31	95.51	91.11	77.75
	T_{err}	0.0668	0.0847	0.0574	0.3695	0.0586	0.4400	0.1562	0.1545	0.6086
FHMM	A_{cc}	99.04	93.63	96.89	89.13	91.71	78.70	92.27	83.22	75.58
	F_1	98.26	97.26	98.21	90.07	97.62	84.53	95.55	92.27	78.09
	T_{err}	0.0920	0.0889	0.0763	0.3709	0.0826	0.5292	0.1557	0.1742	0.5640

Table 7.8 Absolute errors of load energy consumption ratio

Loads	True value T_i (%)	Predicted value D_i (%)		Absolute error E_i (%)	
		FHMM	HSMM	FHMM	HSMM
BME	59.66	54.84	54.67	4.82	4.99
DWE	31.81	36.21	35.64	4.4	3.83
FGE	8.53	8.95	9.69	0.42	1.16
MAE				3.21	3.32

HSMM in state discrimination of the FGE. The HSMM is slightly better than the FHMM in identifying state two fluctuations of the BME. This is because the FHMM uses multi-HMM parallel modeling. The HSMM introduces the concept of time T modeling. Therefore, for different types of loads, it is of great significance to adopt reasonable modeling methods.

References

Aiad M, Lee PH (2018a) Energy disaggregation of overlapping home appliances consumptions using a cluster splitting approach. Sustain Cities Soc 43:487–494. https://doi.org/10.1016/j.scs.2018.08.020

Aiad M, Lee PH (2018b) Modelling and power estimation of continuously varying residential loads using a quantized continuous-state hidden markov model. In: Proceedings of the 2nd international conference on machine learning and soft computing. ACM, pp 180–184

Bengio Y, Frasconi P (1995) An input output HMM architecture. In: Advances in neural information processing systems, pp 427–434

Chen S, Gao F, Liu T (2017) Load identification based on factorial hidden Markov model and online performance analysis. In: 2017 13th IEEE conference on automation science and engineering (CASE). IEEE, pp 1249–1253

Choi K-S, Yang B-W (2016) NILM method using FHMM reflecting correlation between two appliances. Int Inf Inst (Tokyo) Inf 19(2):555

Cole AI, Albicki A (1998) Data extraction for effective non-intrusive identification of residential power loads. In: IMTC/98 conference proceedings. IEEE instrumentation and measurement technology conference. Where instrumentation is going (Cat. No. 98CH36222). IEEE, pp 812–815

Datta R, Hu J, Ray B (2008) On efficient Viterbi decoding for hidden semi-Markov models. In: 2008 19th international conference on pattern recognition. IEEE, pp 1–4

Hagenauer J, Hoeher P (1989) A Viterbi algorithm with soft-decision outputs and its applications. In: 1989 IEEE global telecommunications conference and exhibition' communications technology for the 1990s and beyond. IEEE, pp 1680–1686

Hart GW (1992) Nonintrusive appliance load monitoring. Proc IEEE 80(12):1870–1891

Hartigan JA, Wong MA (1979) Algorithm AS 136: a k-means clustering algorithm. J Roy Stat Soc: Ser C (Appl Stat) 28(1):100–108

Huang X, Jack M (1989a) Unified techniques for vector quantization and hidden Markov modeling using semi-continuous models. In: International conference on acoustics, speech, and signal processing. IEEE, pp 639–642

Huang XD, Jack MA (1989b) Semi-continuous hidden Markov models for speech signals. Comput Speech Lang 3(3):239–251

Kolter JZ, Jaakkola T (2012) Approximate inference in additive factorial HMMs with application to energy disaggregation. In: Artificial intelligence and statistics, pp 1472–1482
Kolter JZ, Johnson MJ (2011) REDD: a public data set for energy disaggregation research. In: Workshop on data mining applications in sustainability (SIGKDD), San Diego, CA, pp 59–62. Citeseer
Kong W, Dong Z, Xu Y, Hill D (2016a) An enhanced bootstrap filtering method for non-intrusive load monitoring. In: 2016 IEEE power and energy society general meeting (PESGM). IEEE, pp 1–5
Kong W, Dong ZY, Hill DJ, Luo F, Xu Y (2016b) Improving nonintrusive load monitoring efficiency via a hybrid programing method. IEEE Trans Industr Inf 12(6):2148–2157
Lee Y, Kassam S (1985) Generalized median filtering and related nonlinear filtering techniques. IEEE Trans Acoust Speech Signal Process 33(3):672–683
Leeb SB, Shaw SR, Kirtley JL (1995) Transient event detection in spectral envelope estimates for nonintrusive load monitoring. IEEE Trans Power Delivery 10(3):1200–1210. https://doi.org/10.1109/61.400897
Liu W, Han W (2010) Improved Viterbi algorithm in continuous speech recognition. In: 2010 international conference on computer application and system modeling (ICCASM 2010). IEEE, pp V7-207–V207-209
Lleida E, Rose RC (2000) Utterance verification in continuous speech recognition: decoding and training procedures. IEEE Trans Speech Audio Process 8(2):126–139
Manmatha R, Feng S (2006) A hierarchical, HMM-based automatic evaluation of OCR accuracy for a digital library of books. In: Proceedings of the 6th ACM/IEEE-CS joint conference on digital libraries (JCDL'06). IEEE, pp 109–118
Moon TK (1996) The expectation-maximization algorithm. IEEE Signal Process Mag 13(6):47–60
Rabiner LR, Juang B-H (1986) An introduction to hidden Markov models. IEEE ASSP Mag 3(1):4–16
Song L, Yao W, Jie T (2018) Non-intrusive load decomposition method based on the factor hidden Markov model. In: 2018 37th Chinese control conference (CCC). IEEE, pp 8994–8999
Wang L, Luo X, Zhang W (2013) Unsupervised energy disaggregation with factorial hidden Markov models based on generalized backfitting algorithm. In: 2013 IEEE international conference of IEEE region 10 (TENCON 2013). IEEE, pp 1–4
Yagou H, Ohtake Y, Belyaev A (2002) Mesh smoothing via mean and median filtering applied to face normals. In: Proceedings of geometric modeling and processing. Theory and applications, GMP 2002. IEEE, pp 124–131
Yu S-Z, Kobayashi H (2003) An efficient forward-backward algorithm for an explicit-duration hidden Markov model. IEEE Signal Process Lett 10(1):11–14
Zhong M, Goddard N, Sutton C (2014) Interleaved factorial non-homogeneous hidden Markov models for energy disaggregation. arXiv:14067665
Zoha A, Gluhak A, Imran M, Rajasegarar S (2012) Non-intrusive load monitoring approaches for disaggregated energy sensing: a survey. Sensors 12(12):16838–16866. https://doi.org/10.3390/s121216838

Chapter 8
Deep Learning Based Appliance Identification

8.1 Introduction

8.1.1 Deep Learning

Deep learning is a method in the field of machine learning and it is also one of the most popular machine learning algorithms currently. By simulating the structure of the human brain, it is intended to make computers work like humans to mine deep features of data. This gives computers the ability to learn, and deep learning has been hugely successful in many fields.

The goal of deep learning is to learn the potential features of data by building a neural network model with a multi-layer structure, so as to improve the accuracy of classifiers or predictors. It abstracts the original features into a form that the computer can understand by converting them layer by layer, which can greatly improve the performance of the model. Compared with the method of artificial constructing features, the features derived using deep learning are more capable of describing the underlying information of big data.

8.1.2 NILM Based on Deep Learning

The deep learning structure has shown its reliable ability to automatically extract advanced features of complex data (LeCun et al. 2015). So it has great potential in NILM. Complex deep learning models can approximate any functions, which provides a new idea for NILM problems. In recent years, different deep learning models have been used to load decomposition problems, such as Recurrent Neural Network (RNN), Convolutional Neural Network (CNN), and Auto-encoder (AE). Lange and Bergés propose a new deep learning method for load decomposition, which can identify the components of power signals in an unsupervised form (Lange and Bergés 2016). The method uses the data of current high-frequency measurement and assumes

H. Liu, *Non-intrusive Load Monitoring*,
https://doi.org/10.1007/978-981-15-1860-7_8

a load model to infer all data. Barsim et al. propose a neural network combination approach to solve the NILM problem (Barsim et al. 2018). The combination of neural networks utilizes the original high-resolution and current-voltage waveform to solve the load identification problem. Long Short-Term Memory (LSTM) networks and convolutional denoising in self-coded deep learning models were proposed by Kelly (2016) to solve the non-intrusive load decomposition problem of low-frequency samples. It is able to separate the energy consumption of the load from the main circuit of the house and achieve better results. The paper trains all different types of loads with the same deep learning structure, which significantly reduces the workload of designing deep learning models. Therefore, designing a deep learning model suitable for all loads has become a potential research direction. And Zhang et al. use CNN to propose a midpoint prediction method for load decomposition, in which the midpoint value of the load window is used as the output of the neural network (Zhang et al. 2018). This method can also reduce the computational cost. The Hidden Markov Model (HMM) and the Deep Neural Network (DNN) method are proposed for load decomposition by Mauch et al. (2016). This method uses two kinds of observation probabilities to train HMM. One of them is used to extract a single load to generate a Gaussian distribution model and the other is to generate a DNN model for the total signal. Although the precision of the DNN-HMM model is better than FHMM, the method only uses a small amount of data for training. A large amount of data is necessary for the training to achieve better generalization precision of the deep learning model.

In addition, deep learning can also be used as feature extractors and classifiers in non-intrusive load identification tasks. In non-intrusive load identification tasks based on event classification, deep learning usually achieves excellent results. De Baets et al. used the Siamese neural network to extract features from the V-I images of the appliances and then use the DBSCAN algorithm to cluster the extracted features (De Baets et al. 2019). This method can identify unknown types of appliances. Huss proposed that extracting the load features by using the CNN network, and then used the features as observations of the Hidden Semi Markov Model (HSMM) (Huss 2015). After capturing the load event of the appliance, Tsai and Lin use k-Nearest Neighbor Rule (k-NNR) and Back-Propagation Artificial Neural Network (BP-ANN) as recognizers, to realize the identification of different electrical appliances. At the same time, the artificial immune algorithm is used to adaptively adjust the feature parameters to improve the identification performance. What's more, the characteristic parameters are optimized by the artificial immune algorithm (Tsai and Lin 2012).

8.2 Appliance Identification Based on End-to-End Decomposition

According to the NILM concept, researches on load identification have developed rapidly in recent years. The classic NILM framework generally consists of five steps, data acquisition, data preprocessing, event detection, feature extraction, and load decomposition, which have achieved rapid progress. The traditional load decomposition algorithm usually uses the hidden Markov algorithm to decompose the load. The study found that the load decomposition algorithm based on deep learning has better effects than the traditional algorithms.

8.2.1 Single Feature Based LSTM Network Load Decomposition

LSTM inherits most of the features of the RNN model, which is a good variant of RNN (Chao et al. 2015). LSTM not only learns the information of the current sequence but also relies on the information of the previous sequence. The special structure of LSTM proposes the problem of information preservation, so it has unique advantages in dealing with time-series problems. For example, LSTM is an ideal model for the problems which have a strong correlation with time series in speech recognition tasks.

In NILM, the load decomposition task is very similar to the speech recognition task. When the appliance is running, the sequences of its power, current, etc. have obvious timing characteristics. It is very suitable to be identified by LSTM. In this section, the active power sequence of the appliance is used as the input of the LSTM. And then each type of electrical appliances to be tested will be modeled one by one to achieve the purpose of load identification. The algorithm flow chart is shown in Fig. 8.1.

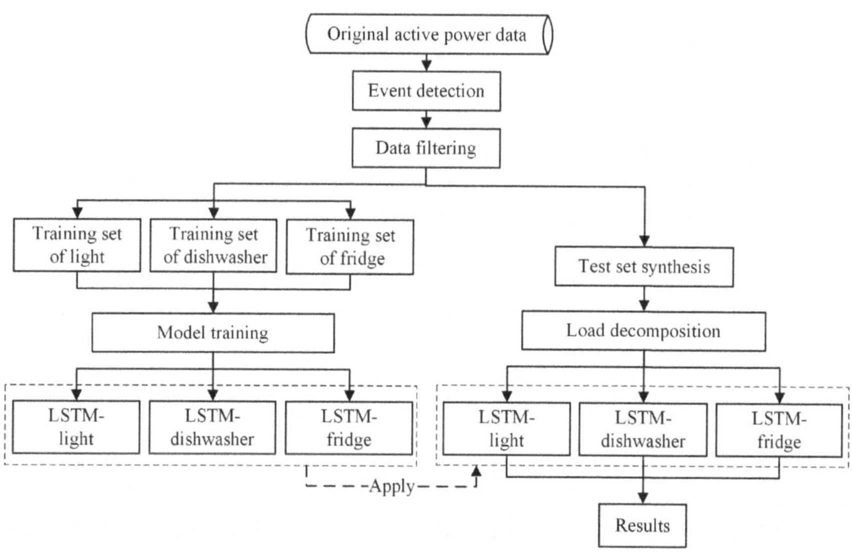

Fig. 8.1 The algorithm flow chart of the LSTM model with single feature

(1) *Data processing*

The original data in this section is derived from the AMPds dataset. The data with the names of BME, DWE, and FGE are selected as the load decomposition object, which corresponds to the electricity data of three kinds of appliances: lighting, dishwasher, and fridge.

It is not difficult to notice that the running time of the lights and the dishwasher is much smaller than the stopping time. If the model is trained by using original data, it will be a serious imbalance between the positive and negative cases, which would have a greater impact on the accuracy of the model. Therefore, this section uses the random rejection method to filter the original active power data of the lighting and the fridge. And delete most of the active power sequences which are the sequence segment when the appliance is closed. This could increase the concentration of positive examples of electrical appliances in the active power sequence. The ratio of the running time in the filtered active power sequence to the stopping time of the appliance is approximately 1:1.

(2) *Establishment and training of the model*

There are only three data types used for load decomposition in this chapter, and the amount of training data is not so large. Building the model with a deeper neural network will not only increase the training time of the model but also may result in under-fitting due to the small amount of data. Therefore, a single-layer LSTM model is used in this section. The number of hidden units in the LSTM layer is 70. And the activation function of the output layer is the softmax function. Establishing the corresponding LSTM models for the three types of electrical appliances that are decomposed. The output of the model is 0 or 1, corresponding to the state of stop and run of the appliance.

A corresponding training set will be built for each type of electrical appliances. The training set includes all combinations of three electrical samples. In order to preserve the timing characteristics of the appliances as much as possible, all samples are not normalized, filtered, etc. The active power sequence, in which a single appliance is running, is directly intercepted from the filtered data. And the active power sequence of a plurality of electrical appliances is obtained by artificial synthesis. The training set for each appliance contains a total sequence length of 300,000. The proportion of each type of sample in the training set is shown in Table 8.1.

Importing datasets to train models. The training situation of the three models is shown in Fig. 8.2.

When the training processes are completed, the training accuracy is about 90% and the model training is successful.

(3) *Experiment*

Three active power sequences have been extracted from the Light, Dishwasher and Fridge that are not in the training set with a length of 40,000, and the testing set are synthesized from these three sequences. This corresponds to the case of the active power sequence when the three loads are operated together. After entering the

Table 8.1 The proportion of samples in the training set

Model	The type of data	The type of appliance		
		Light (%)	Dishwasher (%)	Fridge (%)
LSTM	Light	40	10	10
	Dishwasher	10	40	10
	Fridge	10	10	40
	Light + Dishwasher	10	10	10
	Light + Fridge	10	10	10
	Dishwasher + Fridge	10	10	10
	Light + Dishwasher + Fridge	10	10	10

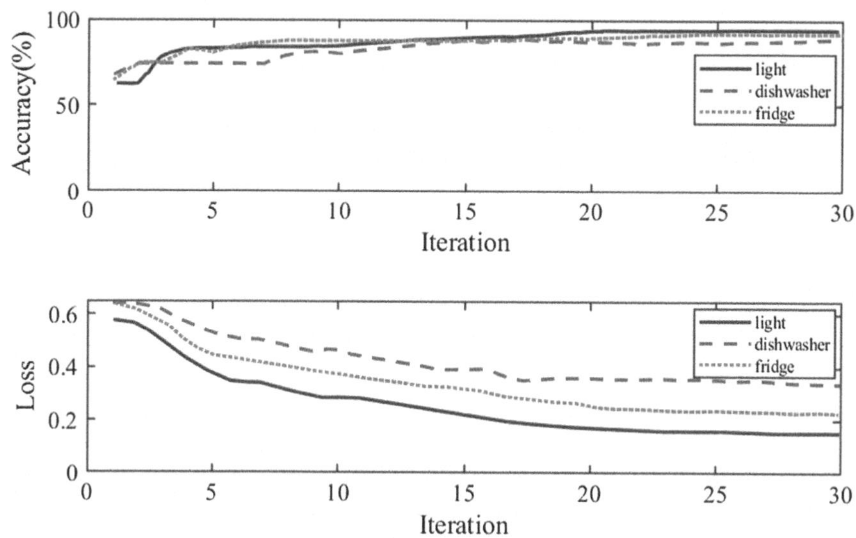

Fig. 8.2 The training process of the three models

testing set data into the three trained models, the output of each model is the working condition of the corresponding electrical appliance.

The prediction results of each model are shown in Table 8.2.

In terms of accuracy, the model has the best identification effect on Light, which reached 88.66% and the worst identification effect on the Fridge is 69.73%. In terms of the F1_Score value, although the accuracy of the dishwasher and Fridge is close, it is obvious that the comprehensive performance of the Dishwasher model is better than the Fridge model. In a nutshell, the identification effect of the model is generally.

Table 8.2 The prediction results of the LSTM model with single feature

Model	Evaluation index	The type of appliance		
		Light	Dishwasher	Fridge
LSTM	Accuracy	0.8866	0.7194	0.6973
	Recall	0.9977	0.6234	0.3957
	Precision	0.8470	0.7236	0.8560
	F1-score	0.9162	0.6698	0.5413

8.2.2 *Multiple Features Based LSTM Network Load Decomposition*

The LSTM model established in the previous section only uses the active power sequence of the appliance as the feature to decompose the load. The performance of the model is general. There are many load features of time-series when the appliance is running. Modeling the load with multiple features can get better results. In this section, the active power sequence of the appliance and the current sequence of the appliance are used as the original feature. At the same time, the original LSTM model is fine-tuned to accommodate the increased data volume. Then the LSTM model is established to correspond to the three electrical appliances. The algorithm flow chart is shown in Fig. 8.3.

(1) *Data processing*

The original data used in this section is derived from the AMPds dataset. The load decomposition object is consistent with the previous section. Three kinds of electrical appliances: lighting, dishwasher, and fridge are still be chosen. The input of the model is the active power sequence and the current sequence of the appliances. The original current sequence is shown in Fig. 8.10.

The current sequence and active power sequence at the corresponding point in time of the filtered data have been extracted in Sect. 8.2.1. Then, a model training set that contains two feature quantities is constructed. The sequence is also not filtered and normalized to preserve the complete feature information of the sequence.

(2) *Model establishment and training*

Compared to the training set constructed with single feature in the previous section, the training set constructed with multiple features is double the amount of data in the former. This means that the number of features included in the training set also increased. Therefore, increasing the number of hidden cells in the original LSTM model to 100 will accommodate the increased dataset and improve the performance of the model. The output function of the model is unchanged and the softmax function is still used. The output value 0 corresponds to the stop state of the appliance and the output value 1 corresponds to the appliance startup state.

The training set consists of seven segments corresponding to seven different combinations of three electrical appliances. The active power sequence and the current

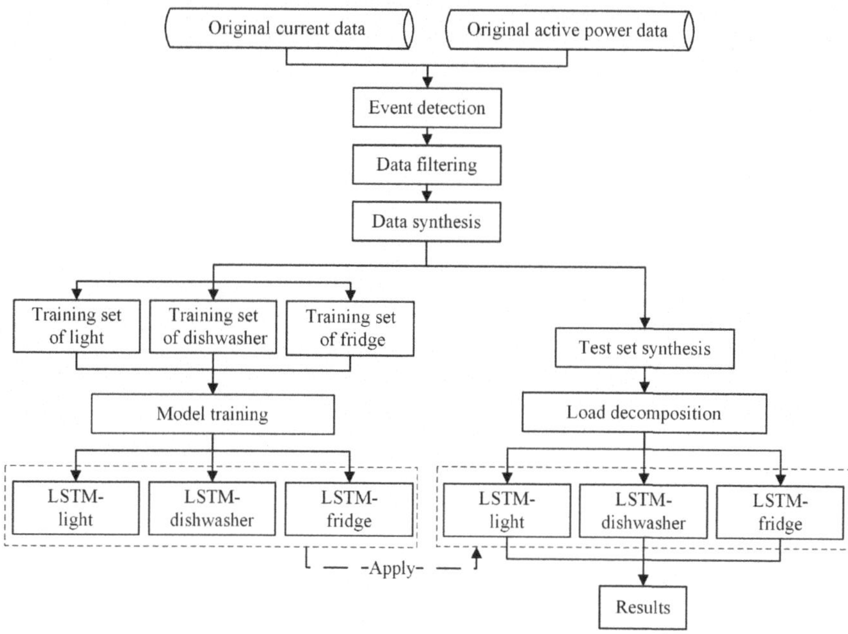

Fig. 8.3 The algorithm flow chart of the LSTM model with multiple features

sequence, in which a single appliance is running, is directly intercepted from the filtered data. The active power sequence and the current sequence of a plurality of electrical appliances are obtained by artificial synthesis. The total length of the training set sequence is 300,000. The storage format of each training sequence is $2 \times N$ including an active power sequence and a current sequence. The time points of the two sequences are consistent. The proportions of each type of samples in the training set are shown in Table 8.3.

Dataset has been imported to train models. The training situation of the three models is shown in Fig. 8.4.

Table 8.3 The proportion of the samples in the training set

Model	The type of data	The type of appliance		
		Light (%)	Dishwasher (%)	Fridge (%)
LSTM	Light	40	10	10
	Dishwasher	10	40	10
	Fridge	10	10	40
	Light + Dishwasher	10	10	10
	Light + Fridge	10	10	10
	Dishwasher + Fridge	10	10	10
	Light + Dishwasher + Fridge	10	10	10

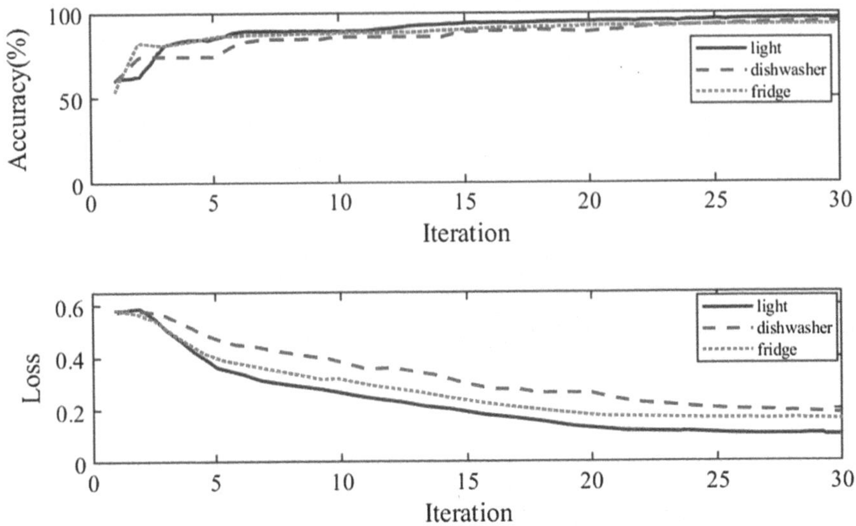

Fig. 8.4 The training process of the three models

After the training process of the model is completed, the training accuracy is above 95%. The training results are good.

(3) *Experiment*

Three active power sequences and the current sequence from the light, dishwasher, and fridge that are not in the training set with a length of 40,000, have been extracted and the testing set is synthesized from these three sequences. This corresponds to the case of the active power sequence when the three loads are operated together, and the testing set data has been entered into the three trained models. The output of each model is the working condition of the corresponding electrical appliance. The prediction results of each model are shown in Table 8.4:

In general, the performance of the improved multi-feature load decomposition model is better than that of a single-feature load decomposition model. The performance of all three appliances has improved. The performance of the dishwasher model is the most improved among them, while the performance improvement of

Table 8.4 The prediction results of the LSTM model with multiple features

LSTM		The type of appliance		
		Light	Dishwasher	Fridge
Evaluation index	Accuracy	0.9110	0.8006	0.7095
	Recall	0.9974	0.6115	0.4747
	Precision	0.8764	0.9266	0.8004
	F1	0.9330	0.7368	0.5959

the other two appliances is not obvious. The improvement of model performance has increased due to the hidden unit of the LSTM network, which makes the model more capable of expressing features. On the other hand, because of the addition of new currents feature, more feature information can be provided for model learning. In terms of improving the effectiveness, the features of the dishwasher contained in the current sequence can be more easily identified than the features of the other two appliances.

8.3 Appliance Identification Based on Appliance Classification

In the event detection based load identification system, the load events of all appliances are first obtained by the event detection algorithm. Then the features of the event are extracted for classification. In the field of feature extraction, traditional artificial feature engineering has dominated for a long time. However, artificial feature engineering requires not only experts with extensive experience in related fields but also a long time to invest. This may result in much project costs, and may also cause the project to progress slowly. In recent years, the use of machine learning models for automated feature learning has been widely developed. The most successful area is a series of neural network models represented by deep learning. They have achieved better results than traditional artificial feature engineering in many fields. Deep learning has great potential in the NILM. The deep neural network can be used as a feature extractor to extract the features of the load event, and as a classifier to identify the extracted load events.

8.3.1 Appliance Identification Based on CNN

Convolutional Neural Network (CNN) is one of the most commonly used deep learning networks (Hadji and Wildes 2018), which performs well in image processing tasks. A complete CNN usually consists of several convolutional layers, pooling layers, and fully connected layers. Usually, the convolutional layer is responsible for extracting features, and the pooling layer selects and filters the extracted features. Finally, the extracted features are nonlinearly combined with the fully connected layer to obtain an output.

In NILM, CNN has been widely used in device classification tasks. This is due to its powerful capabilities of feature classification and the advantages of automatically extracting features. In this section, the V-I images (Hassan et al. 2014) of the appliances are used as the input of the CNN for load identification. The process is shown in Fig. 8.5.

Fig. 8.5 The algorithm flow chart of the CNN model

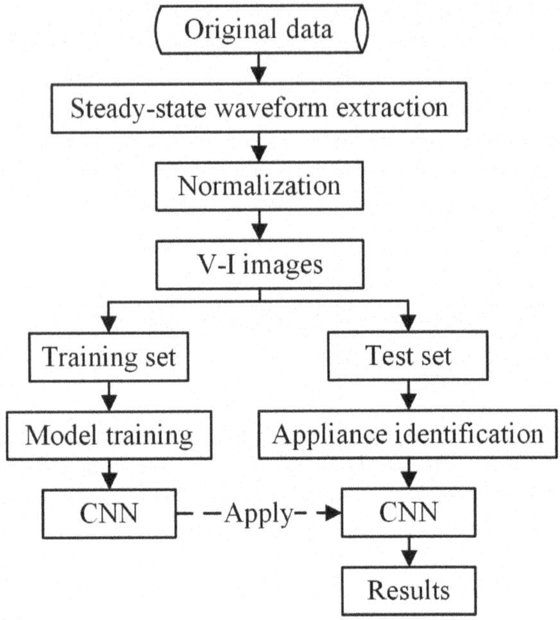

(1) *Data processing*

In this section, the original data used is from the PLAID dataset. The entire PLAID dataset contains a total of 1074 samples with 11 electrical appliances. In order to meet the requirements of the deep learning algorithm for data volume,five types of electrical appliances have been selected, such as the compact fluorescent lamp (hereinafter referred to as CFL), incandescent light bulb (hereinafter referred to as ILB), Microwave, Laptop, and Hairdryer, in the dataset as the identification objects (Fig. 8.6).

The input of the Convolutional Neural Network is the V-I images of each appliance in steady-state operation. 2000 sample points of voltage and current data during steady-state operation of the appliances have been extracted. And the voltage has been taken as the ordinate and the current as of the abscissa, then the VI curve of the appliances is drawn. In order to ensure the stability of the extracted shape of the V-I images, the original voltage and current data have been normalized to $[-1, 1]$ before the curve is drawn. Each electrical sample will be tested twice at different locations in steady-state operation. After removing the error images in the sample, the number of samples of each electrical appliance is finally shown in Table 8.5.

(2) *Model establishment and training*

The structure of the Convolutional Neural Network used in this section is shown in Table 8.6.

The entire network consists of two convolutional layers, one pooling layer and one fully connected layer. The activation function of the output layer is the softmax

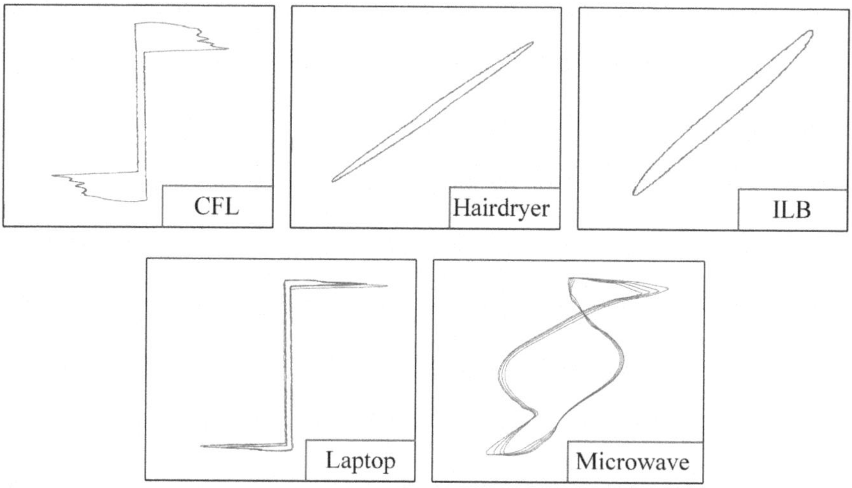

Fig. 8.6 The V-I images of the appliances

Table 8.5 The sample data volume of the CNN model

Categories	CFL	ILB	Microwave	Laptop	Hairdryer
Numbers	341	228	278	344	312

Table 8.6 The structure of the CNN model

Layer 1	Input
Layer 2	CONV 3×3
Layer 3	CONV 3×3/ReLU
Layer 4	MAX POOLING
Layer 5	FULL CONNECT
Layer 6	Output

function. The convolution layer 1 has a total of 32 convolution kernels, and its receiving domain is 3 × 3, and the step size is 2. And the convolution layer 2 has only 16 convolution kernels. It has the same receive domain and step size as convolutional layer 1. What's more, the pooling layer uses the largest pooling and the fully connected layer contains a total of five neurons.

The first 70% of each appliance in the samples is used as a training set and the last 30% is used as a testing set. Then the model is trained by stochastic gradient descent. The training results are shown in Fig. 8.7.

It can be seen from the training process that the accuracy and loss of the model tend to be stable when the number of iterations reaches 70 and the model has been trained.

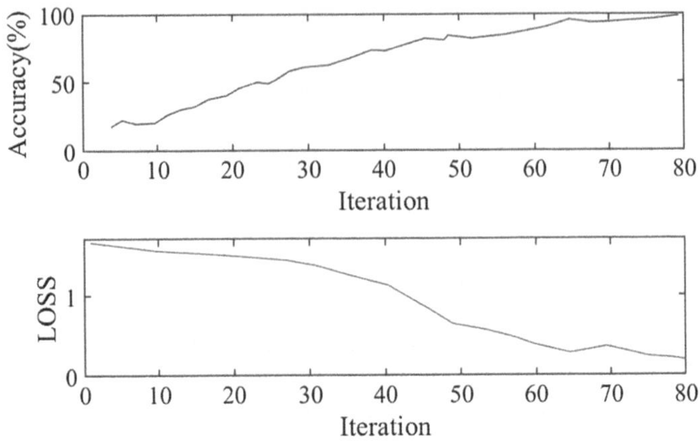

Fig. 8.7 The training process of the CNN model

Table 8.7 The results of the CNN model

CNN		The type of appliance				
		CFL	Hairdryer	ILB	Laptop	Microwave
Evaluation index	Accuracy	0.9889	0.8911	0.9222	0.9511	0.9267
	Recall	0.9608	0.6809	0.8235	0.8544	0.8675
	Precision	0.9899	0.7711	0.7089	0.9263	0.7660
	F1-score	0.9751	0.7232	0.7619	0.8889	0.8136

(3) *Experiment*

The testing set data should be entered into the trained convolution network to get the final appliance identification results. The average accuracy of the CNN model is 84.00%. The identification results of each appliance are shown in Table 8.7, and Fig. 8.8 is a classification confusion matrix of it. In the confusion matrix, the sum of each row is the number of samples for the appliance in the testing set.

According to Table 8.8, among the five types of electrical appliances the model has the best identification effect on CFL. The identification accuracy of CFL reached 98.89% and there are very few false detections. The model has the worst identification effect on the Hairdryer and the identification accuracy is only 89.11%. From the confusion matrix, it can be seen that the most serious confusion is between the Hairdryer and ILB. Both appliances have a few numbers of false detections. This is because the V-I images of the two appliances are similar and the resolution of the model is not high enough. Overall the comprehensive identification accuracy of the model is 84% and the identification effect is acceptable.

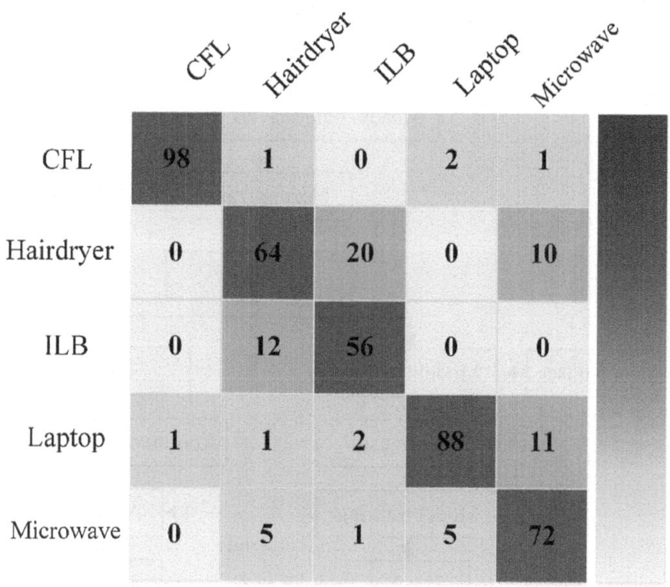

Fig. 8.8 The confusion matrix of the CNN model

Table 8.8 The sample data volume

Categories	CFL	ILB	Microwave	Laptop	Hairdryer
Numbers	341	228	278	344	312

8.3.2 *Appliance Identification Based on AlexNet*

AlexNet was proposed by Krizhevsky et al. (2012). It has eight layers of the network. Compared to CNN in the previous section, AlexNet has a more complex and deeper network structure. At the same time, the model parameters of the AlexNet have also increased. This can improve the ability of the character expression in the model. But when the network gets deeper and deeper, the difficulty of model training has also increased. This makes the model need more data to converge by training. Therefore, the technology of transfer learning is introduced. In this section, the training set and testing set images are consistent with Sect. 8.3.2. For model training, the ImagNet dataset has been used to pre-train the original AlexNet at first. Then the training set works for secondary training. Finally, a deep neural network adapted to the task of load identification is obtained. The flow chart of the algorithm is shown in Fig. 8.9.

(1) *Data processing*

In this section, the method of data processing is consistent with Sect. 8.2.2. The training set and testing set used for model training are also consistent with the previous

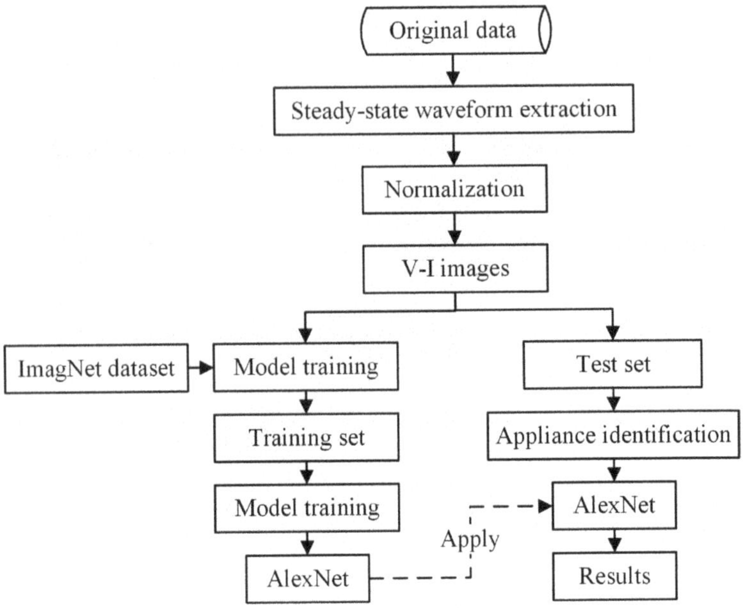

Fig. 8.9 The algorithm flow chart

section. Therefore, it will not be repeated. The number of samples in each electrical appliance is finally shown in Table 8.8.

(2) *Model establishment and training*

AlexNet is consist of 5 convolutional layers and 3 fully connected layers. Among the convolutional layers, three pooling layers are inserted to reduce the parameters in the model. The entire network contains 630 million links, 60 million parameters and 650,000 neurons. The structure of AlexNet is shown in Table 8.9.

As shown in the figure, the input layer of the network is a matrix of 224 × 224 × 3, corresponding to the size of the input image. The first convolutional layer has a total of 96 convolution kernels using a larger convolution and size of 11 × 11 with a step size of 4. The second layer is the LRN layer, followed by a 3 × 3 maximum pooling layer with a step size of 4. The convolutional layer after this is small, usually 5 × 5 or 3 × 3, and the step size is 1. Its purpose is to scan all the pixels. And the maximum pooling layer is still 3 × 3 with a step size of 2. It can be seen that in several convolutional layers in front of the network, although the amount of calculation is large, the parameter amount is small, which is about 1M. Most of the parameters of the model are in the fully connected layer. This is determined by the nature of the convolutional layer sharing weights.

Consistent with the training process in Sect. 8.2.2, the first 70% of each appliance in the sample is used as the training set and the last 30% is used as the testing set. Training the model by stochastic gradient descent. The training results are shown in Fig. 8.10.

Table 8.9 The structure of the AlexNet

Layer 1	Input
Layer 2	CONV 11×11/ReLU
Layer 3	LOCAL CONTRAST NORM
Layer 4	MAX POOLING
Layer 5	CONV 11×11/ReLU
Layer 6	LOCAL CONTRAST NORM
Layer 7	MAX POOLING
Layer 8	CONV 3×3/ReLU
Layer 9	CONV 3×3/ReLU
Layer 10	CONV 3×3/ReLU
Layer 11	MAX POOLING
Layer 12	FULL 4096/ReLU
Layer 13	FULL 4096/ReLU
Layer 14	FULL 4096/ReLU
Layer 15	Output

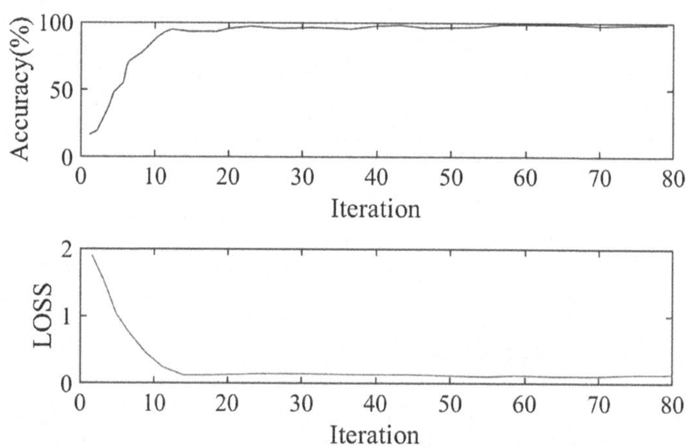

Fig. 8.10 The training process of the AlexNet model

Compared with the CNN network, the convergence speed of the AlexNet network is obviously accelerated during training. When the number of iterations reaches 10, the accuracy and loss of the model will be close to a smooth trend and the training process of the model is completed because the transfer learning method is used to pre-train the network. Most of the parameters in the AlexNet network, which have been adjusted after the network, are trained by the ImagNet dataset. Therefore, when the network is trained by the training set in this section the network only needs to fine-tune a small number of parameters to make it more suitable for the load identification task of this section. This method can not only save the training time of the network, but also reduce the data volume requirements.

(3) *Experiment*

The testing set data has been entered into the trained AlexNet network to get the final appliance identification results. The average accuracy of the model is 95.56%. The identification results of each appliance are shown in Table 8.10 and Fig. 8.11 is the classification confusion matrix of it.

Compared to the CNN model in the previous section, the performance of the AlexNet appliance identification model has been improved. According to Table 8.8, among the five types of electrical appliances, the model has the best identification effect on CFL. The identification accuracy of CFL reached 99.56%. The model has the worst identification effect on the ILB and Microwave but the accuracy has also reached 98.00%. And it can be seen from the confusion matrix that the detection

Table 8.10 The results of the AlexNet model

Model	Evaluation index	The type of appliance				
		CFL	Hairdryer	ILB	Laptop	Microwave
AlexNet	Accuracy	0.9956	0.9867	0.9800	0.9689	0.9800
	Recall	0.9902	1.0000	1.0000	0.8932	0.9036
	Precision	0.9902	0.9400	0.8831	0.9684	0.9868
	F1-score	0.9902	0.9691	0.9379	0.9293	0.9434

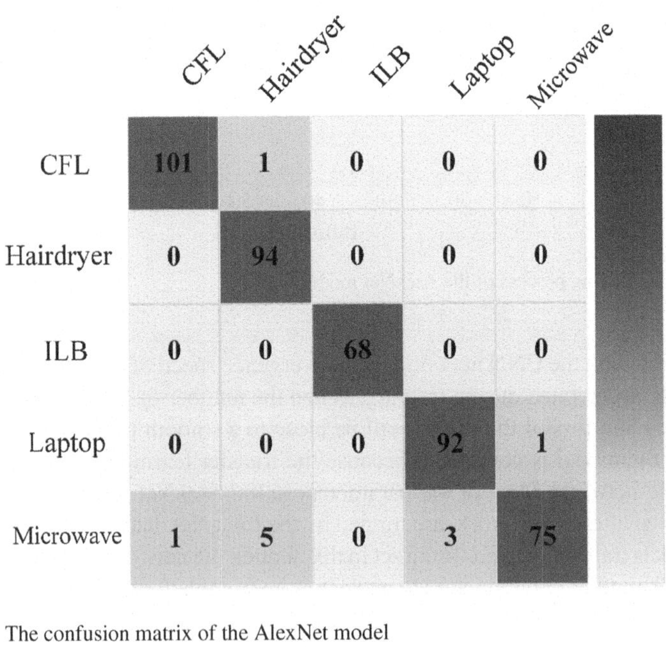

Fig. 8.11 The confusion matrix of the AlexNet model

of the five electrical appliances is only slightly confused. This shows that the resolution of the appliance identification model based on AlexNet is very high and the comprehensive performance of the model is good.

8.3.3 Appliance Identification Based on LeNet-SVM Model

In NILM, the accuracy of appliance identification is closely related to the performance of the feature extraction module. One of the great advantages of deep learning is that it can automatically extract the features of an object. Using deep learning as a feature extractor in NILM tasks, the features of load events can be extracted accurately and conveniently without manual intervention. This can greatly reduce the research threshold of NILM. In this section, the binarized electrical V-I images (Du et al. 2015) are used as the original input data. And the feature extractor for load events is the LeNet which is trained by the MNIST dataset (LeCun et al. 1998). The value of the fully connected layer, which is the penultimate layer of the LeNet, is used as an output feature. Then the SVM is used to classify the features to complete the appliance identification task. The algorithm flow chart is shown in Fig. 8.12 (Fig. 8.13).

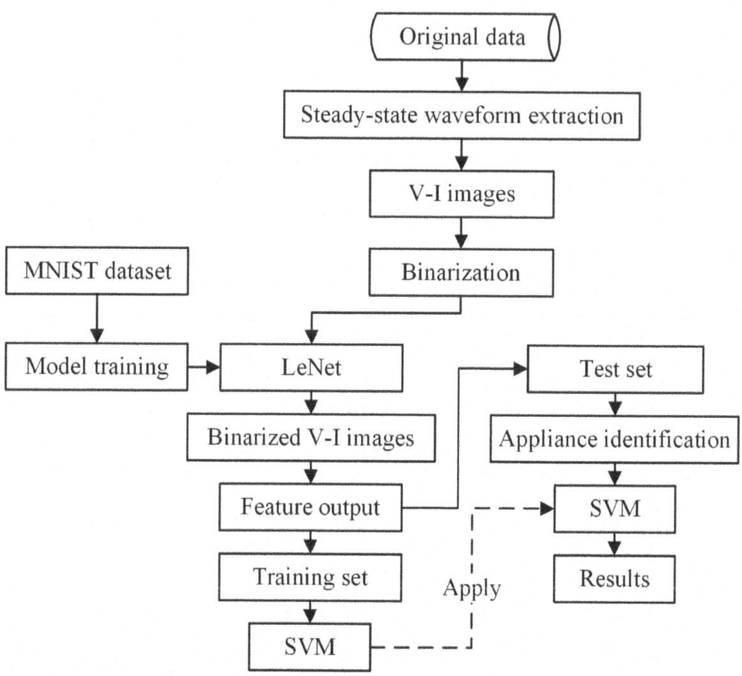

Fig. 8.12 The algorithm flow chart of the LeNet-SVM model

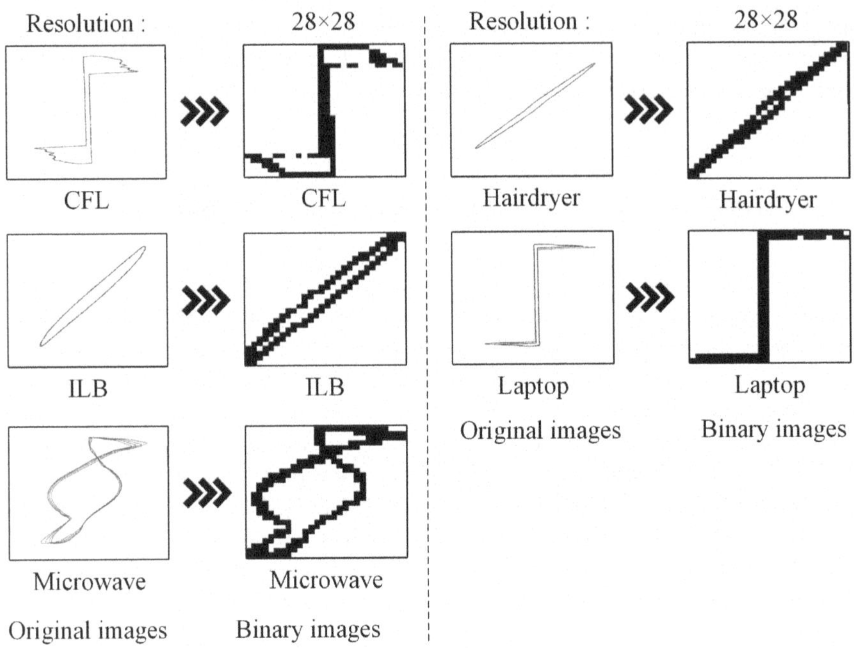

Fig. 8.13 The binarized V-I images

(1) *Data processing*

In this section, the original data come from the PLAID dataset. There are five kinds of electrical appliances: CFL, ILB, Microwave, Laptop, and Hairdryer, which are chosen as identification objects.

The input of LeNet is the binarized V-I images of each appliance in steady-state operation. 2000 sample points of voltage and current data during steady-state operation of the appliance have been extracted to draw the V-I curve. And then the entire V-I image is segmented into a 28 × 28 grid. If there is a curve in the grid, the value is 1, otherwise the value is 0. Processing the V-I trajectory of the appliance by this method can greatly reduce the algorithm's demand for computing resources and the time for load identification to ensure accuracy. Each electrical sample will be tested twice at different locations in the steady-state operation after removing the error image in the sample, the number of samples of each electrical appliance is finally shown in Table 8.11.

(2) *Model establishment and training*

Table 8.11 The sample data volume of the LeNet-SVM model

Categories	CFL	ILB	Microwave	Laptop	Hairdryer
Numbers	341	228	278	344	312

Table 8.12 The structure of the original LeNet model

Layer 1	Input 32 × 32
Layer 2	CONV 5 × 5
Layer 3	POOLING
Layer 4	CONV 5 × 5
Layer 5	POOLING
Layer 6	CONV 5 × 5
Layer 7	FULL CONNECT
Layer 8	Output

LeNet achieved great success in handwritten character recognition. And the input of LeNet is binary images. The binary VI images of the load that needs to be identified are very similar to handwritten character recognition in this section. The binary VI feature images can be treated as special characters. So, LeNet can accurately extract its features. In this section, an adjusted LeNet is used. The output layer of the network is removed and 84 output values of the penultimate layer of the LeNet are used to build the feature vector. The binarized V-I images are edge-filled before inputting in order to match the input layer size. LeNet is mainly composed of 2 convolution layers, 2 pooling layers and 3 fully connected layers. The structure of the original LeNet is as shown in Table 8.12.

As for model training, transfer learning is introduced for the problem of fewer original data samples. First, the MNIST dataset is used to train the original LeNet. Then, the binarized V-I images are entered into LeNet, and the electrical features of the output of the penultimate layer are recorded in the LeNet to train the SVM model. The training set accounts for 70% of all extracted binarized VI images and the remaining 30% are used as the testing set.

Support Vector Machine (SVM) is one of the most commonly used machine learning algorithms. It performs well in the small samples cases. Besides, it also has good robustness and can provide satisfactory classification results. The SVM performs well in load identification tasks (Altrabalsi et al. 2016). In this section, a multi-class SVM classifier is built for load identification. Essentially, SVM is a binary classification model. In order to achieve the identification of the five appliances in this section. The multi-class SVM model was established by the one-versus-one method, which is also known as OVO SVMs. A total of 10 single SVM models have been established, and the kernel functions of all single SVM models are RBF functions.

(3) *Experiment*

The testing set is input into the trained SVM model, and the results of appliance identification are shown in Table 8.13. And Fig. 8.14 is the confusion matrix.

According to the experimental results, the accuracy of the model detection is extremely high and the classification results of the five electrical appliances are

Table 8.13 The results of the LeNet-SVM model

Model	Evaluation index	The type of appliance				
		CFL	Hairdryer	ILB	Laptop	Microwave
LeNet-SVM	Accuracy	1.0000	1.0000	1.0000	1.0000	1.0000
	Recall	1.0000	1.0000	1.0000	1.0000	1.0000
	Precision	1.0000	1.0000	1.0000	1.0000	1.0000
	F1-score	1.0000	1.0000	1.0000	1.0000	1.0000

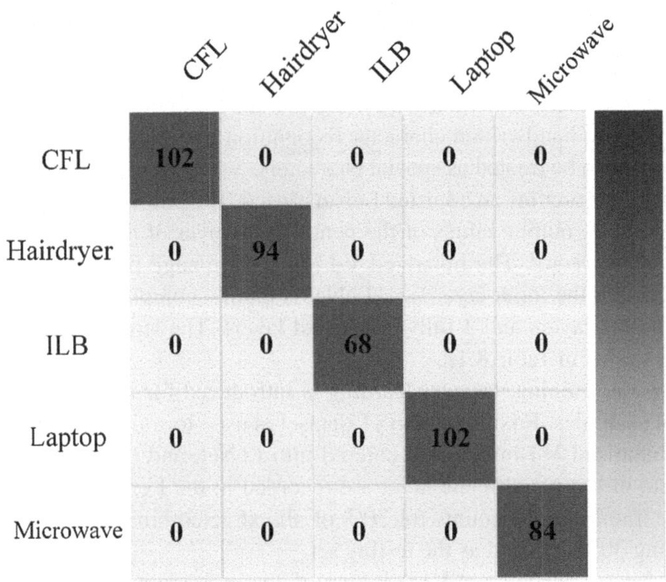

Fig. 8.14 The confusion matrix of the LeNet-SVM model

error-free. This shows that LeNet can accurately extract the characteristics of the electrical appliances in the binarized VI image. On the other hand, the performance of the SVM classifier meets the classification task requirements of the five electrical appliances.

8.4 Experiment Analysis

8.4.1 *Experimental Analysis of End-to-End Decomposition*

The identification results of the two models in Sect. 8.2 are shown in Table 8.14 and Fig. 8.15. Model 1 uses a single feature to decompose the load and Model 2 uses multiple features to decompose the load.

(a) From the results, both models can complete the appliance identification task for three electrical appliances:the light, dishwasher and fridge. Model 2, which

Table 8.14 The results of the two models

LSTM		Evaluation index			
		Accuracy	Recall	Precision	F1-score
Single feature (Model 1)	Light	0.8866	0.9977	0.8470	0.9162
	Dishwasher	0.7194	0.6234	0.7236	0.6698
	Fridge	0.6973	0.3957	0.8650	0.5413
Multiple feature (Model 2)	Light	0.9110	0.9974	0.8764	0.9330
	Dishwasher	0.8006	0.6115	0.9266	0.7368
	Fridge	0.7095	0.4747	0.8004	0.5959

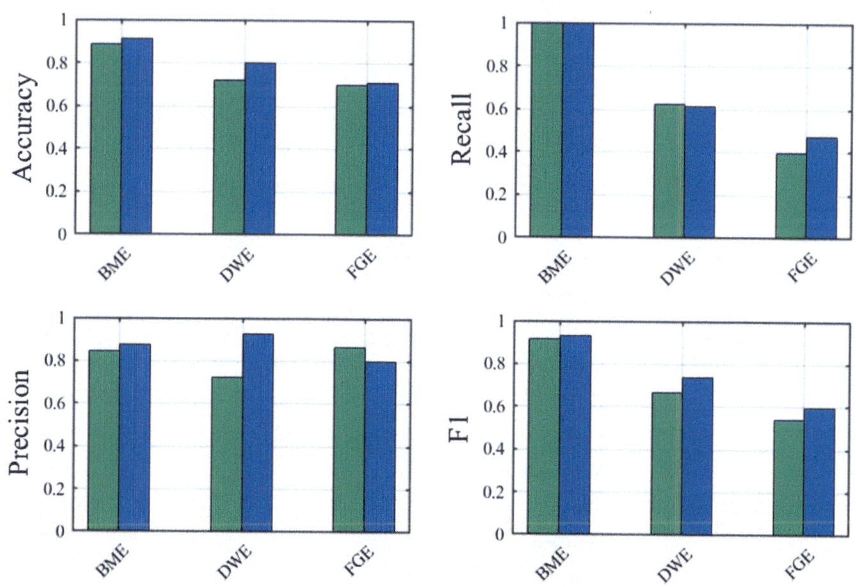

Fig. 8.15 The results of two models (The green corresponds to Model 1, and the blue corresponds to Model 2, and BME, DWE, FGE correspond to light, dishwasher, and fridge)

uses multiple features and the LSTM network with the more hidden unit, has a better effect on the decomposition of the load.

(b) Compared with Model 1, Model 2 has the highest identification performance improvement for dishwasher. Its accuracy and F1 value have increased by about 7%. The identification performance improvement of light is small. Its accuracy and F1 value increase by about 2%. As for the fridge, the identification performance of it has hardly improved. This shows that the newly added current features cannot effectively express the characteristics of the fridge. To further improve the identification performance of the model, more types of feature sequences need to be introduced.

(c) In general, the comprehensive performance of Model 2 is better than that of Model 1. This shows that increasing the complexity of deep neural networks and the types of original feature sequences can effectively improve the performance of load decomposition models.

8.4.2 Experimental Analysis of Appliance Classification

The recognition results of the three models in Sect. 8.3 are shown in Table 8.15 and Fig. 8.16. Model 1 uses CNN for appliance identification and Model 2 uses AlexNet combined with transfer learning technology for appliance identification. Model 3 uses

Table 8.15 The results of the three models

Model		Evaluation index			
		Accuracy	Recall	Precision	F1
CNN (Model 1)	CFL	0.9889	0.9608	0.9899	0.9751
	Hairdryer	0.8911	0.6809	0.7711	0.7232
	ILB	0.9222	0.8235	0.7089	0.7619
	Laptop	0.9511	0.8544	0.9263	0.8889
	Microwave	0.9267	0.8675	0.7660	0.8136
AlexNet (Model 2)	CFL	0.9956	0.9902	0.9902	0.9902
	Hairdryer	0.9867	1.0000	0.9400	0.9691
	ILB	0.9800	1.0000	0.8831	0.9379
	Laptop	0.9689	0.8932	0.9684	0.9293
	Microwave	0.9800	0.9036	0.9868	0.9434
LeNet-SVM (Model 3)	CFL	1.0000	1.0000	1.0000	1.0000
	Hairdryer	1.0000	1.0000	1.0000	1.0000
	ILB	1.0000	1.0000	1.0000	1.0000
	Laptop	1.0000	1.0000	1.0000	1.0000
	Microwave	1.0000	1.0000	1.0000	1.0000

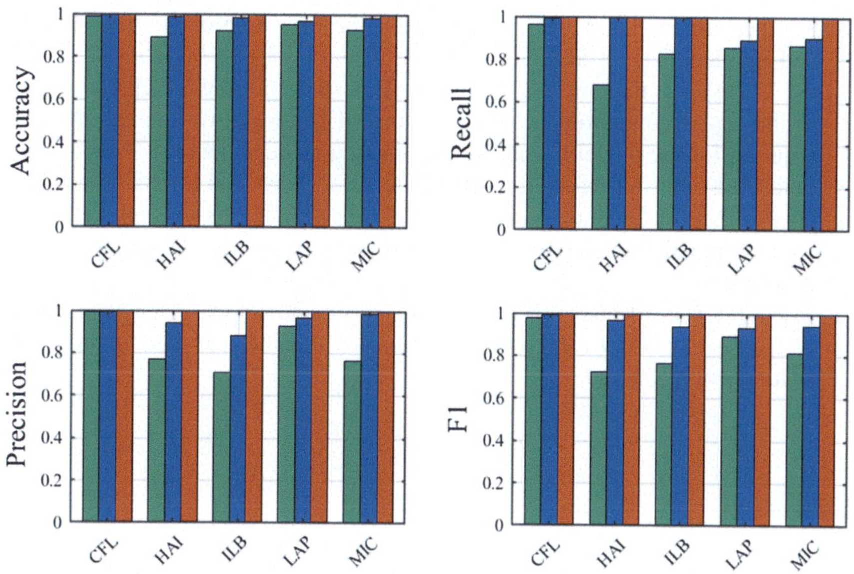

Fig. 8.16 The results of the three models (green, blue, and brown correspond to Model 1, Model 2, Model 3 and CFL, HAI, ILB, LAP. MIC correspond to compact Fluorescent lamp, Hairdryer, Incandescent light bulb, Laptop, and Microwave)

LeNet as a feature extractor and SVM as a classifier to jointly implement appliance identification.

(a) According to the results, the three models can complete the identification tasks of CFL, Hairdryer, ILB, Laptop, and Microwave, and the accuracy is high.

(b) Model 1 has the lowest comprehensive performance with an average accuracy of 84%. By using this model for appliance identification the main errors are concentrated on the Hairdryer and ILB, which have similar V-I images. The former is partially confused with ILB and Microwave and the latter is only confused with the Hairdryer. because the reason is that the depth of the CNN network in Model 1 is shallow, which will result in an insufficient resolution of the model. This can lead to a decline in the performance of the appliance identification model.

(c) The performance of Model 2 is greatly improved compared with Model 1 and its average accuracy is 95%. There are very few errors in the identification of the appliance because the AlexNet network used in Model 2 is deeper than the CNN network used in Model 1. Therefore, Model 2 is more capable of abstract features. On the other hand, the introduction of transfer learning methods to train deep neural networks expands the training data volume of the network and speeds up the training speed of the network. This makes the Model 2 network more comprehensive than model 1.

(d) The performance of Model 3 takes the highest position of the three models and its identification accuracy is 100%. The model first binarizes the original V-I images, to compress the image. This simplifies the V-I characteristics of the appliance and reduces the workload of feature extraction and the time required for the model to run. Model 3 also uses deep neural networks combined with transfer learning to extract features from load events. The difference is that Model 3 uses a better performance SVM classifier instead of the softmax function to classify the extracted features, so it has better recognition performance.

(e) In summary, the deep neural network is very suitable for extracting the features of load events. And the higher performance classifier can also improve the accuracy of the model.

References

Altrabalsi H, Stankovic V, Liao J, Stankovic L (2016) Low-complexity energy disaggregation using appliance load modelling. AIMS Energy 4(1):884–905

Barsim KS, Mauch L, Yang B (2018) Neural network ensembles to real-time identification of plug-level appliance measurements. arXiv:06963

Chao L, Tao J, Yang M, Li Y, Wen Z (2015) Long short term memory recurrent neural network based multimodal dimensional emotion recognition. In: 5th international workshop on audio/visual emotion challenge

De Baets L, Develder C, Dhaene T, Deschrijver D (2019) Detection of unidentified appliances in non-intrusive load monitoring using siamese neural networks. Int J Electr Power Energy Syst 104:645–653. https://doi.org/10.1016/j.ijepes.2018.07.026

Du L, He D, Harley RG, Habetler TG (2015) Electric load classification by binary voltage–current trajectory mapping. IEEE Trans Smart Grid 7(1):358–365

Hadji I, Wildes RP (2018) What do we understand about convolutional networks? arXiv:08834

Hassan T, Javed F, Arshad N (2014) An empirical investigation of V-I trajectory based load signatures for non-intrusive load monitoring. IEEE Trans Smart Grid 5(2):870–878

Huss A (2015) Hybrid model approach to appliance load disaggregation: Expressive appliance modelling by combining convolutional neural networks and hidden semi Markov models

Kelly D (2016) Disaggregation of domestic smart meter energy data

Krizhevsky A, Sutskever I, Hinton GE (2012) Imagenet classification with deep convolutional neural networks. In: Advances in neural information processing systems, pp 1097–1105

Lange H, Bergés M (2016) The neural energy decoder: energy disaggregation by combining binary subcomponents. In: Proceedings of the 3rd international workshop on non-intrusive load monitoring

LeCun Y, Bottou L, Bengio Y, Haffner P (1998) Gradient-based learning applied to document recognition. Proc IEEE 86(11):2278–2324

LeCun Y, Bengio Y, Hinton GJn (2015) Deep learning. Nature 521 (7553):436

Mauch L, Barsim KS, Yang B (2016) How well can HMM model load signals. In: Proceeding of the 3rd international workshop on non-intrusive load monitoring (NILM 2016)

Tsai MS, Lin YHJAE (2012) Modern development of an adaptive non-intrusive appliance load monitoring system in electricity energy conservation. Appl Energy 96(8):55–73

Zhang C, Zhong M, Wang Z, Goddard N, Sutton C (2018) Sequence-to-point learning with neural networks for non-intrusive load monitoring. In: Thirty-second AAAI conference on artificial intelligence

Chapter 9
Deterministic Prediction of Electric Load Time Series

9.1 Introduction

9.1.1 Background

The serious shortage of traditional energy and the growing environmental pollution are the biggest challenges facing the sustainable development of human society (Ehrenfeld 2010). In order to solve the energy source crisis and environmental problems, various low-carbon technologies such as energy efficiency technology, renewable energy technology, and smart transportation technology have developed rapidly and are applied widely (Migendt 2017; Ge et al. 2017; Cooper 2012). The large-scale application of various low-carbon technologies is mainly focused on renewable energy generation and terminal user service optimization (Zeng et al. 2014). The relationship between the supply side and the demand side of the traditional power grid has undergone major changes, and it has brought new challenges to the development and safe operation of the transmission and distribution. Under such a development background, the concept of smart grid emerged as the times require. And smart grid system has been widely applied in the whole world, becoming the common development trend of the world power industry (Mengelkamp et al. 2018). The construction of smart grid concept can stimulate the market to upgrade manufactory products, thereby increasing the country's international economic competitiveness (Zhang et al. 2015).

As one of the core technologies of the smart grid, electric load forecasting plays an increasingly important role (Kang and Xia 2004). Electric load forecasting takes an important part in power system dispatching, power consumption, planning, and other management departments. The improvement of load forecasting technology can be benefited from the power management plan, which is the reasonable arrangement of grid operation mode and unit maintenance plan. The research of load forecasting technology is deepening with the rapid development of modern science and technology. Various load forecasting methods have been explored, such as classical unit consumption method, elastic coefficient method, statistical analysis method, gray

H. Liu, *Non-intrusive Load Monitoring*,
https://doi.org/10.1007/978-981-15-1860-7_9

prediction method, and expert system methods (Feinberg and Genethliou 2005). Fuzzy mathematics, neural network methods, and deep learning methods are used more and more widely (Markovié and Fraissler 2010; Nalcaci et al. 2018; Chen et al. 2018).

In recent years, academic researches on load forecasting have focused on the use of intelligence models such as neural networks and deep learning algorithms. A series of optimization algorithms are used to optimize the performance of the predictor, such as Wavelet Decomposition, Variational Mode Decomposition, and Particle Swarm Optimization. Shi et al. proposed a new Pool-based Deep Recurrent Neural Network (PDRNN) to predict household electrical load (Shi et al. 2018). This model could solve the overfitting by increasing the diversity and quantity of data. So it has a higher prediction accuracy than other comparable models in-home load prediction. Xiao et al. developed a hybrid model composed of the Multi-objective Optimization Algorithm (MOA) and Multi-objective Flower Pollination Algorithm (MOFPA) for short-term load forecasting (Xiao et al. 2017). In this hybrid model, MOFPA was used to optimize the weight of a single model in order to achieve high precision and high stability. At the same time, data preprocessing techniques were proposed including Fast Ensemble Experience Mode Decomposition (FEEMD) and multiple seasonal patterns to achieve model optimization. Li et al. proposed a Data-driven Linear Clustering (DLC) method to achieve long-term urban load forecasting (Yiyan et al. 2017). They preprocessed the large-scale substation long-term load dataset by linear clustering method and then constructed the optimal ARIMA model to achieve long-term load forecasting.

9.1.2 Advance Prediction Strategies

In practical engineering applications, single-step prediction often fails to meet actual needs. It is necessary to achieve advanced multi-step prediction based on historical data, so as to achieve more long-term predictions in the future. Therefore, the importance and practicability of multi-step prediction often exceed single-step prediction (Chatfield and Weigend 1994). In the field of time series prediction, recursive strategies and multiple input multiple output strategies are generally used to achieve advanced multi-step prediction.

Recursive strategy, also known as a direct multi-step forecasting strategy, is the most common advanced multi-step predictive strategy. The recursive strategy uses a historical data point of length M as the predictive model input to make a 1-step prediction, and the model output is single output. Then, the prediction result is taken as historical data, and the input model is used to obtain the prediction result of the advanced 2-step, and the prediction result of the advanced n-step is obtained by analogy. Since the recursive strategy has the problem of error accumulation, the prediction error is passed to the next step. Thus, the prediction accuracy is easy to decrease as the number of prediction steps increase. A schematic diagram of the

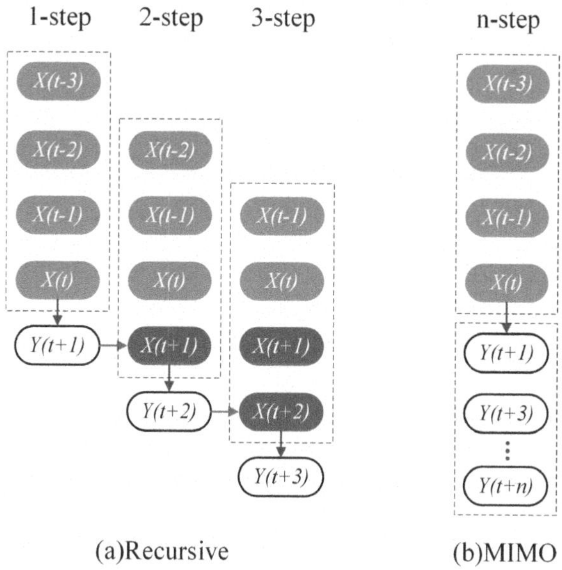

Fig. 9.1 Schematic diagram of advanced prediction strategies

recursive strategy prediction process is shown in Fig. 9.1a. The length of the prediction model's input for the Multi-Input Multi-Output (MIMO) strategy is M, and all n prediction results are output at one time. Due to the large number of output elements, the complexity of the prediction model increases as the prediction step increases, and the computational space increases at the same time. However, this strategy does not have the problem of error accumulation, so it is suitable for occasions where the number of partial prediction steps is low. A schematic diagram of the strategy is shown in Fig. 9.1b. The recursive strategy is utilized in the chapter.

9.1.3 Original Electric Load Time Series

The time series in the chapter is obtained from the US PJM company, which summarizes the GW-hour net energy for load consumed from different electric distribution companies within PJM. These data characterize the actual power usage in the company's power supply areas. The sampling period of the original load time series is 1 h. The load original time series with a sample length of 1200 in the CE area is selected as the short-term electric load time series. The chapter uses $\{X_S\}$ to indicate short-term electric load time series. By equal interval resampling, the chapter obtains a long-term electric load time series with a sample length of 1200 from the original load series in the CE area, whose sampling frequency is 1 day. The chapter uses $\{X_L\}$ to indicate long-term electric load time series.

In the modeling process of the load time series prediction model, the sample data needs to be divided into a training set for model construction and a test set for model

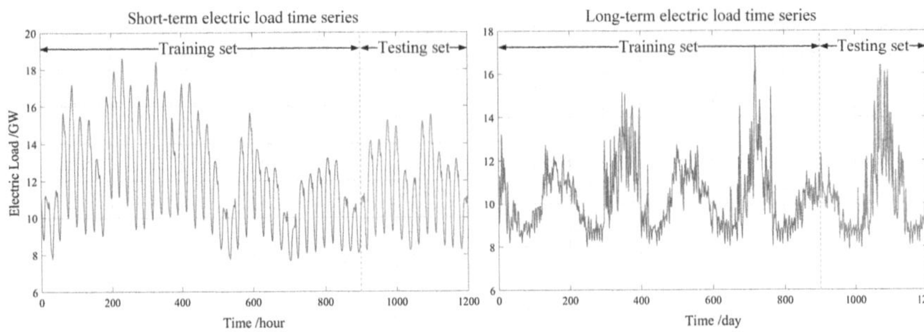

Fig. 9.2 Separation of electric load time series

performance verification. The chapter utilizes the 1st to the 900th sample points of two electric load time series as training samples, and the 901st to the 1200th sample points as test samples. The sample separation results are shown in Fig. 9.2.

By visually analyzing the two types of load time series in Fig. 9.2, it can be seen that both of them have strong instability and high-frequency volatility. The difference is that the long-term electric load series has a significantly long period of time that is not available in the short-term electric load series. The fluctuation period is basically stable at around 365 days, and the changes in the peaks and valleys are basically consistent with the season changes during the year. This phenomenon indicates that the long-term electric load series can represent the power usage habits of the entire community in the area to a certain extent. At the same time, the long-term electric load series has stronger high-frequency volatility than the short-term electric load series. Excessive fluctuation frequencies may affect the predictive performance of load deterministic prediction models.

9.2 Load Forecasting Based on ARIMA Model

9.2.1 Model Framework

Figure 9.3 shows the model framework for predicting the electric load series with the ARIMA model and EMD-ARIMA hybrid model.

9.2.2 Theoretical Basis of ARIMA

For the time series $\{X_t, t \in T\}$ composed of random variables, its nature cannot be obtained intuitively. It needs to be indirectly inferred by studying the nature of its observation sequence $\{x_t, t = 1, 2, \ldots, n\}$. The random time series has the following related concepts:

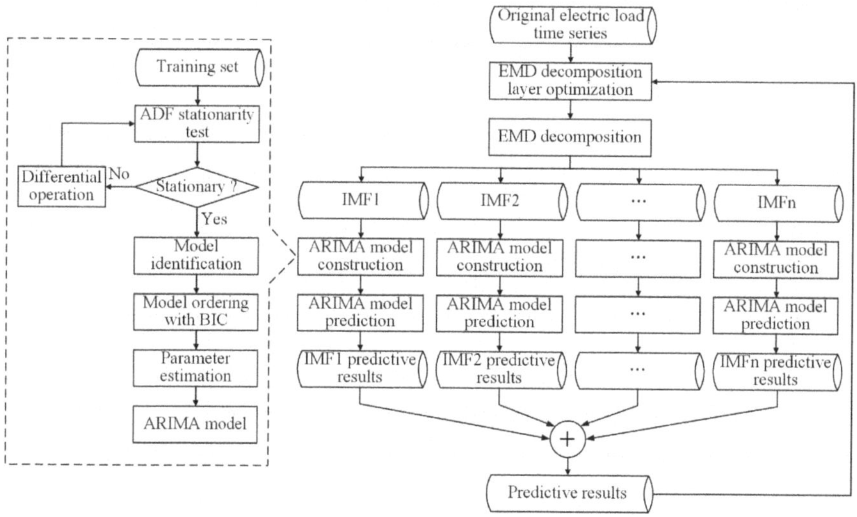

Fig. 9.3 Modeling flow chart of ARIMA and EMD-ARIMA model

(a) Stationarity

The stationarity of the time series is based on the evaluation of random variables. It can be divided into two types: wide stationarity and strict stationarity (Madsen 2007).

The condition of wide stationarity is relatively loose, and the following conditions must be met. If the mean and variance of a sequence random variable are independent of time, the sequence satisfies the condition of wide stationarity. The definition of wide stationarity has an implicit premise that the mean and variance of the time series are both exist.

If a time series satisfies the strict stationarity, the joint distribution of the random variable $\{X_{t_1}, X_{t_1-1}, \ldots, X_{t_1-p}\}$ of this time series is the same as the joint distribution of the random variable $\{X_{t_2}, X_{t_2-1}, \ldots, X_{t_2-p}\}$. To a strict stationarity time series, the joint probability distribution of any subset is independent of time.

Wide stationarity has certain requirements for time series statistical indicators, and strict stationarity has certain requirements for the joint distribution of time series. So strict stationarity is more difficult to verify than wide stationarity. In the chapter, stationarity refers to the wide stationarity.

(b) Linear

The ARIMA model assumes that the time series is linear. However, the linear representation of the sequence can also achieve high-precision approximations for non-linearities, thus enabling the prediction of time series. If a time series is linear, it satisfies the following equation:

$$X_t = a + \sum_{p=0}^{\infty} \theta_p \varepsilon_{t-p} \tag{9.1}$$

where a is a constant, $\theta_0 = 1$ and θ_p are constants. ε_{t-p} is an independent and identically distributed random vector, whose mean is 0 and variance is σ^2. As can be seen from the above equation, X_t is a function of the random variable $\{\varepsilon_{t-p}, p = 0 \sim \infty\}$. Therefore, when the time is $t - 1$, ε_t is the only unknown variable. Therefore, this unknown variable is called Innovation, which is the newly added information from the previous moment to the next moment.

(c) Reversibility

If a time series X_t is a function of the hysteresis value X_{t-k} and the Innovation at time t, then this time series is considered to be reversible (Diks et al. 1995). It can be expressed as follows:

$$X_t = b + \varepsilon_{t-k} + \sum_{k=1}^{\infty} \rho_k \varepsilon_{t-k} \tag{9.2}$$

where b is a constant, ρ_k is the coefficient of the lag term. For a reversible time series, when $k \to \infty$, $\rho_k \to 0$.

Generally speaking, time series models can be divided into four categories: Auto Regressive Model, Moving Average Model, Auto Regressive Moving Average Model, and Auto Regressive Integrated Moving Average Model.

(a) Auto Regressive Model

If the response of a time series model at time t is directly related to its past response X_{t-k} and Innovation ε_t, and has no direct relationship with the perturbation of the past time $\{\varepsilon_{t-p}, p = 1 \sim \infty\}$, it will be called Auto Regressive model, abbreviated as AR(p) model.

The Auto Regressive Model can be expressed as follows:

$$\begin{cases} X_t = a_0 + a_1 X_{t-1} + a_2 X_{t-2} + \cdots + a_p X_{t-p} + \varepsilon_t \\ E(\varepsilon_t) = 0,\ E(\varepsilon_t, \varepsilon_i) = 0,\ E(\varepsilon_t, X_i) = 0,\ t \neq i \\ Var(\varepsilon_t) = \sigma^2 \end{cases} \tag{9.3}$$

where, $\{a_1, a_2, \ldots, a_p\}$ represents the autoregressive coefficient and p is the model order. In particular, when $a_0 = 0$, the model AR(p) is called the centralized model.

(b) Moving Average Model

If the response of a time series model at time t has no direct relationship with its past response X_{t-k} and is directly related to the perturbation of the past time $\{\varepsilon_{t-p}, p = 1 \sim \infty\}$, it will be called Moving Average model, abbreviated as MA(q) model.

The MA model can be expressed as follows:

$$\begin{cases} X_t = b_0 + \varepsilon_t - b_1\varepsilon_{t-1} - b_2\varepsilon_{t-2} - \cdots - b_q\varepsilon_{t-q} \\ E(\varepsilon_t) = 0,\ E(\varepsilon_t, \varepsilon_i) = 0,\ t \neq i \\ Var(\varepsilon_t) = \sigma_\varepsilon^2 \end{cases} \tag{9.4}$$

where, $\{b_1, b_2, \ldots, b_p\}$ represents the moving coefficient and q is the model order. In particular, when $b_0 = 0$, the model MA(q) is called the centralized model.

(c) Auto Regressive Moving Average Model

If the response of a time series model at time t not only has a direct relationship with its past response X_{t-k} but also is directly related to the perturbation of the past time $\{\varepsilon_{t-p}, p = 1 \sim \infty\}$, it is called Auto Regressive Moving Average model, abbreviated as ARMA(q) model.

The ARMA(q) model can be expressed as follows:

$$\begin{cases} X_t = c_0 + a_1X_{t-1} + a_2X_{t-2} + \cdots + a_pX_{t-p} + \varepsilon_t - b_1\varepsilon_{t-1} - b_2\varepsilon_{t-2} - \cdots - b_q\varepsilon_{t-q} \\ E(\varepsilon_t) = 0,\ E(\varepsilon_t, \varepsilon_i) = 0,\ t \neq i \\ Var(\varepsilon_t) = \sigma_\varepsilon^2 \end{cases} \tag{9.5}$$

in particular, when $c_0 = 0$, the model ARMA(q) is called the centralized model.

(d) Auto Regressive Integrated Moving Average Model

In real situations, the time series obtained is usually non-stationary. However, the traditional linear model requires that the time series satisfy the stationarity. It is necessary to find new ways to solve such problems. Some methods are utilized to simplify and convert non-stationary time series to stationary time series, and then use the ARMA model to fit them, the above problem can be solved. The differential sequences of many non-stationary sequences show the nature of stationary sequences. The ARIMA model uses d-order difference to convert non-stationary time series into stationary time series. The model is abbreviated as ARIMA(p, d, q). Among them, p, d, q represent the autoregressive order, the different order, and the moving average order, respectively.

9.2.3 Modeling Steps of ARIMA Predictive Model

(1) *Preprocessing and stationary testing*

Before applying the ARMA model, it is important to test the stationarity of each time series to avoid the occurrence of false regressions. Time series stationarity can be obtained by observing the sequence diagram, but the method is subject to subjective impression and has strong uncertainty. In order to improve the inspection accuracy, this chapter uses the ADF test to verify the stationarity of the training set of electric load time series. It is one of the most widely used statistical methods for time series

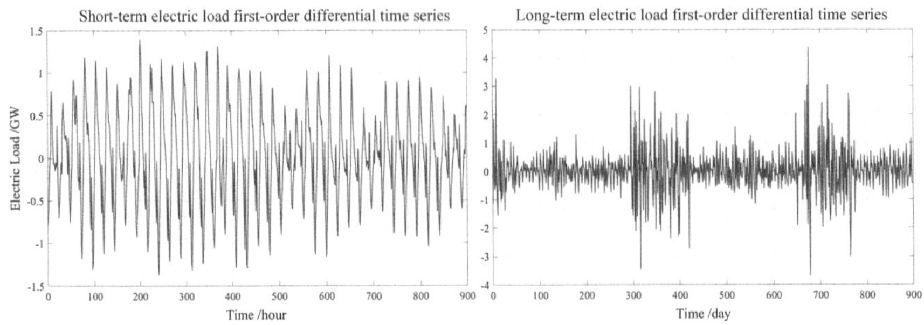

Fig. 9.4 Electric load time series first-order differential series

stationarity (Amin and Alam 2018). After the ADF test, both types of electric load time series are non-stationary time series and cannot be directly fitted by the ARMA model. Therefore, two types of electric load time series need to be differentially processed to improve their stationarity. The time series after the 1-order difference are shown in Fig. 9.4. The differential time series are represented by $\{X_S'\}$ and $\{X_L'\}$ respectively.

It can be drawn from the figure that $\{X_S'\}$ and $\{X_L'\}$ are obviously more stationary than the original electric load time series $\{X_S\}$ and $\{X_L\}$. The differential time series is basically stable near the mean of 0, fluctuating at a lower amplitude. $\{X_L'\}$ still has a higher fluctuation frequency than $\{X_S'\}$. The ADF test shows that both types of differential electric load time series have stationarity, which satisfies the premise of the ARMA model.

(2) *Ordering ARIMA model*

The Autocorrelation Coefficients and Partial Autocorrelation Coefficients are often used for model selection and parameter ordering of ARMA models (Inoue 2008). The applicable optimal model type can be obtained by performing calculation analysis on the autocorrelation coefficient and the partial correlation coefficient of the time series. The characteristics of the autocorrelation coefficient and the partial correlation coefficient are shown in Table 9.1.

The Autocorrelation Coefficients and Partial Autocorrelation Coefficients of the first 20 samples for two types of 1-order differential electric load series are shown in Fig. 9.5. It can be seen from figure that the Autocorrelation Coefficient and Partial Autocorrelation Coefficients of the two types of 1-order differential electric load series exhibit the trailing characteristics. Therefore, it is preliminary judged that the ARMA(p, q) model is selected for the 1-order differential series of the two types of electric load time series.

The method of ranking the model by Autocorrelation Coefficient and Partial Autocorrelation Coefficient is often very subjective. Therefore, after selecting the model type, this chapter uses the Bayesian Information Criterion (BIC) to achieve the order of the ARMA(p, q) model (Watanabe 2013).

Table 9.1 Correlation coefficient characteristics of ARIMA model

First-order differential series	Models	AR(p)	MA(q)	ARMA(p, q)
$\{X_S'\}$	Autocorrelation coefficient	Trailing	q-order censored	Trailing
	Partial autocorrelation coefficient	p-order censored	Trailing	Trailing
$\{X_L'\}$	Autocorrelation coefficient	Trailing	q-order censored	Trailing
	Partial autocorrelation coefficient	p-order censored	Trailing	Trailing

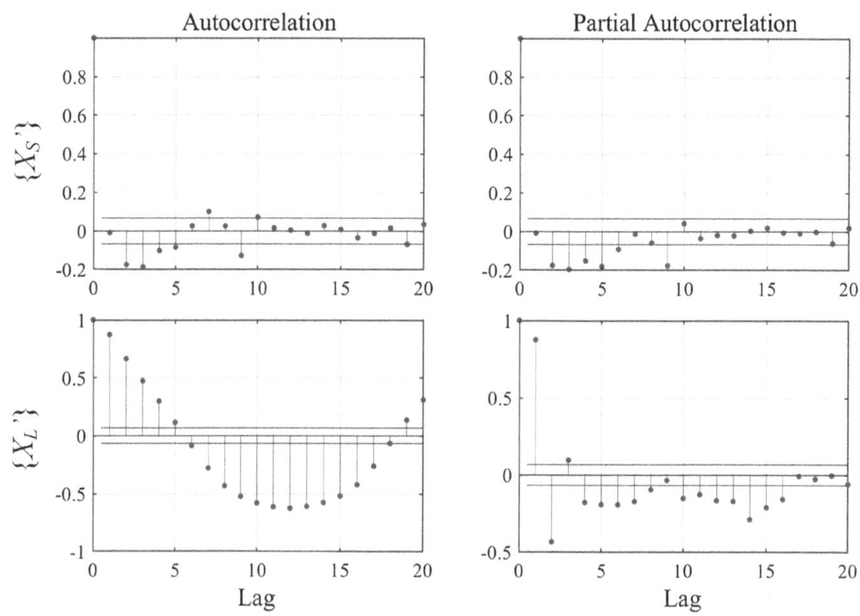

Fig. 9.5 Autocorrelation coefficient and partial autocorrelation coefficient

The parameter optimization results by the BIC criterion are shown in Fig. 9.6. It can be drawn from the figure that the BIC values of $\{X_L'\}$ are far greater than the BIC values of $\{X_S'\}$ in general. At the same time, in the process of searching for the optimal parameters, the BIC values of $\{X_S'\}$ show strong numerical differences for different combinations of p and q. And under some parameter selection conditions, the BIC value is much smaller than 0. It is proved that the ARMA model composed of the corresponding parameters has a good fitting ability for $\{X_S'\}$. Conversely, for different combinations of p and q, the BIC values of $\{X_L'\}$ are basically stable, lacking

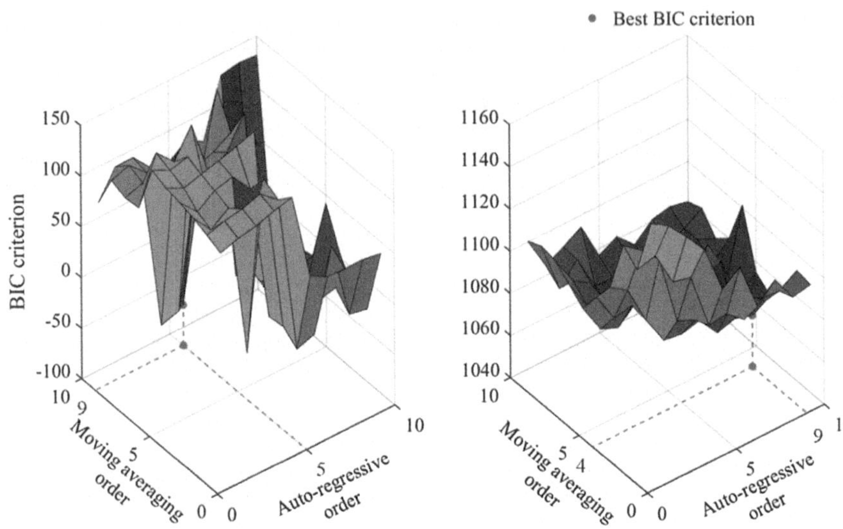

(a)Short-term electric load time series (b)Long-term electric load time series

Fig. 9.6 BIC criterion parameter optimization results

in variability, and generally greater than 1000. It indicates that the BIC criterion parameter optimization process is not effective for $\{X_L'\}$. Finally, the model selected for the short-term electric load time series is ARIMA(5, 1, 9), and the model selected for the long-term electric load time series is ARIMA(9, 1, 4). The corresponding BIC values are -60 and 1064, respectively.

9.2.4 Predictive Results

The predictive results of the ARIMA model for short-term electric load time series and long-term electric load time series are shown in Fig. 9.7.

The evaluation indices of ARIMA predictive results for short-term electric load time series and long-term electric load time series are shown in Tables 9.2 and 9.3.

From the prediction results and the calculation results of the evaluation indices, the following conclusions can be drawn:

(a) As the number of advance prediction steps increases, the prediction accuracy and evaluation index values of the ARIMA model for the short-term electric load time series and the long-term electric load time series are significantly degraded. For example, when performing a 1-step prediction on the short-term electric load time series, it can be clearly seen that the prediction result of the testing set can fit the true value well, and there is no obvious prediction deviation among them. Meanwhile, the evaluation indices MAE, MAPE, RMSE, SDE, and R are

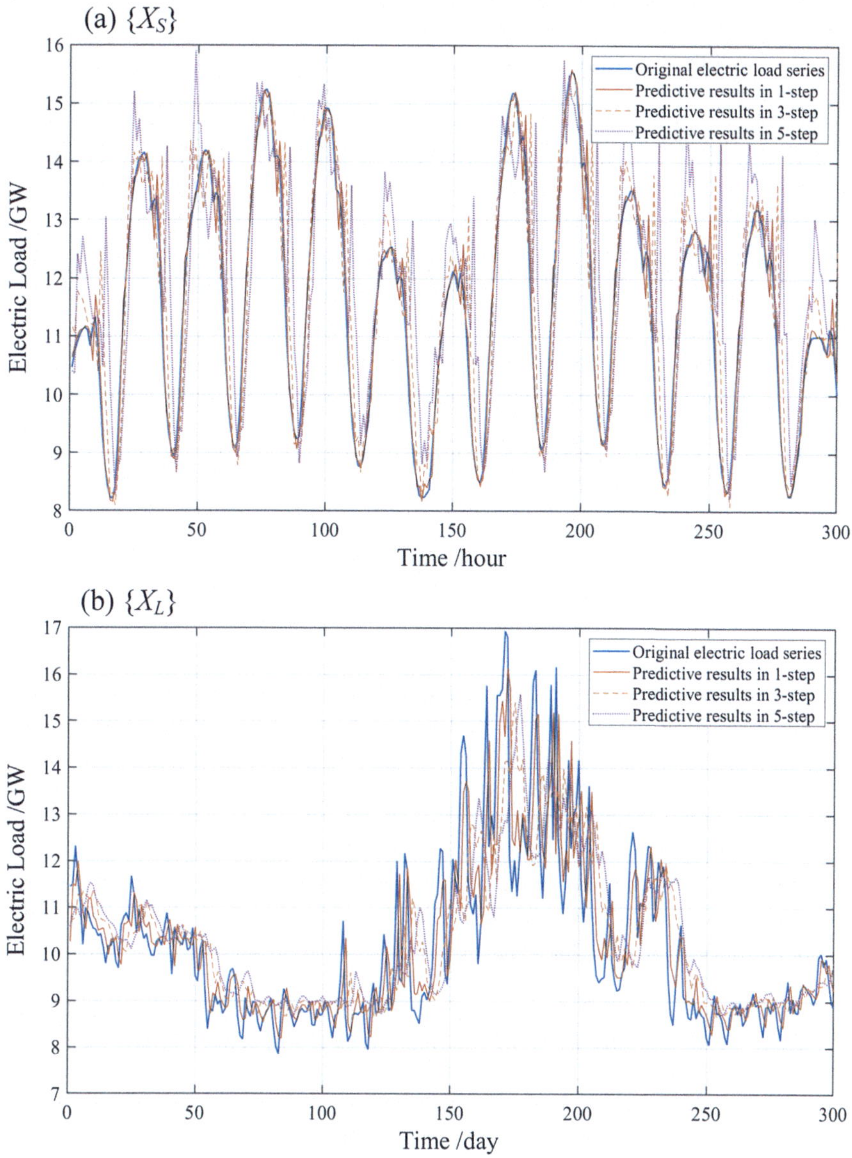

Fig. 9.7 The predictive results of by ARIMA

0.1759 (GW), 1.5221%, 0.2506 (GW), 0.2475 (GW) and 0.9923, respectively. It indicates that the model prediction results have extremely low relative and absolute errors, and the model has excellent prediction performance. However, when performing a 5-step prediction, it can be clearly seen that there are a significant prediction bias and prediction delay between the prediction result

Table 9.2 The predictive performance indices of ARIMA model for $\{X_S\}$

Model	Advance step	MAE (GW)	MAPE (%)	RMSE (GW)	SDE (GW)	R
ARIMA	1	0.1759	1.5221	0.2506	0.2475	0.9923
	2	0.4051	3.5523	0.5513	0.5372	0.9639
	3	0.6082	5.5161	0.8482	0.8134	0.9160
	4	0.7661	7.1609	1.1127	1.0513	0.8565
	5	0.9641	9.1842	1.3641	1.2653	0.7846

Table 9.3 The predictive performance indices of ARIMA model for $\{X_L\}$

Model	Advance step	MAE (GW)	MAPE (%)	RMSE (GW)	SDE (GW)	R
ARIMA	1	0.5956	5.4681	0.9099	0.9085	0.8637
	2	0.7986	7.4004	1.1589	1.1550	0.7698
	3	0.8552	7.9146	1.2448	1.2385	0.7296
	4	0.8751	8.1270	1.2546	1.2464	0.7250
	5	0.9051	8.4155	1.2957	1.2857	0.7053

and the true value. The predicted values of extreme points have large deviations. The evaluation indices MAE, MAPE, RMSE, SDE, and R are 0.9641 (GW), 9.1842%, 1.3641 (GW), 1.2653 (GW) and 0.7846, respectively. It indicates that the model prediction results have relatively high relative and absolute errors, and the model prediction accuracy is significantly degraded as the number of advance prediction steps increases.

(b) When the same numbers of advanced prediction steps are selected, the prediction accuracy of the ARIMA model for the short-term electric load time series is generally higher than that of the long-term electric load time series. For example, when performing a 3-step prediction on the short-term electric load time series, the evaluation indexes MAE, MAPE, RMSE, SDE, and R are 0.6082 (GW), 5.5161%, 0.8482 (GW), 0.8134 (GW) and 0.9160, respectively. When performing a 3-step prediction on the long-term electric load time series, the evaluation indexes MAE, MAPE, RMSE, SDE, and R are 0.8552 (GW), 7.9146%, 1.2448 (GW), 1.2385 (GW) and 0.7296, respectively. The calculation results of the latter's predictive evaluation indices are significantly higher than the former. At the same time, through the comparative analysis of Figs. 9.10 and 9.11, the prediction result of short-term electric load time series using the ARIMA model can better fit the fluctuation of the true value, no matter how many the steps are. However, when the number of predicted steps is high, the prediction result of the long-term electric load time series using the ARIMA model can only follow the true value well in the overall trend, and the effective fitting of the extreme points cannot be achieved. This is mainly due to the fact that the latter has higher fluctuation frequencies and stronger high-frequency uncertainty than the former.

9.2.5 The Theoretical Basis of EMD

Empirical Mode Decomposition (EMD) was first proposed by Huang et al. 1998. The basic idea of the EMD is to separate the time series into a series of oscillation functions, namely the Intrinsic Mode Function (IMF).

The first condition ensures that the Intrinsic Mode Function has only one vibration mode in each vibration cycle without other complex distortions. The second condition uses the mean of the upper and lower envelopes instead of the local mean of the data set to ensure the symmetry of the function. The resulting instantaneous frequency does not cause interference fluctuations due to some asymmetric waves. Unlike the general harmonic component, the Intrinsic Mode Function can be non-stationary, and it can be a frequency-modulated and amplitude-modulated wave.

The EMD decomposition essentially separates the signal into several IMFs and a residual. Different IMFs have different frequency components and widths. In general, the most important information in a signal tends to concentrate on the first several IMFs.

The EMD has a lot of advantages compared to traditional Fourier decomposition and wavelet decomposition. It can be better explained, and the actual process can be achieved in a simple step. At the same time, the window truncation processing of the traditional analysis method is abandoned, which has lower requirements for parameter selection and stronger adaptability. Due to the above characteristics, the EMD decomposition is more suitable for time-frequency analysis of non-stationary and nonlinear signals than analysis methods such as wavelet transform.

When the noise contained in the original data is not white noise, some time scales will be lost during the decomposition process, which will cause Mode Mixing (Tong et al. 2012). Mode Mixing refers to an IMF that includes signals with large differences in time scales, or a signal of similar magnitude appears in different IMFs. Mode Mixing easily leads to a lack of physical meaning for a single IMF. All the data, in reality, is a combination of signal and noise, which is ubiquitous, so the Mode Mixing phenomenon is inevitable. This is one of the main shortcomings of the EMD decomposition.

9.2.6 Optimization of EMD Decomposition Layers

As an important parameter in the EMD decomposition process, the decomposition layer has an important influence on the final decomposition result and the prediction effect of the hybrid prediction model. It is of great significance to take certain measures to optimize the number of decomposition layers to reduce the influence of Mode Mixing on the decomposition results and to improve the prediction accuracy of the final hybrid model. In the chapter, the optimization method based on the hybrid model prediction accuracy index is used to select the optimal EMD decomposition layer number. Figures 9.8, 9.9, 9.10, 9.11 show the EMD decomposition results for the short-term electric load time series and the long-term electric load time series when the number of decomposition layers is 3 and 8, respectively.

The following conclusions can be drawn from Figs. 9.8, 9.9, 9.10 and 9.11. When the number of decomposition layers is extremely, the decomposition results have a serious phenomenon of Mode Mixing. When the number of decomposition layers is 3, some IMFs contain information on different time scales. The EMD decomposition fails to adequately extract valid information from different time scales of the original time series. Conversely, when the number of decomposition layers is 8, time-scale information of adjacent features is apparently included in different IMF components. Trend information at the same time scale of the original time series is excessively stripped into different IMFs.

Figures 9.12 and 9.13 shows the advance 1-step and 5-step predictive evaluation indices of the electric load series using the EMD-ARIMA hybrid model under different EMD decomposition layers.

According to Figs. 9.12 and 9.13, it can be concluded that when the number of decomposition layers is low, the prediction error of the EMD-ARIMA model for both types of electric load series is relatively large. With the number of decomposition layers increases, the prediction accuracy is gradually improved. When the number of decomposition layers reaches 7 layers, the prediction accuracy gradually becomes

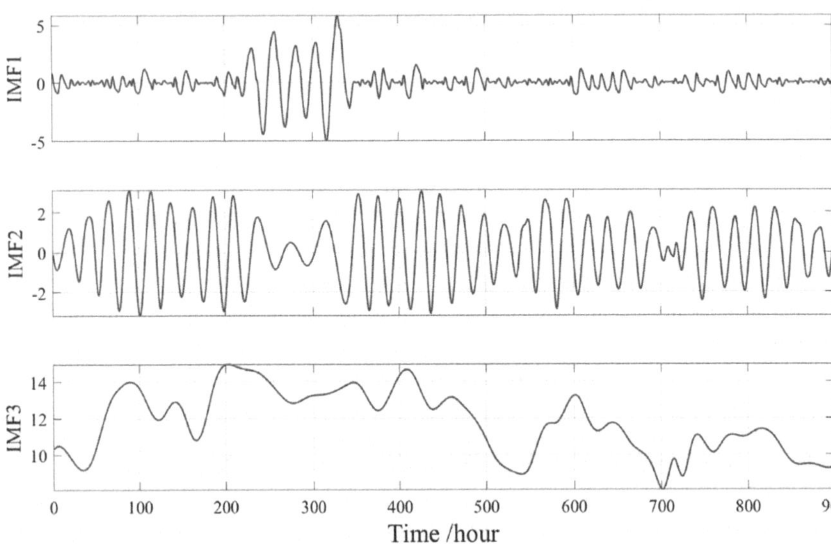

Fig. 9.8 The 3-layer EMD decomposition results of $\{X_S\}$

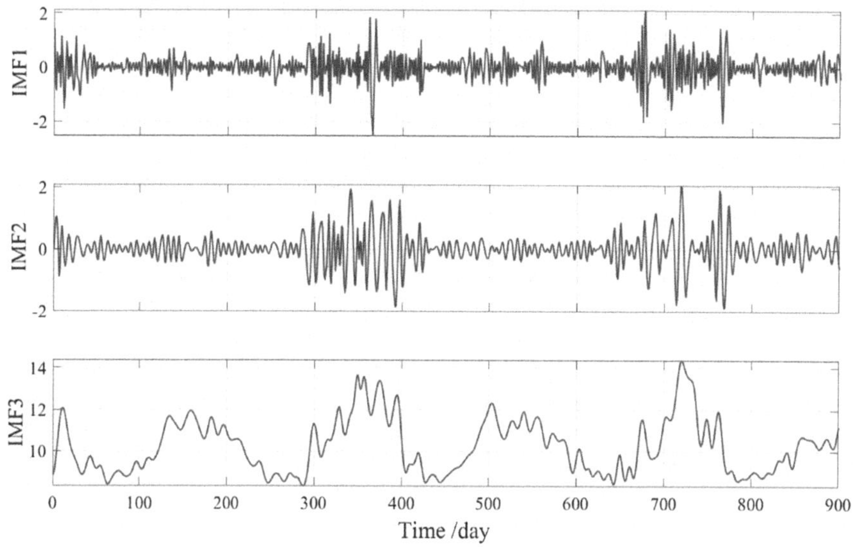

Fig. 9.9 The 3-layer EMD decomposition results of $\{X_L\}$

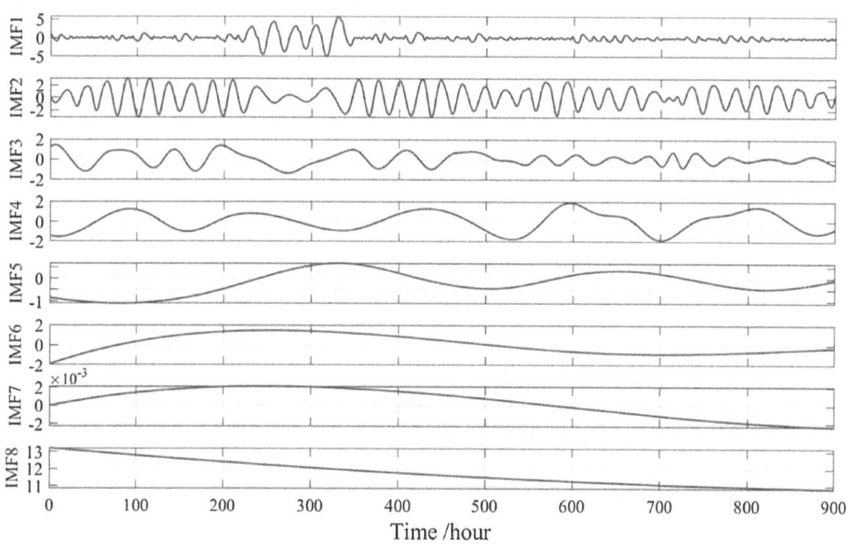

Fig. 9.10 The 8-layer EMD decomposition results of $\{X_S\}$

stable. The above phenomenon indicates that for the EMD-ARIMA hybrid model when the number of decomposition layers is low, the Mode Mixing phenomenon of the EMD decomposition has a greater influence on the prediction accuracy. However, the high decomposition layer has a less influence on the prediction accuracy. Through

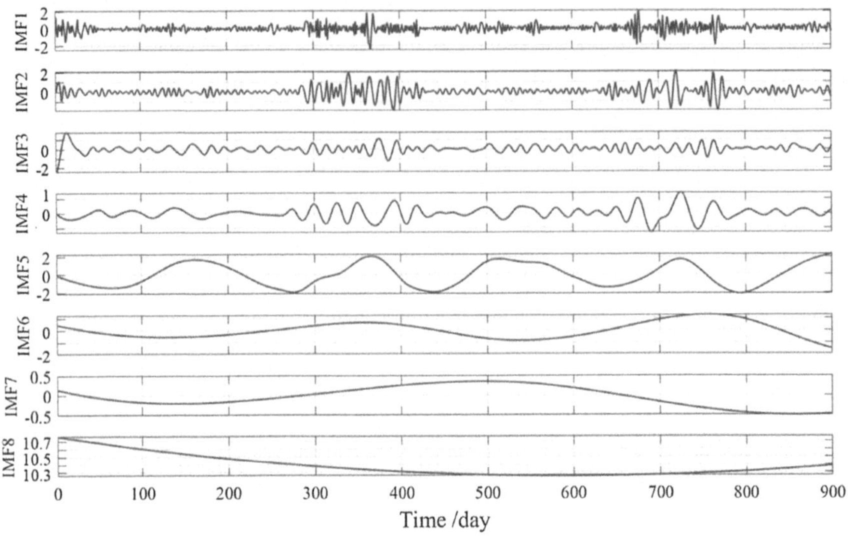

Fig. 9.11 The 8-layer EMD decomposition results of $\{X_L\}$

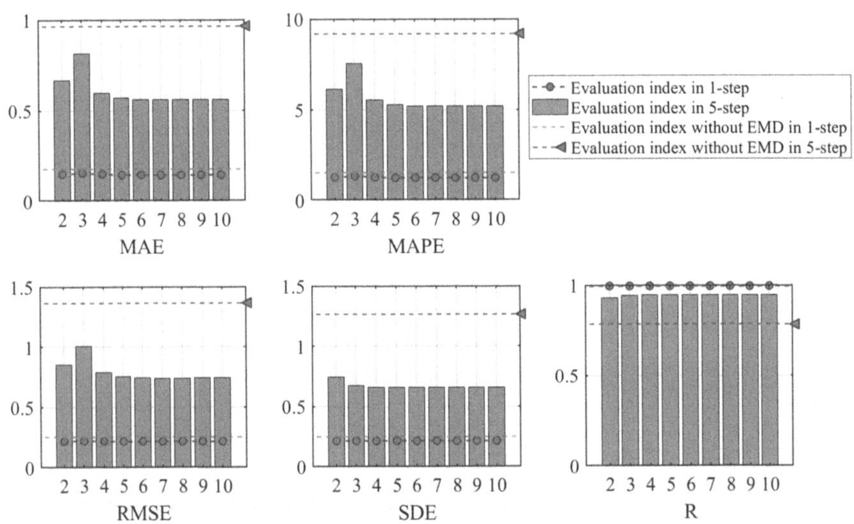

Fig. 9.12 Evaluation indexes for $\{X_S\}$ by EMD-ARIMA model under different decomposition layers

comparative analysis, when using the EMD-ARIMA hybrid model for prediction, the optimal EMD decomposition layers of short-term electric load time series and long-term electric load time series are both set as 8 layers.

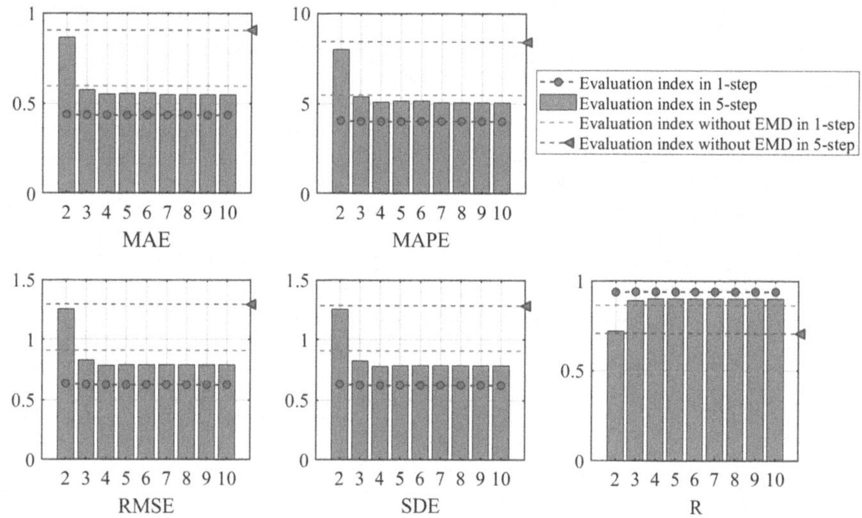

Fig. 9.13 Evaluation indexes for $\{X_L\}$ by EMD-ARIMA model under different decomposition layers

9.2.7 Predictive Results

The predictive results of the EMD-ARIMA hybrid model optimal decomposition parameter for short-term electric load time series and long-term electric load time series are shown in Fig. 9.14.

The evaluation indices of EMD-ARIMA predictive results for short-term electric load time series and long-term electric load time series are shown in Tables 9.4 and 9.5.

From the prediction results and the calculation results of the evaluation indicators, the following conclusions can be drawn:

(a) Similar to the single ARIMA model, the prediction accuracy of the EMD-ARIMA hybrid model is significantly increased, both for the short-term electric load time series and the long-term electric load time series as the number of advance prediction steps increases. When performing a 1-step prediction on the short-term electric load time series, the evaluation indexes MAE, MAPE, RMSE, SDE, and R are 0.1436 (GW), 1.2334%, 0.2144 (GW), 0.2131 (GW) and 0.9943, respectively. For advance 5-step prediction, the evaluation indexes MAE, MAPE, RMSE, SDE and R are 0.5598 (GW), 5.1719%, 0.7422 (GW), 0.6565 (GW) and 0.9456, respectively. The relative and absolute error indicators have improved significantly. However, unlike the single ARIMA model, the EMD-ARIMA hybrid model still has a relatively good predictive performance for the short-term electric load time series when the number of predicted steps is high. The deviation and the time delay between the predicted value and the

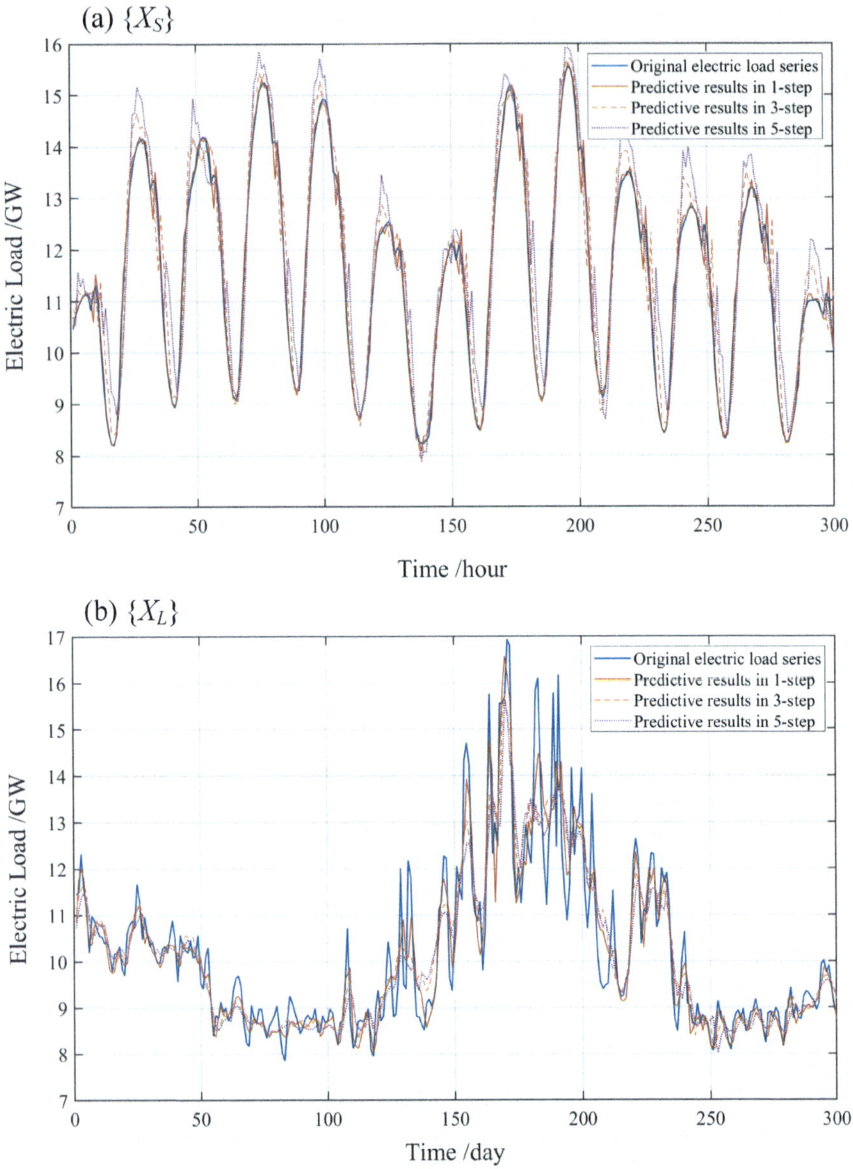

Fig. 9.14 The predictive results by EMD-ARIMA hybrid model

Table 9.4 The predictive performance indices of EMD-ARIMA model for $\{X_S\}$

Model	Advance step	MAE (GW)	MAPE (%)	RMSE (GW)	SDE (GW)	R
EMD-ARIMA	1	0.1436	1.2334	0.2144	0.2131	0.9943
	2	0.3090	2.6658	0.4142	0.4002	0.9802
	3	0.4242	3.7591	0.5527	0.5154	0.9671
	4	0.4877	4.4334	0.6500	0.5883	0.9569
	5	0.5598	5.1719	0.7422	0.6565	0.9456

Table 9.5 The predictive performance indices of EMD-ARIMA model for $\{X_L\}$

Model	Advance step	MAE (GW)	MAPE (%)	RMSE (GW)	SDE (GW)	R
EMD-ARIMA	1	0.4348	4.0140	0.6246	0.6221	0.9381
	2	0.4640	4.3196	0.6791	0.6780	0.9260
	3	0.5088	4.7226	0.7308	0.7307	0.9137
	4	0.5163	4.7917	0.7423	0.7418	0.9111
	5	0.5474	5.0465	0.7876	0.7835	0.9000

true value are small, and there are significant prediction biases only at the partial extreme points. For the long-term electric load time series, the prediction accuracy of the EMD-ARIMA hybrid model is not significantly degraded as the number of advance prediction steps increase. The model still retains certain prediction accuracy when the number of high prediction step is high. The evaluation index R of the EMD-ARIMA hybrid model is generally higher than 0.9, indicating that the correlation between the predicted value and the true value is generally higher.

(b) Same as the single ARIMA model, the prediction accuracy of the EMD-ARIMA hybrid model for short-term electric load time series is generally higher than that for long-term electric load time series when the same numbers of advanced prediction steps are selected. For example, when performing a 3-step prediction on the short-term electric load time series, the evaluation indexes MAE, MAPE, RMSE, SDE, and R are 0.4242 (GW), 3.7599%, 0.5527 (GW), 0.5154 (GW) and 0.9671, respectively. When the 3-step prediction is performed on the long-term electric load time series, the evaluation indexes MAE, MAPE, RMSE, SDE, and R are 0.5088 (GW), 4.7626%, 0.7308 (GW), 0.7307 (GW) and 0.9137, respectively. The calculation results of the latter's predictive evaluation indexes are significantly higher than the former. At the same time, through the comparative analysis of Fig. 9.14, the short-term electric load time series using the ARIMA model can better fit the fluctuation of the true value, no matter how many the steps are. However, when the number of predicted step is high, there is a significant prediction bias in the prediction results of the long-term electric load time series using the EMD-ARIMA hybrid model.

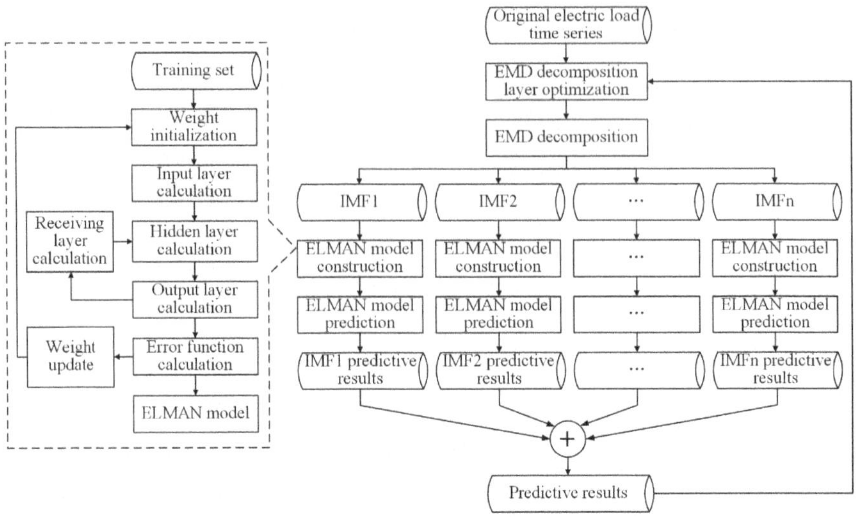

Fig. 9.15 Modeling flow chart of Elman and EMD-Elman predictive model

9.3 Load Forecasting Based on Elman Neural Network

9.3.1 Model Framework

Figure 9.15 shows the model framework of predicting the electric load series with Elman predictive model and the EMD-Elman hybrid predictive model.

9.3.2 Steps of Modeling

The Elman neural network is a dynamic recurrent neural network. Compared with the traditional BP neural network, it adds a receiving layer in the hidden layer. Thereby, a network structure based on the input layer, the hidden layer, the receiving layer, and the output layer is formed. The output at a moment on the hidden layer is not only transmitted to the output layer but also transmitted to the receiving layer for memory and storage. It acts as a delay operator with a certain delay effect, and again enters the input of the hidden layer after a certain delay through the form of internal feedback. This structure increases the sensitivity of the Elman neural network to the historical state, which makes the network has a better processing ability for dynamic information and achieves the effect of dynamic modeling (Köker 2005).

The calculation method of the Elman neural network can be expressed as follows:

$$h(t) = f_h(\omega_1 \cdot x(t) + \omega_2 \cdot h(t-1) + \delta) \tag{9.6}$$

$$y(t) = f_o(\omega_3 \cdot h(t) + \theta) \tag{9.7}$$

where $x(t)$ is the historical input data, $h(t)$ is the hidden layer output data, and $y(t)$ is the output data. ω_1, ω_2, and ω_3 are the weight vectors between the hidden layer and the input layer, between the receiving layer and the hidden layer, as well as between the hidden layer and the output layer, respectively. δ and θ are the threshold vectors of the hidden layer and the output layer respectively. $f_h(\cdot)$ and $f_o(\cdot)$ are the transfer functions of the hidden layer and the output layer, respectively.

9.3.3 Predictive Results

The predictive results of the Elman model for short-term electric load time series and long-term electric load time series are shown in Fig. 9.16.

The evaluation indices of Elman predictive results for short-term electric load time series and long-term electric load time series are shown in Tables 9.6 and 9.7.

From the prediction results and the calculation results of the evaluation indices, the following conclusions can be drawn:

(a) Like the ARIMA model, the prediction accuracy of the Elman model is also significantly reduced as the number of advance prediction steps increases. When performing a 5-step prediction, there is a significant prediction bias between the predicted values and the true values for both two types of electric load time series. Especially for the long-term electric load time series, the predicted values of some sample intervals are completely out of the true value, and the model lacks feasibility and practicability.
(b) Like the ARIMA model, the prediction accuracy of Elman model for the short-term electric load time series is generally higher than the long-term electric load time series when the same number of advanced prediction steps are selected. The former MAE indices are generally lower than 0.62 (GW), and the MAPE indices are generally lower than 5.5%. The latter's MAE indices are all higher than 0.65 (GW), and the MAPE indices are all higher than 6.0%. Therefore, whether it is relative error or absolute error, the error level of the Elman model for the former is significantly higher than the latter. At the same time, by comparing and analyzing Fig. 9.16, the prediction results of the short-term electric load time series using the Elman model can better fit the fluctuation of the true value regardless of the number of predicted steps. On the contrary, regardless of the number of predicted steps, the prediction results of the long-term electric load time series have obvious prediction bias and prediction delay.

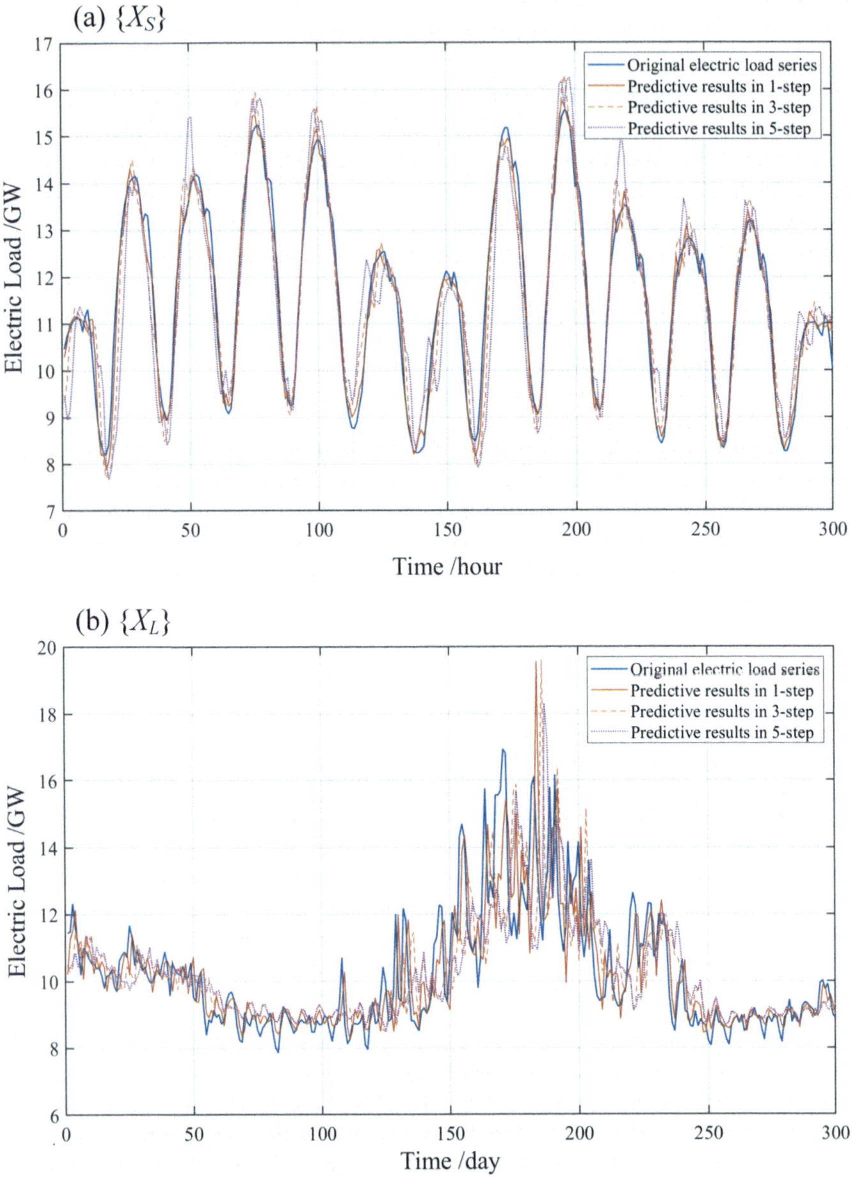

Fig. 9.16 The predictive results by Elman

Table 9.6 The predictive performance indices of Elman model for $\{X_S\}$

Model	Advance step	MAE (GW)	MAPE (%)	RMSE (GW)	SDE (GW)	R
ELMAN	1	0.2638	2.3362	0.3337	0.3335	0.9860
	2	0.3962	3.5029	0.5120	0.5116	0.9671
	3	0.4786	4.2456	0.6104	0.6090	0.9534
	4	0.5508	4.8720	0.7130	0.7087	0.9369
	5	0.6117	5.3292	0.8250	0.8172	0.9164

Table 9.7 The predictive performance indices of Elman model for $\{X_L\}$

Model	Advance step	MAE (GW)	MAPE (%)	RMSE (GW)	SDE (GW)	R
ELMAN	1	0.6669	6.0331	1.0892	1.0890	0.7968
	2	0.8493	7.6906	1.3329	1.3322	0.6809
	3	0.9061	8.1864	1.4036	1.4013	0.6324
	4	0.9160	8.2807	1.4012	1.3980	0.6312
	5	0.9302	8.4400	1.3767	1.3722	0.6449

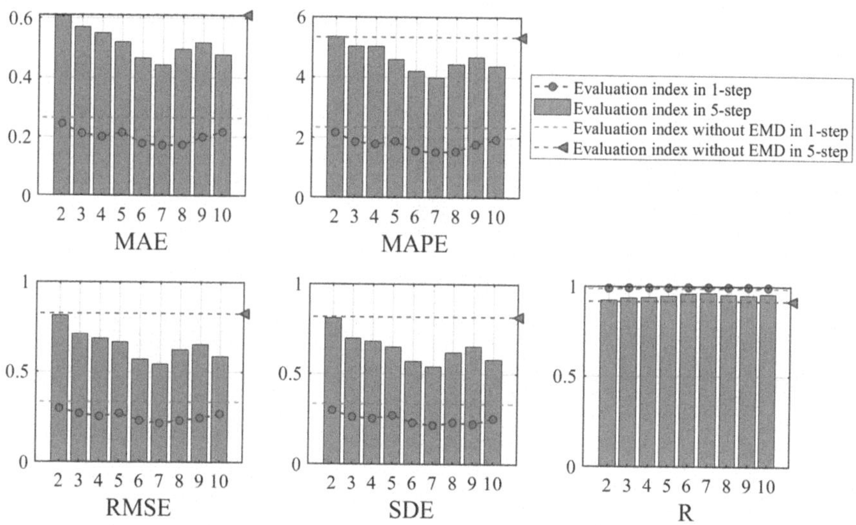

Fig. 9.17 Evaluation indexes for $\{X_S\}$ by EMD-Elman model under different decomposition layers

9.3.4 Optimization of EMD Decomposition Layers

Figures 9.17 and 9.18 show the advance 1-step and 5-step predictive evaluation indices of the electric load series using the EMD-Elman hybrid model under different decomposition layers.

According to Figs. 9.17 and 9.18, it can be found that when the number of decomposition layers is low, the prediction error of the EMD-Elman model for both types of electric load series is large relatively. The prediction accuracy is gradually improved as the number of decomposition layers increases. When the number of decomposition layers reaches the optimal number of layers, the prediction accuracy tends to decrease. At the same time, by comparing Figs. 9.17 and 9.18, it can be concluded that the number of decomposition layers has a stronger influence on the EMD-Elman hybrid model than the EMD-ARIMA hybrid model. The above phenomenon indicates that for the EMD-Elman hybrid model, when the number of decomposition layers is too low or too high, the Mode Mixing phenomenon of EMD decomposition has a great influence on the prediction accuracy. The effect of optimizing the EMD decomposition layer is obvious. Through comparative analysis, when using the EMD-Elman hybrid model for prediction, the optimal EMD decomposition layers of short-term electric load time series and long-term electric load time series are 7 and 5 layers, respectively.

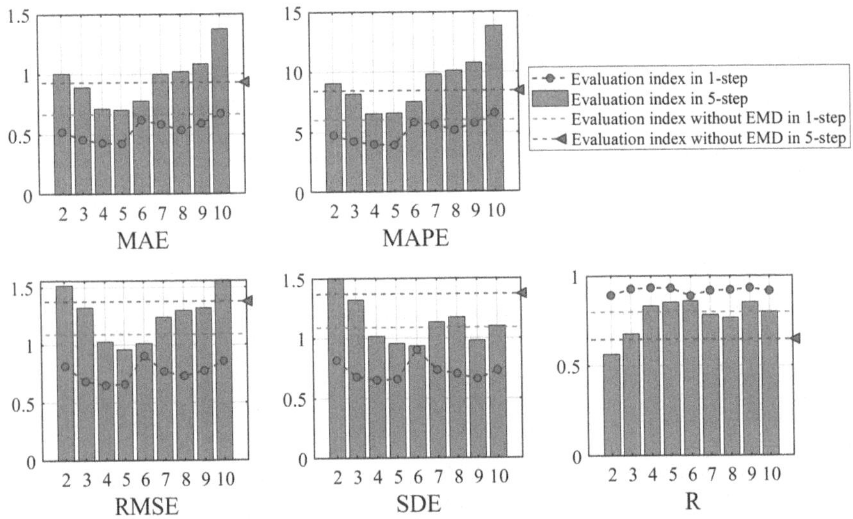

Fig. 9.18 Evaluation indexes for $\{X_L\}$ by EMD-Elman model under different decomposition layers

9.3.5 *Predictive Results*

The predictive results of the EMD-Elman hybrid model optimal decomposition parameter for short-term electric load time series and long-term electric load time series are shown in Fig. 9.19.

The evaluation indices of the EMD-Elman predictive results for short-term electric load time series and long-term electric load time series are shown in Tables 9.8 and 9.9.

According to Tables 9.8 and 9.9, the following conclusions can be drawn:

(a) Similar to the EMD-ARIMA hybrid model, the prediction accuracy of the EMD-Elman model is also significantly reduced as the number of advance prediction steps increases. However, when the number of predicted step is high, the EMD-Elman hybrid model still has a relatively good predictive performance for the short-term electric load time series. The deviation and the time delay between the predicted value and the true value are small, and there are significant prediction biases only at the partial extreme points. For the long-term electric load time series, the EMD-Elman hybrid model still has a relatively good predictive performance when the number of advanced prediction step is high. The evaluation index R of the EMD-Elman hybrid model is generally higher than 0.85, so the predicted time series and the original time series show relatively high trend following ability and fluctuation correlation.

(b) Like the EMD-ARIMA model, the prediction accuracy of the Elman model of the short-term electric load time series is generally higher than the long-term electric load time series when the same numbers of advanced prediction steps are selected. The MAE indices of the former are generally lower than 0.45 (GW), and the MAPE indices are generally lower than 4.0%. The latter MAE indices are all higher than 0.42 (GW), and the MAPE indices are all higher than 3.9%. Therefore, limited by the strong high-frequency volatility of the latter, the EMD-Elman hybrid model has better predictive performance for the former than the latter. The same conclusion can be drawn by a comparative analysis of Fig. 9.16.

9.4 Experiment Analysis

9.4.1 *Comparative Analysis of Predictive Performance*

Tables 9.10 and 9.11 show the comparison of the prediction performance of the short-term electric load time series and long-term electric load time series in the four types of models in the chapter. The section uses four comparative evaluation indices to characterize the comparison between Model 1 and Model 2. The calculation formulas are shown as follows:

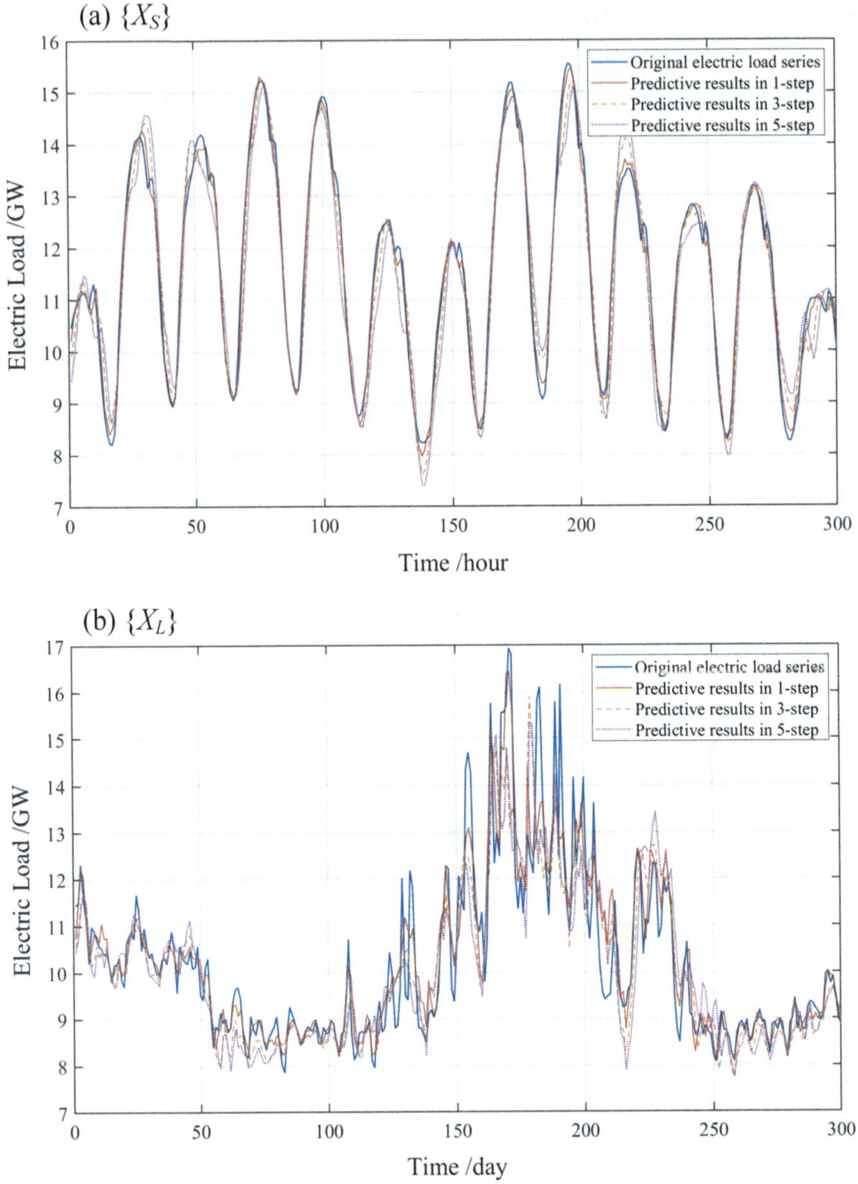

Fig. 9.19 The predictive results by EMD-Elman hybrid model

Table 9.8 The predictive performance indices of EMD-Elman model for $\{X_S\}$

Model	Advance step	MAE (GW)	MAPE (%)	RMSE (GW)	SDE (GW)	R
EMD-ELMAN	1	0.1713	1.5212	0.2167	0.2154	0.9941
	2	0.2678	2.4081	0.3393	0.3302	0.9862
	3	0.3321	3.0154	0.4080	0.4060	0.9790
	4	0.3880	3.5280	0.4768	0.4767	0.9709
	5	0.4409	3.9726	0.5445	0.5399	0.9624

Table 9.9 The predictive performance indices of EMD-Elman model for $\{X_L\}$

Model	Advance step	MAE (GW)	MAPE (%)	RMSE (GW)	SDE (GW)	R
EMD-ELMAN	1	0.4256	3.9456	0.6605	0.6584	0.9304
	2	0.5026	4.6941	0.7380	0.7289	0.9139
	3	0.5840	5.5484	0.8191	0.7917	0.8985
	4	0.6629	6.3718	0.8940	0.8441	0.8844
	5	0.7039	6.5729	0.9568	0.9567	0.8509

$$\mathrm{P}_{MAE} = \frac{MAE_2 - MAE_1}{MAE_2} * 100\% \tag{9.8}$$

$$\mathrm{P}_{MAPE} = \frac{MAPE_2 - MAPE_1}{MAPE_2} * 100\% \tag{9.9}$$

$$\mathrm{P}_{RMSE} = \frac{RMSE_2 - RMSE_1}{RMSE_2} * 100\% \tag{9.10}$$

$$\mathrm{P}_{SDE} = \frac{SDE_2 - SDE_1}{SDE_2} * 100\% \tag{9.11}$$

By analyzing Tables 9.10 and 9.11, the following conclusions can be drawn:

(a) For the short-term electric load time series, the single ARIMA model exhibits better predictive performance than the single Elman model only when performing a 1-step prediction. When performing a multi-step prediction, as the number of prediction steps increases, the advantages of the single Elman model relative to the prediction performance of a single ARIMA model are gradually expanded. At the same time, the above rules are also satisfied between the EMD-Elman hybrid model and the EMD-ARIMA hybrid model. The above phenomenon proves that for the short-term electric load time series, the error accumulation caused by the recursive strategy has much less impact on the Elman model than the ARIMA model.

(b) For the long-term electric load time series, the single ARIMA model exhibits better predictive performance than the single Elman model, regardless of the

Table 9.10 The predictive performance comparison of different model for $\{X_S\}$

Model advance step	Advance step	P_{MAPE} (%)	P_{MAE} (%)	P_{RMSE} (%)	P_{SDE} (%)
ARIMA versus ELMAN	1-step	−49.9947	−53.4886	−33.1573	−34.7606
	2-step	2.2011	1.3884	7.1224	4.7715
	3-step	21.3142	23.0334	28.0318	25.1281
	4-step	28.1095	31.9634	35.9204	32.5895
	5-step	36.5571	41.9742	39.5203	35.4160
EMD-ARIMA versus EMD-ELMAN	1-step	−19.2957	−23.3286	−1.0965	−1.0849
	2-step	13.3119	9.6702	18.0890	17.4936
	3-step	21.7110	19.7822	26.1776	21.2281
	4-step	20.4435	20.4216	26.6420	18.9579
	5-step	21.2430	23.1878	26.6329	17.7619
ARIMA versus EMD-ARIMA	1-step	18.3421	18.9649	14.4549	13.8940
	2-step	23.7281	24.9534	24.8703	25.5084
	3-step	30.2602	31.8532	34.8380	36.6343
	4-step	36.3485	38.0894	41.5863	44.0456
	5-step	41.9348	43.6869	45.5941	48.1183
ELMAN versus EMD-ELMAN	1-step	35.0548	34.8880	35.0520	35.4113
	2-step	32.3933	31.2561	33.7414	35.4602
	3-step	30.6119	28.9747	33.1591	33.3336
	4-step	29.5611	27.5868	33.1283	32.7307
	5-step	27.9188	25.4549	34.0009	33.9363

number of advance prediction steps. At the same time, the prediction accuracy of the EMD-ARIMA hybrid model is also generally higher than the EMD-Elman model. The above phenomenon shows that the ARIMA algorithm has better fitting ability than the Elman neural network, because of the higher frequency fluctuation and the weaker high-frequency stability. The Elman neural network is more sensitive to high-frequency fluctuations than the ARIMA algorithm.

(c) Whether it is the ARIMA model or the Elman model, after adding the EMD decomposition algorithm, the prediction accuracy of the model for both types of electric load time series is greatly improved. Most of the evaluation indices have increased by more than 20%. This phenomenon shows that EMD decomposition algorithm can extract effective information of different time scales from the original time series after selecting appropriate decomposition layers. Finally, the prediction performance of single prediction algorithm can be improved.

Table 9.11 The predictive performance comparison of different model for $\{X_L\}$

Model advance step	Advance step	P_{MAPE} (%)	P_{MAE} (%)	P_{RMSE} (%)	P_{SDE} (%)
ARIMA versus ELMAN	1-step	−11.9652	−10.3345	−19.7004	−19.8679
	2-step	−6.3431	−3.9212	−15.0211	−15.3417
	3-step	−5.9575	−3.4342	−12.7581	−13.1428
	4-step	−4.6666	−1.8911	−11.6835	−12.1669
	5-step	−2.7671	−0.2913	−6.2511	−6.7266
EMD-ARIMA versus EMD-ELMAN	1-step	2.1272	1.7057	−5.7466	−5.8292
	2-step	−8.3220	−8.6688	−8.6727	−7.4987
	3-step	−14.7769	−17.4881	−12.0726	−8.3510
	4-step	−28.4078	−32.9757	−20.4372	−13.7900
	5-step	−28.5781	−30.2471	−21.4812	−22.1104
ARIMA versus EMD-ARIMA	1-step	26.9928	26.5912	31.3580	31.5199
	2-step	41.9011	41.6299	41.3951	41.2970
	3-step	40.5048	40.3310	41.2858	41.0030
	4-step	41.0061	41.0393	40.8360	40.4829
	5-step	39.5220	40.0332	39.2150	39.0596
ELMAN versus EMD-ELMAN	1-step	36.1817	34.6019	39.3598	39.5401
	2-step	40.8199	38.9633	44.6297	45.2887
	3-step	35.5527	32.2236	41.6427	43.5016
	4-step	27.6247	23.0518	36.1987	39.6216
	5-step	24.3324	22.1219	30.5020	30.2755

9.5 Conclusion

The chapter uses two electric load time series to carry out a series of experimental analysis and comparative analysis, and reached the following conclusions:

(a) Due to the accumulation of errors in recursive strategies, the prediction accuracy will gradually decrease as the number of advance prediction steps increases. The prediction accuracy and evaluation indexes of the four models including ARIMA, Elman, EMD-ARIMA, EMD-Elman used in this experiment are obviously degraded.

(b) The prediction accuracy of the four models for the short-term electric load time series is generally higher than that of the long-term electric load time series. The prediction results of short-term electric load time series can better fit the fluctuation of real values, no matter how many steps are predicted. However, when the number of predicted step is high, the prediction result of the long-term electric load time series can only follow the true value well in the overall trend. The effective fitting of the extreme points cannot be achieved. This is mainly due to the fact that the latter has higher fluctuation frequencies and stronger high-frequency uncertainty than the former.

(c) The number of decomposition layers is an important parameter in the EMD decomposition process, which has an important influence on the final decomposition result and the prediction effect of the hybrid prediction model. For the two types of electric load time series, when the number of decomposition layers is too high or too low, the decomposition results have a serious phenomenon of Mode Mixing. When the number of decomposition layers is too low, the EMD decomposition fails to adequately extract the effective information of different time scales of the original time series. Conversely, when the number of decomposition layers is too high, the trend information at the same time scale of the original time series is excessively stripped into different IMFs. For the EMD-ARIMA hybrid model, the Mode Mixing phenomenon of the EMD decomposition has a greater influence on the prediction accuracy with a low decomposition layer, while the high decomposition layer has less influence on the prediction accuracy. For the EMD-Elman hybrid model, when the number of decomposition layers is too low or too high, the Mode Mixing phenomenon of the EMD decomposition has a great influence on the prediction accuracy. The EMD decomposition layer number optimization effect is more obvious.

(d) For the short-term electric load time series, the error accumulation produced by the recursive strategy has a much smaller impact on the Elman model than the ARIMA model. As the number of prediction steps increases, the advantages of the Elman model in predicting performance are gradually expanding. For long-term electric load time series with higher fluctuating frequency and lower frequency stability, the ARIMA algorithm has better fitting ability than Elman neural network. Elman neural network is more sensitive to high-frequency fluctuations.

(e) Regardless of the ARIMA model or the Elman model, after adding the EMD decomposition algorithm, the prediction accuracy of the model for both types of electric load time series is greatly improved. The EMD decomposition algorithm can effectively improve the prediction performance of a single prediction algorithm.

References

Amin SB, Alam T (2018) The relationship between energy consumption and sectoral output in Bangladesh: An empirical analysis. J Dev Areas 52(3):39–54

Chatfield C, Weigend AS (1994) Time series prediction: forecasting the future and understanding the past. In: Weigend AS, Gershenfeld NA (eds) Future of time series. Addison-Wesley, Reading, MA, pp 1–70. Int J Forecast 10(1):161–163

Chen K, Chen K, Qin W, He Z, He J (2018) Short-term load forecasting with deep residual networks. IEEE Trans Smart Grid (99):1–1

Cooper DE (2012) Intelligent transportation systems for smart cities: a progress review. Sci China (Inf Sci) 55(12):2908–2914

Diks C, Houwelingen JCV, Takens F, Degoede J (1995) Reversibility as a criterion for discriminating time series. Phys Lett A 201(2–3):221–228

Ehrenfeld D (2010) Sustainability: living with the imperfections. Conserv Biol 19(1):33–35

Feinberg EA, Genethliou D (2005) Load forecasting. In: Applied mathematics for restructured electric power systems. Springer

Ge X, Yang J, Gharavi H, Sun Y (2017) Energy efficiency challenges of 5G small cell networks. IEEE Commun Mag 55(5):184–191

Huang NE, Zheng S, Long SR, Wu MC, Shih HH, Zheng Q, Yen NC, Chi CT, Liu HH (1998) The empirical mode decomposition and the Hilbert spectrum for nonlinear and non-stationary time series analysis. Proc Math Phys Eng Sci 454(1971):903–995

Inoue A (2008) AR and MA representation of partial autocorrelation functions, with applications. Probab Theory Relat Fields 140(3–4):523–551

Kang C, Xia Q (2004) Review of power system load forecasting and its development. Autom Electr Power Syst 28(17):1–11

Köker R (2005) Reliability-based approach to the inverse kinematics solution of robots using Elman's networks. Eng Appl Artif Intell 18(6):685–693

Madsen H (2007) Time series analysis. Chapman & Hall/CRC Boca Raton Fl 27 (1864):103–121

Markovié ML, Fraissler WF (2010) Short-term load forecast by plausibility checking of announced demand: an expert-system approach. Int Trans Electr Energy Syst 3(5):353–358

Mengelkamp E, Notheisen B, Beer C, Dauer D, Weinhardt C (2018) A blockchain-based smart grid: towards sustainable local energy markets. Comput Sci Res Dev 33(1–2):207–214

Migendt M (2017) public policy influence on renewable energy investments—a panel data study across OECD countries. In: ULB institutional repository. Springer Gabler, Wiesbaden

Nalcaci G, Özmen A, Weber GW (2018) Long-term load forecasting: models based on MARS, ANN and LR methods. CEJOR 2:1–17

Shi H, Xu M, Ran L (2018) Deep learning for household load forecasting—a novel pooling deep RNN. IEEE Trans Smart Grid 9(5):5271–5280

Tong W, Zhang M, Yu Q, Zhang H (2012) Comparing the applications of EMD and EEMD on time–frequency analysis of seismic signal. J Appl Geophys 83(29–34):29–34

Watanabe S (2013) A widely applicable Bayesian information criterion. J Mach Learn Res 14(1):867–897

Xiao L, Wei S, Yu M, Jing M, Jin C (2017) Research and application of a combined model based on multi-objective optimization for electrical load forecasting. Energy 119:1057–1074

Yiyan LI, Han D, Yan Z (2017) Long-term system load forecasting based on data-driven linear clustering method. J Modern Power Syst Clean Energy 17:1–11

Zeng B, Zhang J, Yang X, Wang J, Dong J, Zhang Y (2014) Integrated planning for transition to low-carbon distribution system with renewable energy generation and demand response. IEEE Trans Power Syst 29(3):1153–1165

Zhang D, Miao X, Liu L, Zhang Y, Liu K (2015) Research on development strategy for smart grid big data. Proc Csee 35(1):2–12

Chapter 10
Interval Prediction of Electric Load Time Series

10.1 Introduction

With the continuous growth of the power demand, the modeling of the electric load time series is becoming an important component in the operation of electric energy. The modeling and forecasting of the electric load time series are beneficial to improve the operation efficiency of the power systems. As explained in (Shepero et al. 2018), significant savings can be achieved even with 1% improvement of mean absolute percentage error in load forecasting.

The electric load forecasting can be divided into different types according to the forecasting horizon. It has been widely accepted that the long-term is utilized to achieve years ahead load forecasting, the medium-term is involved to predict the load months to a year ahead, the short-term provides the load information a day to weeks ahead and the ultra short-term load forecasting refers to minutes to hours ahead (Dedinec et al. 2016). Besides, the electric load forecasting can be divided into different types according to the forecasting scenarios. For instance, the aggregate load forecasting is a hot topic in smart energy management (Alobaidi et al. 2018). Li et al. proposed an novel structure to model the short term load based on the ISO New England data (Li et al. 2016). Dudek achieved hourly national load forecasting in Poland (Dudek 2016). Taylor et al. evaluated their models on dataset from ten European countries (Taylor and McSharry 2007). It has been proved that the aggregate load forecasting can achieve satisfactory results by considering the historical data, the time of the day, and the temperature.

With the installation of the smart meters, it is becoming much easier to get more information from the household sector. Much more attentions have been paid to the residential sector to manage the power system at a household level. As a result, the household level load forecasting is put forward recently. Yu et al. proposed an individual household load forecasting model based on the sparse coding technique (Yu et al. 2016). The experimental results based on 5000 households proved that the sparse coding technique was effective in improving forecasting accuracy. Alobaidi et al. constructed a household energy consumption forecasting framework based on

H. Liu, *Non-intrusive Load Monitoring*,
https://doi.org/10.1007/978-981-15-1860-7_10

ensemble learning, in which the two-stage resampling techniques were utilized to train the Artificial Neuron Network (ANN) models (Alobaidi et al. 2018). Shi et al. provided a deep recurrent neural network to carry out modeling the uncertainty in the household load time series (Shi et al. 2018). In the proposed deep recurrent neural network a pooling technique was designed to prevent over-fitting. The forecasting results demonstrated that the proposed pooling based deep recurrent neural network could provide better performance than the conventional deep recurrent neural network.

The input features play an important role in load forecasting. Son et al. provided a short-term electricity demand forecasting model based on the Support Vector Regression (SVR) and feature selection (Son and Kim 2017). In their work, 19 features including 14 variables and 5 social variables were taken into consideration. The experimental results indicated that the mean temperature, the maximum temperature, the wind speed, the daylight time, the global solar radiation, the cooling degree days, the total number of cooling degree days, the vapor pressure, the consumer price index, and the real electricity price were the top 10 variables that affect the monthly electricity demand.

Kong et al. carried out a detailed analysis of the difference between the aggregate load and the house household load based on the clustering method named Density-Based Spatial Clustering of Application with Noise (Kong et al. 2019). The clustering results indicated that the household load varied greatly from different houses and time, while the aggregated load formed only one cluster and had no outlier. The possible reason is that the uncertainty of the house household load is offset during the aggregation. Compared to the aggregate load, the household load is more non-stationary and stochastic, which makes it hard to get an accurate point prediction. Compared to the point prediction, the interval prediction can provide more information about the forecasting results in the sight of probability. The interval prediction is regarded as an effective way to model the household load. In the chapter, the interval prediction of the household load will be discussed comprehensively with different input features.

In the Chapter, the household load interval forecasting experiment is carried out based on the AMPds dataset (Makonin et al. 2016). The details of the AMPds dataset are given in Sect. 2.2.

10.2 Interval Prediction Based on Quantile Regression

10.2.1 The Performance Evaluation Metrics

As described in Sect. 9.2, the Mean Absolute Error, the Mean Absolute Percentage Error, the Root Mean Square Error and the Standard Deviation of Error are utilized to make a comprehensive evaluation of the point forecasting. Similarly, some specific metrics are utilized to judge the performance of the interval predictions. The details of the metrics are given as follows (Khosravi et al. 2011):

(a) Prediction interval coverage probability

The coverage probability of the interval predictions is utilized to evaluate the probability that the real target points lie within the constructed intervals. A large coverage probability indicates that more real target points lie within the constructed intervals. Namely, the constructed intervals provide better performance of reliability. The detailed definition of the prediction interval coverage probability is given as follows (Khosravi et al. 2011):

$$PICP = \frac{1}{m}\sum_{i=1}^{m}\varepsilon_i \tag{10.1}$$

where m is the number of the samples in the testing set, ε_i is Boolean variable which can be defined as follow:

$$\varepsilon_i = \begin{cases} 1, \text{ if } y_i \in [L_i, U_i]; \\ 0, \text{ if } y_i \notin [L_i, U_i]. \end{cases} \tag{10.2}$$

where the y_i is the target value of the sample i, the L_i is the lower boundary of the prediction interval and U_i is the upper boundary of the prediction interval. It is obvious that if all the target values fall into the constructed interval, the $PICP = 1$, and inversely, if all the target values fall out of the constructed interval, the $PICP = 0$. In real application, the Prediction Interval Coverage Probability (PICP) is expected to be close to the Prediction Interval Nominal Confidence (PINC) $1 - \alpha$, namely, minimizing the average coverage error. The average coverage error is defined as follows (Sideratos and Hatziargyriou 2012):

$$ACE = PICP - PINC \tag{10.3}$$

(b) Width of the prediction interval

As explained in the above section, Prediction Interval Normalized Average Width (PINAW) is an important index to evaluate the performance of the interval prediction results. But the PICP is not enough to make a comprehensive judgment of the performance. That is because it can always get good PICP value as the width of the prediction interval is big enough, in which the interval prediction results make no sense for decision making. So the width of the prediction interval is another important parameter to judge the performance of the interval predictions. Prediction Interval Average Width (PIAW) is one of the most widely used prediction interval width evaluation metric, the definition of the average width is given as follows (Khosravi et al. 2011):

$$PIAW = \frac{1}{m}\sum_{i=1}^{m}(U_i - L_i) \tag{10.4}$$

where m is the number of the samples in the testing set, the L_i is the lower boundary of the prediction interval and U_i is the upper boundary of the prediction interval. It can be found that the PINAW is an absolute width evaluation metric. Based on the PIAW, the Prediction Interval Normalized Average Width (PINAW), which is a relative width evaluation metric can be defined as follow (Khosravi et al. 2011):

$$PINAW = \frac{1}{mw}\sum_{i=1}^{m}(U_i - L_i) \tag{10.5}$$

where w is the range of the underlying targets which can be defined as follow:

$$w = \max(y_i) - \min(y_i) \tag{10.6}$$

where the y_i is the target value of the sample i. Similar to the Root Mean Square Error in the point prediction, which is more sensitive to large error, the prediction interval normalized root-mean-square width is defined to magnify and recognize the effect of the wider intervals. The definition of the root-mean-square width is given as follows (Khosravi et al. 2011):

$$PINRW = \frac{1}{w}\sqrt{\frac{1}{m}\sum_{i=1}^{m}(U_i - L_i)^2} \tag{10.7}$$

(c) Comprehensive evaluation of coverage probability and interval width

It can be found that the coverage probability and the interval width are two important aspects of the prediction intervals. In the real application of the prediction intervals, it is important to get both large coverage probability and narrow prediction intervals which can give more useful information. However, larger interval width tends to result in higher coverage probability. It is hard to get optimal solutions for both metrics at the same time. New metrics should be defined to balance the conflict and make a comprehensive evaluation of coverage probability and interval width. The coverage width-based criterion is such a metric that can be defined as follows (Khosravi et al. 2011):

$$CWC = PINAW\left(1 + \eta(PICP)e^{-\gamma(PICP-\mu)}\right) \tag{10.8}$$

where μ is the nominal confidence level which can be set as the prescribed probability $1-\alpha$, γ is a variable coefficient (Khosravi et al. 2011):

$$\eta(PICP) = \begin{cases} 1, \text{ if } PICP - \mu < 0; \\ 0, \text{ if } PICP - \mu \geq 0. \end{cases} \tag{10.9}$$

It is easy to find that when $PICP \geq \mu$, the *CWC* is equal to the *PINAW*, when $PICP < \mu$, an exponential penalty term is added to the *CWC*. As a result, the *CWC* is larger than the *PINAW* significantly.

Besides, the interval score is also defined to make a comprehensive evaluation of coverage probability and interval width, the interval score is expressed as follows (Wan et al. 2014):

$$Sc = \frac{1}{m}\sum_{i=1}^{m} Sc_i \tag{10.10}$$

where Sc_i is the interval score of the sample point i that can be defined as follows (Wan et al. 2014):

$$Sc_i = \begin{cases} -2\alpha(U_i - L_i) - 4(L_i - y_i), \text{ if } y_i - L_i < 0; \\ -2\alpha(U_i - L_i), \text{ if } y_i \in [L_i, U_i], \\ -2\alpha(U_i - L_i) - 4(y_i - U_i), \text{ if } y_i - U_i > 0. \end{cases} \tag{10.11}$$

where the L_i is the lower boundary of the prediction interval and U_i is the upper boundary of the prediction interval. y_i is the target value of the sample i.

10.2.2 Original Sequence for Modeling

The total electric load time series, as well as the monitored loads from the sub-meters, are utilized to carry out the interval prediction experiments. In the study, the original electric load time series is hourly averaged. Figure 10.1 displays the total electric load time series, the electric load time series from the utility, the furnace, the clothes washer, and the home office, respectively. As shown in Fig. 10.1, there are 2000 sample points in each time series. In order to construct the interval prediction models and evaluate the performance of the constructed models, each of the electric load time series should be grouped. In the study, each of the electric load time series is divided into the training set and the testing set. Specifically, the 1st–1800th sample points of each times series are utilized as the training data to carry out the model construction process, while the 1801st–2000th sample points are utilized as testing samples to carry out the model evaluation process, in which the sample points are fed into the constructed models and the output values of the models are evaluated based on the performance evaluation metrics.

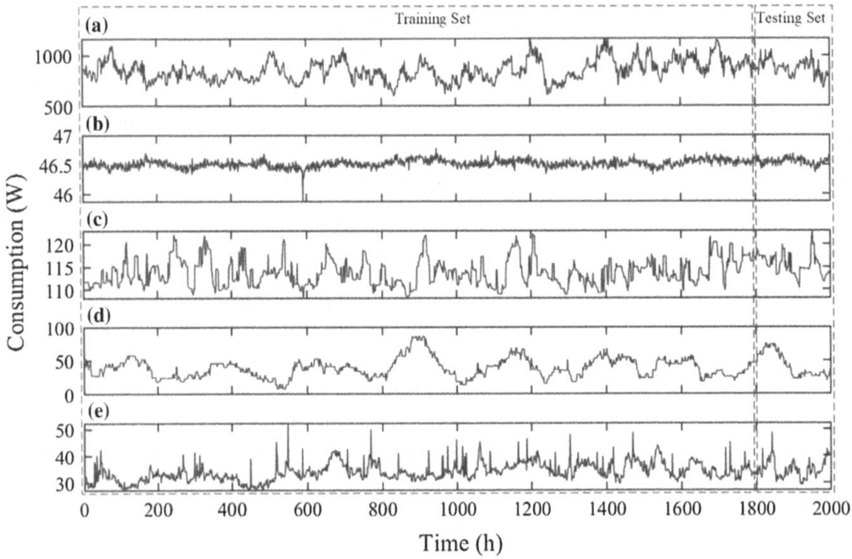

Fig. 10.1 Modeling flow chart of QR prediction model driven by single data

10.2.3 The Theoretical Basis of Quantile Regression

The Quantile Regression (QR) proposed by Koenker and Bassett is an extension of the traditional least squares expectation estimation (Koenker and Bassett 1978). The τ quantile is defined as the solution for the following problem:

$$\min_{b \in R^K} \left(\sum_{y_i - x_i b \geq 0} \tau |y_i - x_i b| + \sum_{y_i - x_i b < 0} (1 - \tau) |y_i - x_i b| \right) \tag{10.12}$$

The equation describes a minimization optimization problem that can be solved by linear programming techniques. The target value corresponds to the τ quantile is larger than τ proportion of all values, where $\tau \in [0, 1]$. If conducting an interval prediction base on the quantile regression, it is necessary to find out two quantiles. For instance, given prediction interval nominal confidence of 0.99, it needs to find out the 0.05 quantile and 0.995 quantile, and the intervals between the two quantiles are the target prediction intervals (Khan et al. 2019).

10.2.4 *Quantile Regression Based on the Total Electric Load Time Series*

In the section, the household load interval prediction task is carried out under the time series forecasting framework. In the framework, the historical total electric load information is utilized as the input of the model to predict the interval of the current point. Figure 10.2 shows the modeling flow chart of the quantile regression based on the total electric load (QR-TEL) time series, as can be seen from Fig. 10.2, the modeling process is as follows:

(a) Arranging the original total electric load time series into training data and testing data. More specifically, the original total electric load time series is reorganized into a large sum of samples, and each sample consists of input data and target data. The last 200 samples are utilized as the testing set and the rest are training set.
(b) Normalizing all data of the training set as well as the input data of the testing set. More specifically, the upper quantile is calculated as follows:

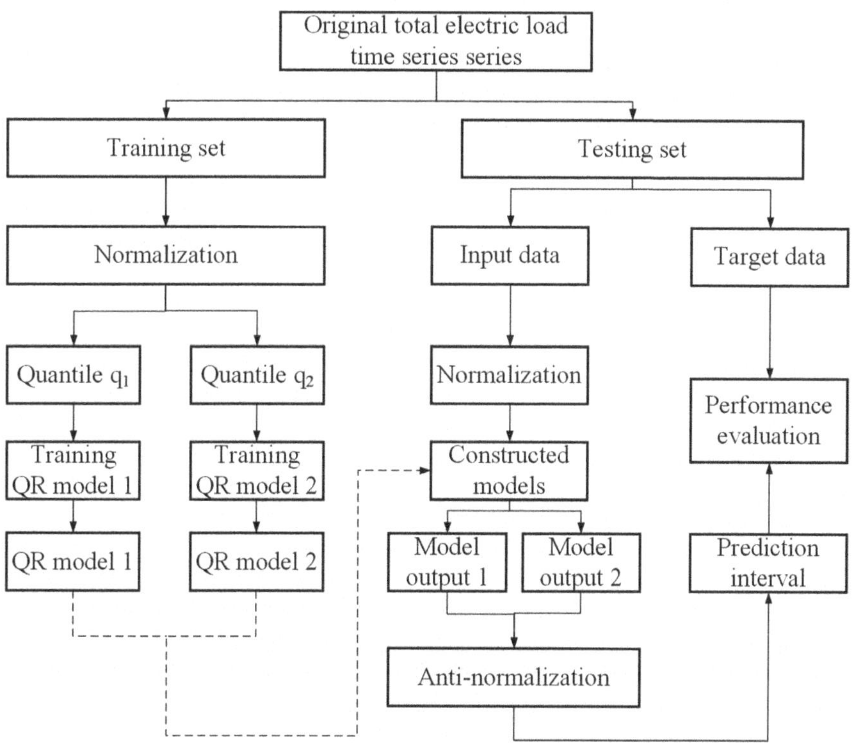

Fig. 10.2 Modeling flow chart of the proposed QR-TEL model

$$q_1 = 1 - \frac{1}{2}PINC \tag{10.13}$$

The lower quantile is calculated as follows

$$q_2 = \frac{1}{2}PINC \tag{10.14}$$

(c) Training the QR models under the determined quantiles based on the training set, putting the normalized input data of the testing set to the constructed model, and getting the output of the two models, carrying out the anti-normalize to get the final prediction intervals.
(d) Comparing the prediction intervals with the actual output data in the testing set, evaluating the performance according to the PICP, ACE, PIAW, PINAW, CWC, and Sc.

The intervals prediction results with nominal confidence 99%, 95% and 90% obtained by the proposed QR-TEL model are given in Figs. 10.3, 10.4 and 10.5. The performance evaluation indices of the proposed QR-TEL model are shown in Table 10.1.

As can be seen in Figs. 10.3, 10.4 and 10.5 and Table 10.1, it can be concluded that:

(a) The proposed QR-TEL model can accomplish the interval prediction with a certain accuracy. The prediction intervals can capture the trend and variation of the total electric load time series. Most of the measured data are perfectly covered

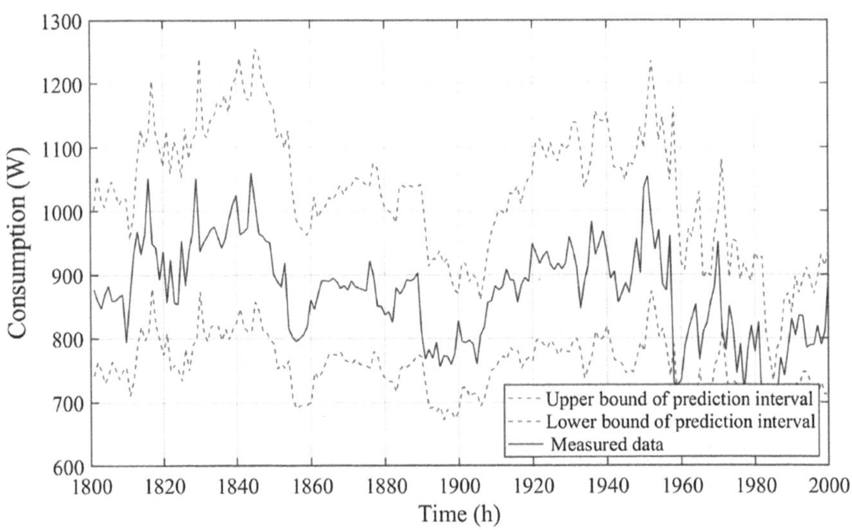

Fig. 10.3 Intervals prediction results with nominal confidence 99% obtained by the proposed QR-TEL model

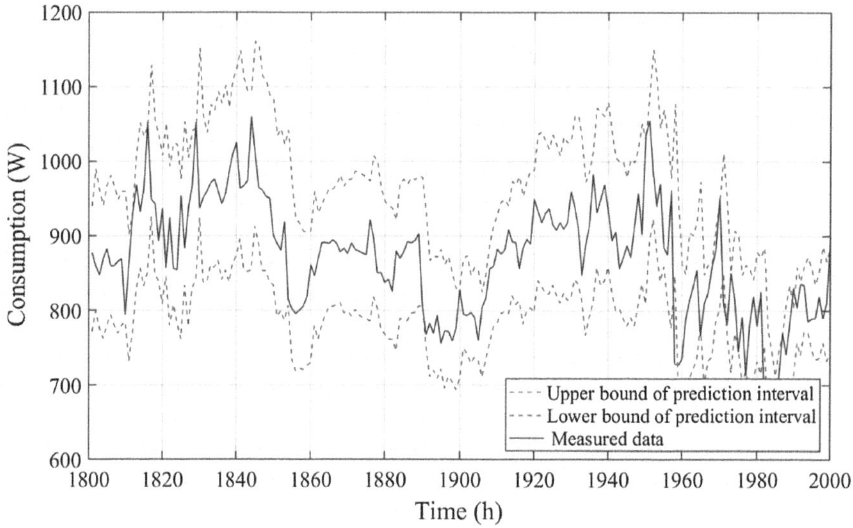

Fig. 10.4 Intervals prediction results with nominal confidence 95% obtained by the proposed QR-TEL model

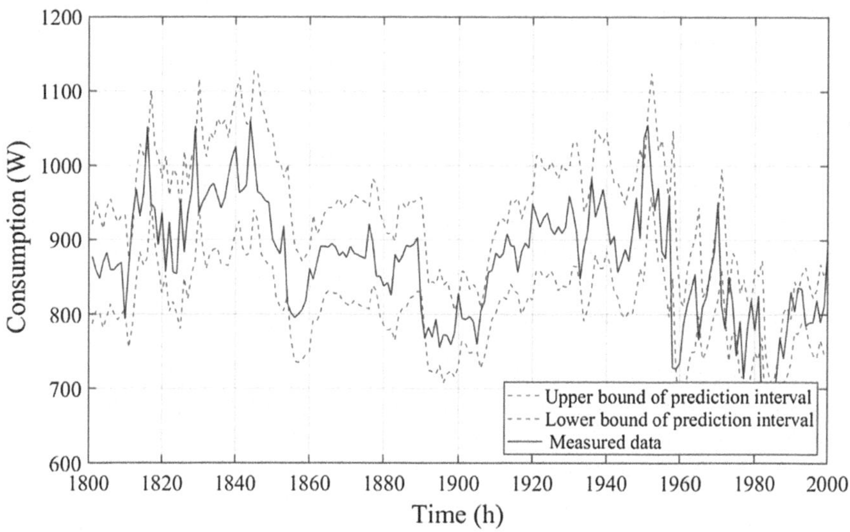

Fig. 10.5 Intervals prediction results with nominal confidence 90% obtained by the proposed QR-TEL model

Table 10.1 The performance evaluation indices of the proposed QR-TEL model

Model	PINC	PICP	ACE	PIAW	PINAW	CWC	Sc
QR-TEL	0.9900	0.9900	0.0000	276.3926	0.7158	0.7158	−7.8528
	0.9500	0.9350	−0.0150	181.7943	0.4708	1.4676	−24.8104
	0.9000	0.8450	−0.0550	135.2866	0.3504	5.8312	−40.9927

by the upper and lower bound of prediction intervals. However, a deviation is also observed around the dramatic change points such as the 1960th sample point and 1980th sample point.

(b) With the decrease of the PINC, the values of the PICP, PIAW and PINAW decrease continuously, the absolute values of the ACE, CWC and Sc increase continuously. The prediction intervals become narrower with small PINC, but at the same time, there are more samples exceeding the prediction intervals in small nominal confidence. As a result, the overall prediction performance becomes worse.

10.2.5 Quantile Regression Based on Additional Time and Date Information

It is well known that the household load highly depends on human activity. The model house belongs to a family of three-person, in which the male and the daughter are full-time students while the female is self-employed. There is a reason to believe that their activities will affect by the day of week. For instance, there may be more persons in the house during the weekend and result in higher total electricity consumption. Besides, the number of hours from midnight is another import factor that affects human activity. For instance, the total electricity consumption before dawn may be lower as people are in sleep and fewer appliances are in operation.

In the section, the household load interval prediction task is carried out by taking the time and date information into consideration. In the section, the historical total electric load information, as well as the time and date information, are utilized as the input of the model to predict the interval of the current point. Figure 10.6 shows the modeling flow chart of the quantile regression based on the total electric load time series, the number of hours from midnight and the day of week (QR-TEL-H-W), as can be seen from Fig. 10.6, the modeling process is as follows:

(a) Preparing the original dataset. The original dataset consists of three parts: the original total electric load time series, the number of hours from midnight and the day of the week. Among the three parts, the day of week is a discontinuous variable and should be encoded as dummy variables, so that every kind of variable has the same distance from each other.

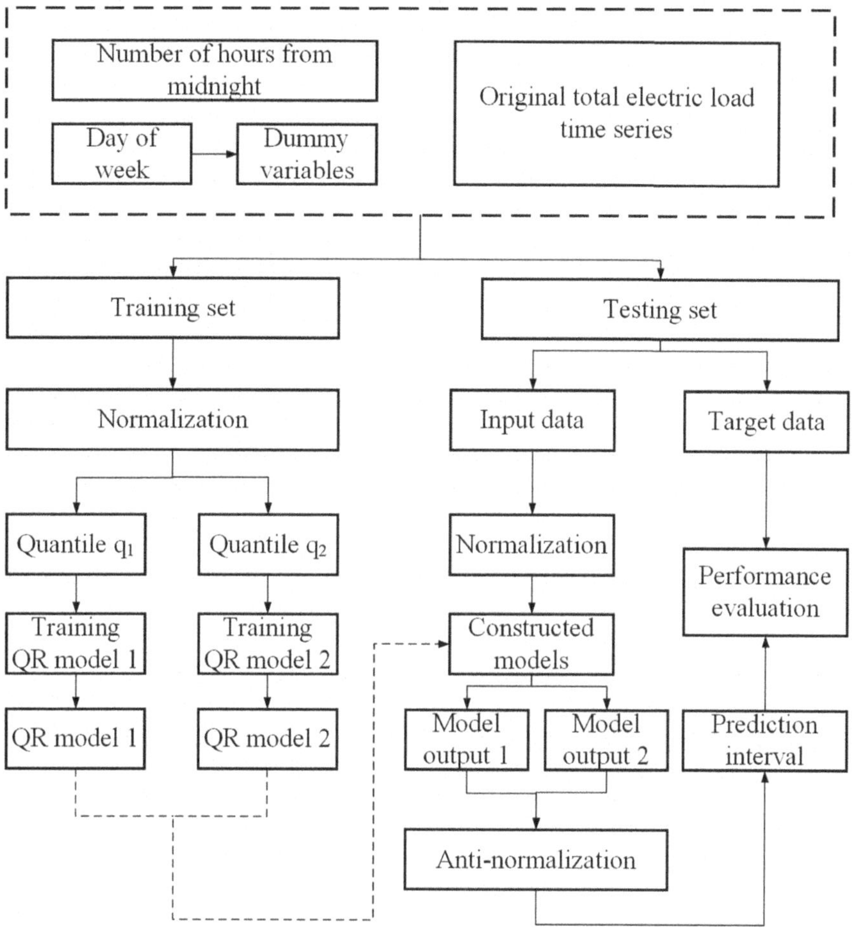

Fig. 10.6 Modeling flow chart of the proposed QR-TEL-H-W model

(b) Arranging the original dataset into training data and testing data. Normalizing all the training set as well as the input data of the testing set. Determining the quantiles according to the PINC.

(c) Training the QR models under the determined quantiles based on the training set, putting the normalized input data of the testing set to the constructed model, getting the output of the two models, and carry out the anti-normalize to get the final prediction intervals. Comparing the prediction intervals with the actual output data in the testing set and evaluating the performance according to the PICP, ACE, PIAW, PINAW, CWC, and Sc.

In the section, the quantile regression based on the total electric load time series, the number of hours from midnight (QR-TEL-H), the quantile regression based on the total electric load time series and the day of week (QR-TEL-W), and the

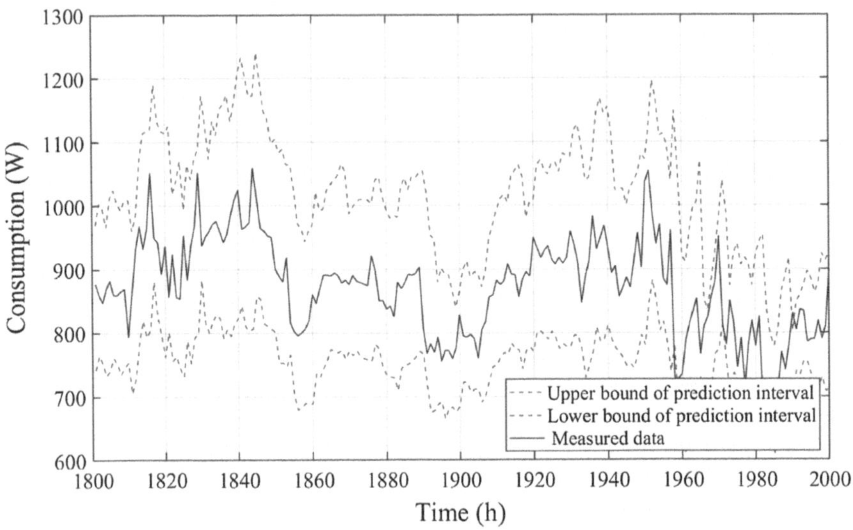

Fig. 10.7 Intervals prediction results with nominal confidence 99% obtained by the proposed QR-TEL-H model

quantile regression based on the total electric load time series, the number of hours from midnight and the day of week (QR-TEL-H-W) are established. The intervals prediction results with nominal confidence 99% obtained by the proposed QR-TEL-H model, the QR-TEL-W model, and the QR-TEL-H-W model are given in Figs. 10.7, 10.8 and 10.9. The performance evaluation indices of these models are shown in Table 10.2.

As can be seen in Figs. 10.7, 10.8 and 10.9 and Table 10.2, it can be concluded that:

(a) All the proposed models can accomplish the interval prediction with a certain accuracy. Most of the measured data are perfectly covered by the upper and lower bounds of prediction intervals. The prediction intervals can capture the trend and variation of the total electric load time series.
(b) The QR-TEL-W model provides the best overall performance among the proposed QR-TEL-H model, QR-TEL-W model, and QR-TEL-H-W model. For instance, the QR-TEL-H model provides the best ACE result with nominal confidence 99, 95, and 90%. The QR-TEL-H model also provides the best CWC with nominal confidence 99 and 95%. Besides, the QR-TEL-H model also provides the best Sc results with nominal confidence 99 and 90%.
(c) The performance of the QR-TEL-W model is better than that of the QR-TEL model, the date of week information is effective in improving the interval prediction performance.
(d) The performance of the QR-TEL-H model is worse than that of the QR-TEL model, the number of hours from midnight information seems not to be effective in improving the interval prediction performance.

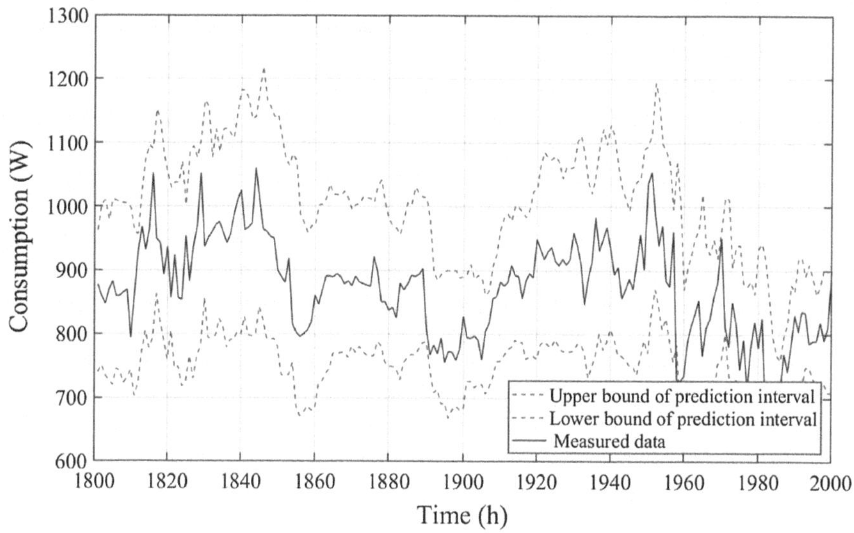

Fig. 10.8 Intervals prediction results with nominal confidence 99% obtained by the proposed QR-TEL-W model

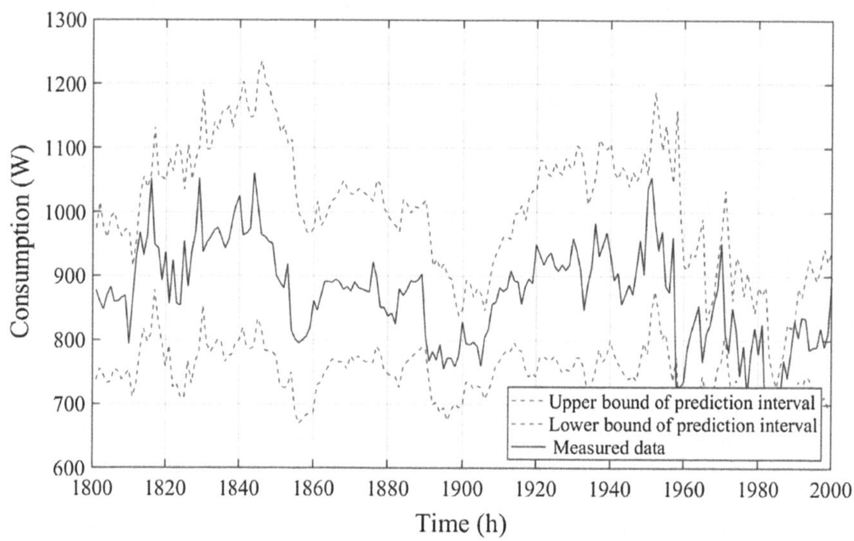

Fig. 10.9 Intervals prediction results with nominal confidence 99% obtained by the proposed QR-TEL-H-W model

Table 10.2 The performance evaluation indices of the proposed QR-TEL-H model, QR-TEL-W model, and QR-TEL-H-W model

Model	PINC	PICP	ACE	PIAW	PINAW	CWC	Sc
QR-TEL-H	0.9900	0.9800	−0.0100	265.4953	0.6876	1.8213	−7.8579
	0.9500	0.9250	−0.0250	174.7873	0.4527	2.0327	−24.9798
	0.9000	0.8600	−0.0400	132.8495	0.3441	2.8864	−42.1779
QR-TEL-W	0.9900	0.9900	0.0000	258.8454	0.6704	0.6704	−6.9477
	0.9500	0.9500	0.0000	184.3273	0.4774	0.4774	−24.1180
	0.9000	0.8600	−0.0400	134.5683	0.3485	2.9237	−40.4305
QR-TEL-H-W	0.9900	0.9850	−0.0050	259.1315	0.6711	1.5329	−7.1917
	0.9500	0.9450	−0.0050	181.4388	0.4699	1.0733	−23.8663
	0.9000	0.8300	−0.0700	130.0989	0.3369	11.4949	−40.5556

10.2.6 Quantile Regression Based on Information from Sub-meters

As well known, the total electric load consists of the load of all appliances in the house. In the section, the household load interval prediction task is carried out by taking the load consumption information of the individual appliances. Namely, the information from sub-meters is utilized as the input of the model to predict the interval of the current point. As explained in Sect. 2.2, the electric load of main house in the AMPds dataset consists of data from 19 sub-meters. In the section, quantile regression model based on all information from sub-meters (QR-SUB-A), quantile regression model based on information from sub-meters and correction analysis (QR-SUB-C), and quantile regression model based on information from sub-meters and feature selection method (QR-SUB-F) are constructed to carry out the interval prediction. Figure 10.10 shows the modeling flow chart of the proposed QR-SUB-A model, QR-SUB-C model, and QR-SUB-F model. As can be seen from Fig. 10.10, the modeling process is as follows:

(a) Preparing the original dataset. The original dataset consists of data from 19 sub-meters. For the QR-SUB-A model, all of the data is utilized as the input of the model. Specifically, all sub-meter data of current time point is adopted to carry out the interval prediction of the next time point.
(b) For the QR-SUB-C model, the correlation analysis is applied to judge the linear correction between the sub-meter data time series and the original total electric load time series. Pearson's correlation coefficients between the sub-meter data time series and the original total electric load time series are given in Table 10.3. In the study, the data with small correlation coefficients, which are from the DNE, EQE, OUE, THE, and B1E, are removed from the input data.
(c) For the QR-SUB-F model, the Minimum Redundancy Maximum Relevance (mRMR) feature selection method is utilized to rank the importance of all the sub-meter data. The results are given in Table 10.3. As can be seen, the data

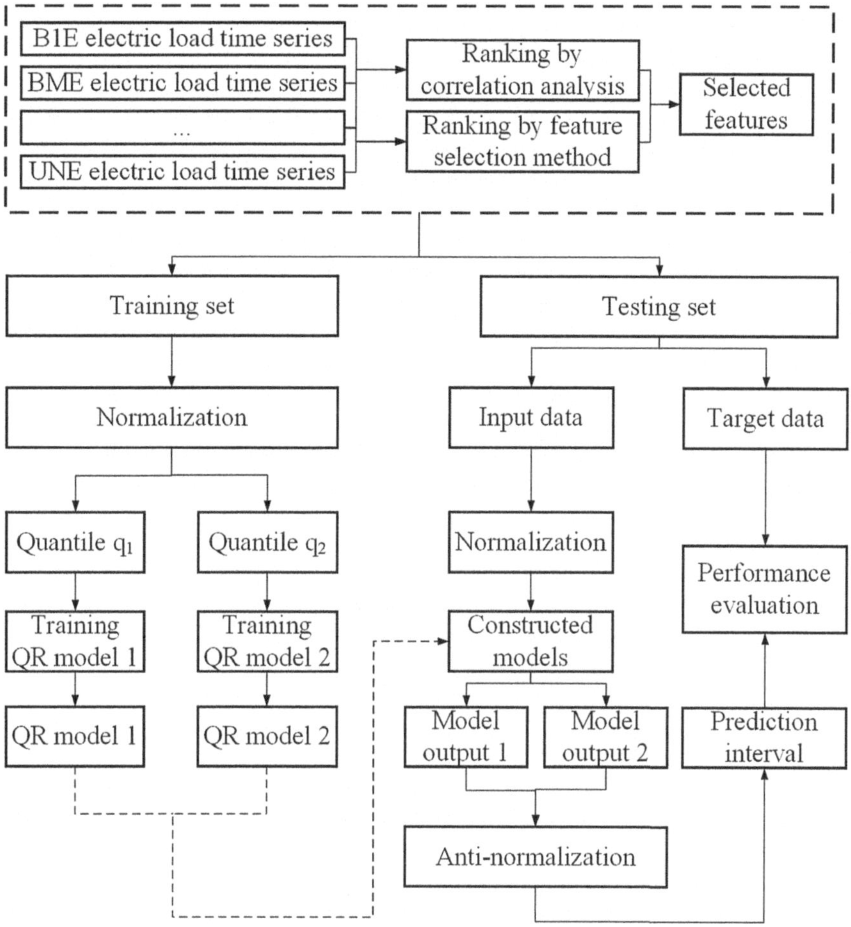

Fig. 10.10 Modeling flow chart of the proposed QR-SUB-A model, QR-SUB-C model and QR-SUB-F model

from the hot water sub-meter is regarded as the most important feature in the total electric load interval prediction. Similar to the strategy of the QR-SUB-C model, the data from the BME, DNE, UNE, FGE, and EBE, which are regarded as the least important features, are removed from the input data.

(d) The rest steps are similar to that of the QR-TEL model, which includes dividing the training data and testing data, normalizing the relevant data, training the QR model and performance evaluation of the prediction intervals.

The intervals prediction results with nominal confidence 99% obtained by the proposed QR-SUB-A model, the QR-SUB-C model and the QR-SUB-F model are given in Figs. 10.11, 10.12 and 10.13. The performance evaluation indices of these models are shown in Table 10.4.

Table 10.3 The correlation results and the mRMR ranking results of the appliances

Abbreviation	Description	Pearson's correlation coefficient	mRMR ranking
CDE	Clothes dryer	0.6509	9
HPE	Heat pump	0.6353	14
UNE	Unmetered	0.4849	17
FRE	Furnace	0.3095	11
OFE	Home office	0.1632	8
BME	Basement	0.1509	18
TVE	TV, PVR, etc.	0.1237	13
WOE	Wall oven	0.1085	2
DWE	Dishwasher	0.1023	6
CWE	Clothes washer	0.0624	12
B2E	Bedrooms	0.0526	10
UTE	Utility	0.0511	5
EBE	Workbench	0.0436	15
FGE	Fridge	0.0387	16
DNE	Dining room	0.0176	17
EQE	Equipment	0.0144	4
OUE	Outside	0.0085	3
HTE	Hot water	0.0043	1
B1E	Bedroom	−0.0185	7

As can be seen in Figs. 10.11, 10.12 and 10.13 and Table 10.4, it can be concluded that:

(a) Both the proposed QR-SUB-A model and the proposed QR-SUB-F model can accomplish the interval prediction with a certain accuracy. Most of the measured data are perfectly covered by the upper and lower bound of prediction intervals. The prediction intervals can capture the trend and variation of the total electric load time series.
(b) The performance of the QR-SUB-C model is worse than that of the QR-SUB-A model and the QR-SUB-F model. For instance, with nominal confidences 99, 95 and 90%, the CWC results of the QR-SUB-C model are higher than that of the QR-SUB-A model and the QR-SUB-F significantly. The features selected by Pearson's correlation coefficients cannot describe the intervals well.
(c) The QR-SUB-F model provides the best interval prediction results with nominal confidences 99% while the QR-SUB-A model provides the best results with nominal confidences 95% and 90%. The features selection method can capture most of the important features in the sub-meter data.

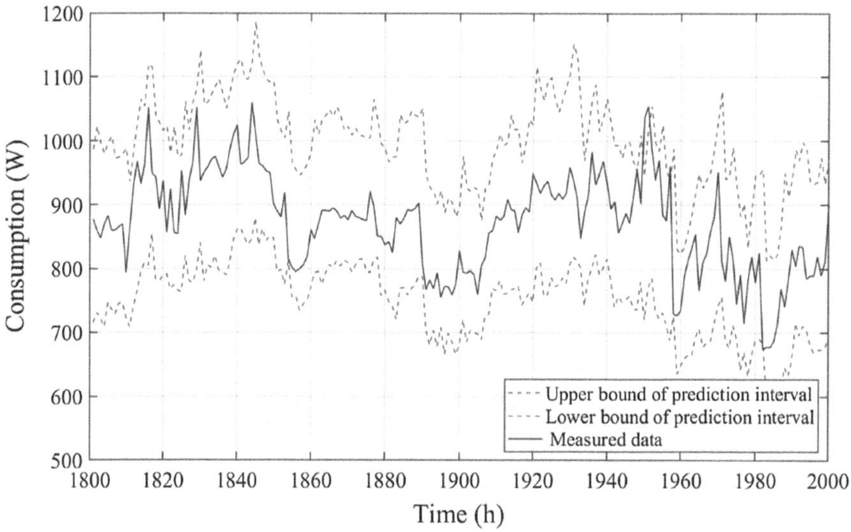

Fig. 10.11 Intervals prediction results with nominal confidence 99% obtained by the proposed QR-SUB model

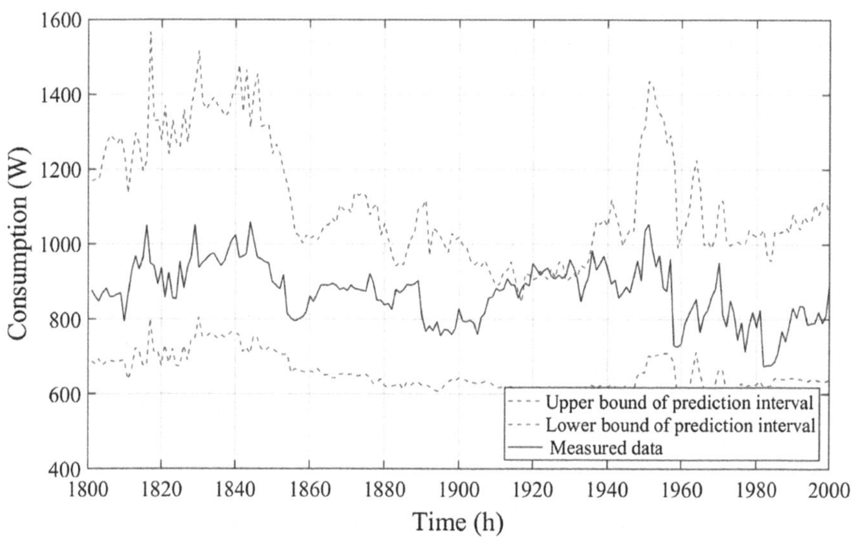

Fig. 10.12 Intervals prediction results with nominal confidence 99% obtained by the proposed QR-SUB-C model

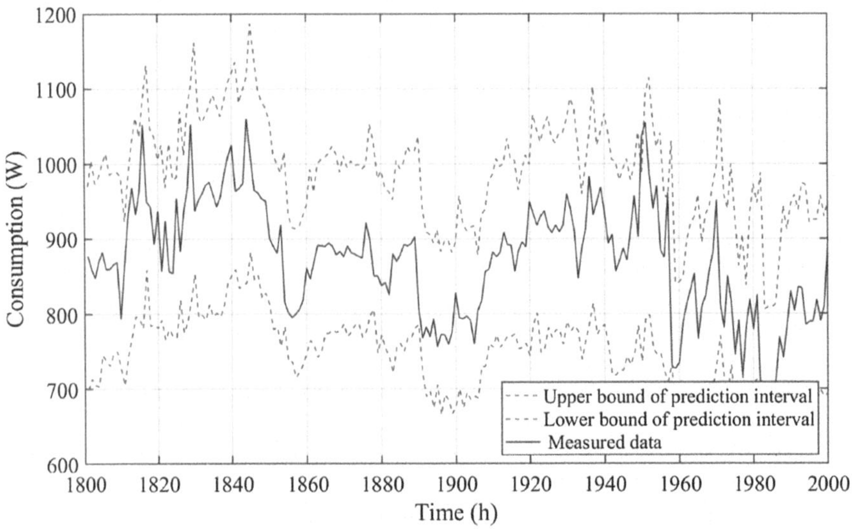

Fig. 10.13 Prediction results with nominal confidence 99% obtained by the proposed QR-SUB-F model

Table 10.4 The performance evaluation indices of the proposed QR-SUB-A model, QR-SUB-C model, and QR-SUB-F model

Model	PINC	PICP	ACE	PIAW	PINAW	CWC	Sc
QR-SUB-A	0.9900	0.9750	−0.0150	247.7723	0.6417	2.0002	−8.2355
	0.9500	0.9500	0.0000	165.0591	0.4275	0.4275	−23.0412
	0.9000	0.8750	−0.0250	134.3508	0.3480	1.5624	−38.8356
QR-SUB-C	0.9900	0.9600	−0.0300	457.3159	1.1844	6.4925	−13.7791
	0.9500	0.8450	−0.1050	313.3850	0.8116	155.4815	−60.6325
	0.9000	0.8100	−0.0900	261.8130	0.6781	61.7157	−93.0945
QR-SUB-F	0.9900	0.9800	−0.0100	246.2613	0.6378	1.6893	−6.8507
	0.9500	0.9450	−0.0050	162.6976	0.4214	0.9624	−23.4755
	0.9000	0.8700	−0.0300	137.5894	0.3563	1.9534	−39.6156

10.3 Interval Prediction Based on Gaussian Process Filtering

10.3.1 The Theoretical Basis of Gaussian Process Regression

The Gaussian Process Regression (GPR) is an extension of the linear Bayesian regression, in which the kernel is utilized to express the similarity of input vectors (Koenker and Bassett 1978). The squared exponential kernel, the exponential kernel, the Matern

3/2 kernel, the Matern 5/2 kernel, and the rational quadratic kernel are widely used kernels (Rasmussen 2004). The definition of the squared exponential kernel is given as follows:

$$K_1(x_i, x_j) = \sigma_f^2 \exp\left(\frac{-(x_i - x_j)^{\mathrm{T}}(x_i - x_j)}{2\sigma_l^2}\right) \tag{10.15}$$

where x_i and x_j are input vectors, σ_f and σ_l are parameters of the kernel. The definition of the exponential kernel is given as follows (Wilson et al. 2015):

$$K_2(x_i, x_j) = \sigma_f^2 \exp\left(\frac{-\sqrt{(x_i - x_j)^{\mathrm{T}}(x_i - x_j)}}{\sigma_l}\right) \tag{10.16}$$

The definition of the Matern 3/2 kernel is given as follows (Melkumyan and Ramos 2011):

$$K_3(x_i, x_j) = \sigma_f^2\left(1 + \frac{r}{\sigma_l}\right)\exp\left(\frac{-r}{\sigma_l}\right) \tag{10.17}$$

where $r = \sqrt{3(x_i - x_j)^{\mathrm{T}}(x_i - x_j)}$. The definition of the Matern 5/2 kernel is given as follows (Melkumyan and Ramos 2011):

$$K_4(x_i, x_j) = \sigma_f^2\left(1 + \frac{r}{\sigma_l} + \frac{r^2}{3\sigma_l^2}\right)\exp\left(\frac{-r}{\sigma_l}\right) \tag{10.18}$$

where $r = \sqrt{5(x_i - x_j)^{\mathrm{T}}(x_i - x_j)}$. The definition of the rational quadratic kernel is given as follows (Wilson et al. 2015):

$$K_5(x_i, x_j) = \sigma_f^2\left(1 + \frac{-(x_i - x_j)^{\mathrm{T}}(x_i - x_j)}{2\alpha\sigma_l^2}\right)^{-\alpha} \tag{10.19}$$

where α is the scale-mixture parameter.

10.3.2 Gaussian Process Regression Based on the Total Electric Load Time Series

In the section, the historical total electric load information is utilized as the input of the model to predict the interval of the current point. Figure 10.14 shows a modeling flow chart of the Gaussian process regression based on the total electric load time

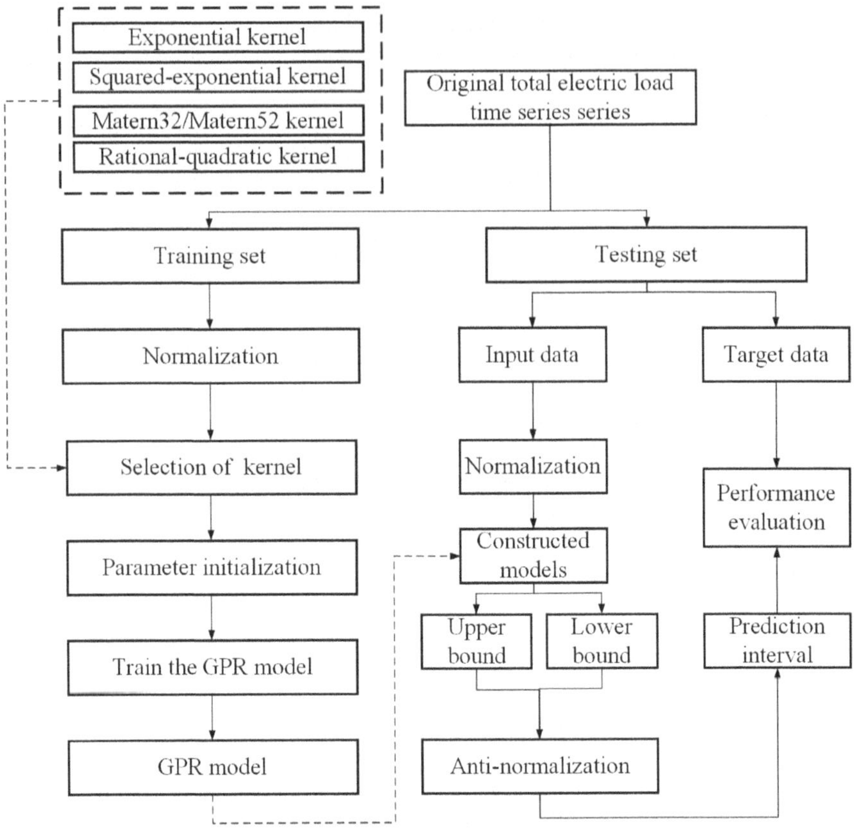

Fig. 10.14 Modeling flow chart of the proposed GPR-TEL model

series (GPR-TEL), as can be seen from Fig. 10.14, the main steps of the modeling are as follows:

(a) Preparing the kernel functions. As the kernel function is an important part of the GPR, the commonly used kernel functions are prepared. All of the kernel functions will be utilized to construct the interval prediction models, and the performance of the kernel functions will be compared in the section.
(b) Getting the training data and testing data from the original total electric load time series. Normalizing all the training set as well as the input data of the testing set.
(c) Selecting the kernels and initializing the input features of the selected kernels. In the section, the σ_f is initialized as the standard deviation of the target value in the training data. The σ_l is initialized as 10.
(d) Training the GPR models under the determined parameters based on the training set.

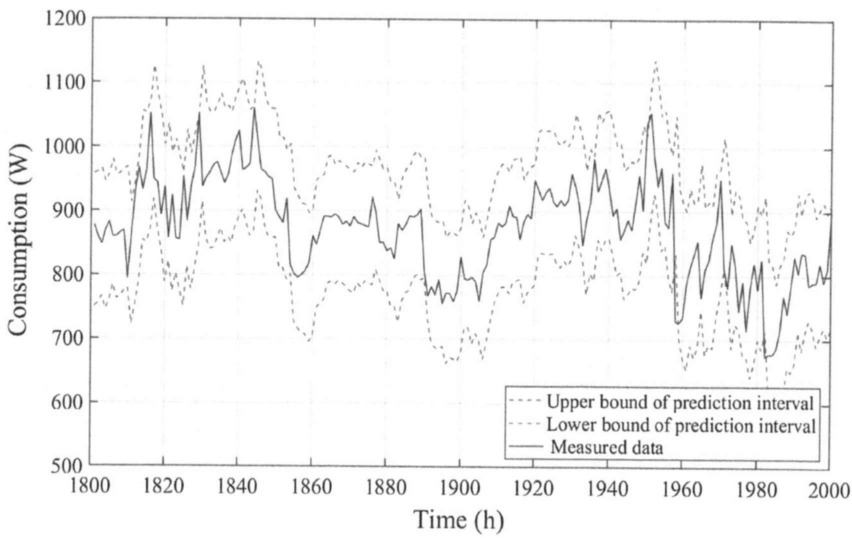

Fig. 10.15 Intervals prediction results with nominal confidence 99% obtained by the proposed GPR-TEL model with the Exponential kernel

(e) Putting the normalized input data of the testing set to the constructed model, and getting the upper bound and lower bound of the interval with a certain nominal confidence, carrying out the anti-normalize to get the final prediction intervals.
(f) Calculating performance evaluating metrics and comparing the performance of the GPR models with different kernels.

The interval prediction results with nominal confidence 99, 95 and 90% obtained by the proposed GPR-TEL models with different kernels are shown in Figs. 10.15, 10.16, 10.17, 10.18 and 10.19. The performance evaluation indices of the proposed GPR-TEL models are given in Table 10.5.

As shown in Figs. 10.15, 10.16, 10.17, 10.18 and 10.19 and Table 10.5, it can be concluded that:

(a) All the GPR-TEL models with different kernels can accomplish the interval prediction with a certain accuracy. The prediction intervals can capture the trend and variation of the total electric load time series. Most of the measured data are perfectly covered by the upper and lower bound of prediction intervals. With the decrease of the PINC, the values of the PICP, PIAW, and PINAW decrease continuously. The prediction intervals become narrower with small PINC, but at the same time, there are more samples exceeding the prediction intervals in small nominal confidence.
(b) The exponential kernel provides the best interval prediction results with nominal confidences 99%. For instance, the ACE and the CWC value of the exponential kernel-based model are −0.01 and 1.3822, respectively, which are significantly better than that of other models.

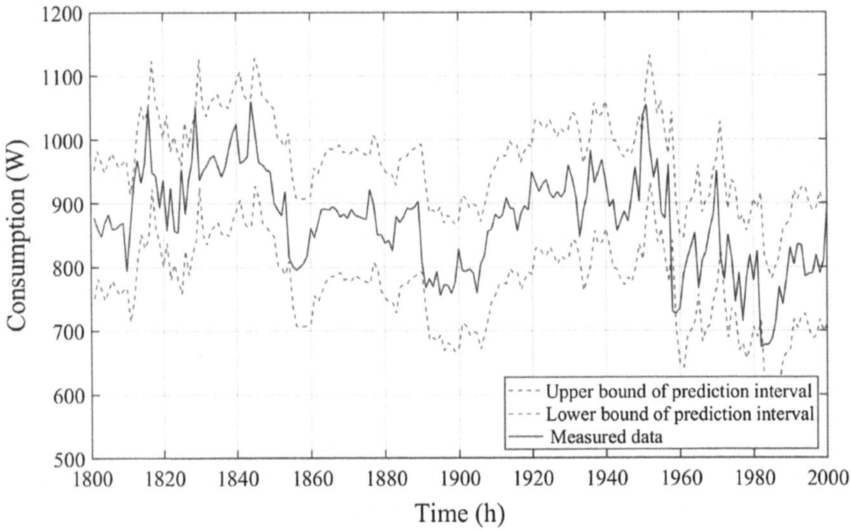

Fig. 10.16 Intervals prediction results with nominal confidence 99% obtained by the proposed GPR-TEL model with the Squared-exponential kernel

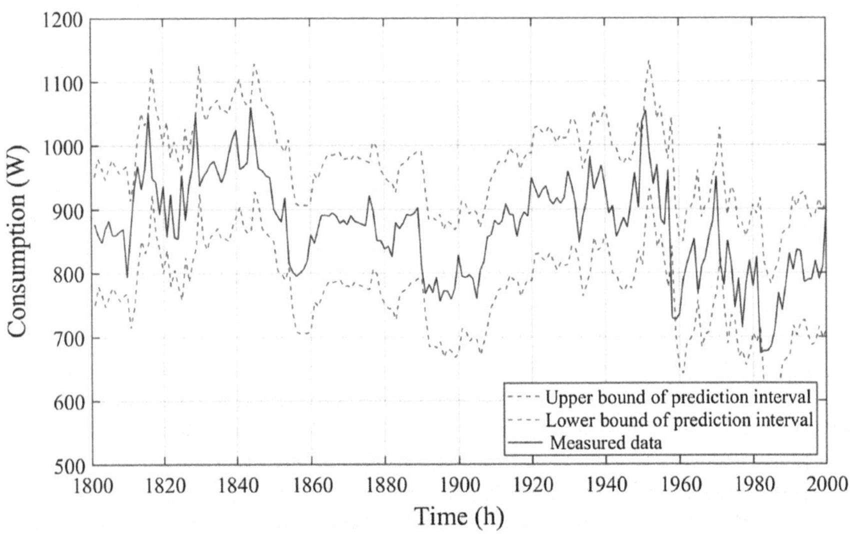

Fig. 10.17 Intervals prediction results with nominal confidence 99% obtained by the proposed GPR-TEL model with the Matern32 kernel

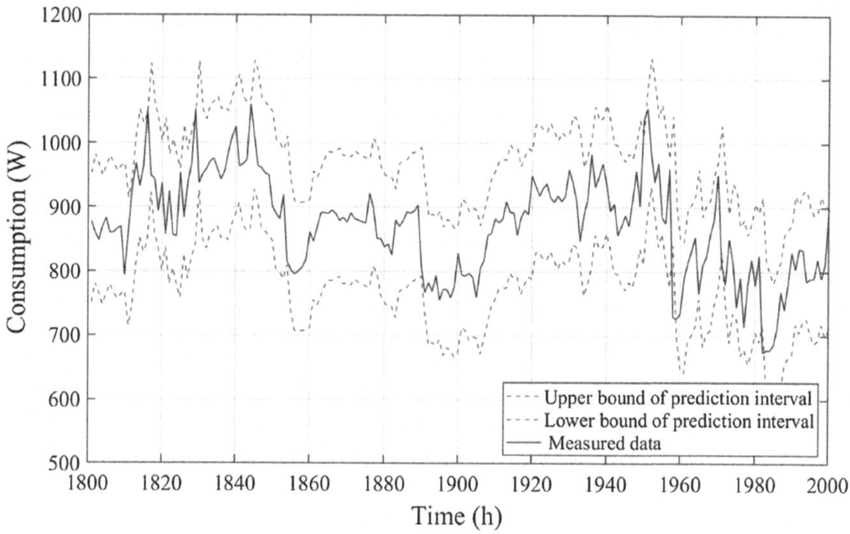

Fig. 10.18 Intervals prediction results with nominal confidence 99% obtained by the proposed GPR-TEL model with the Matern52 kernel

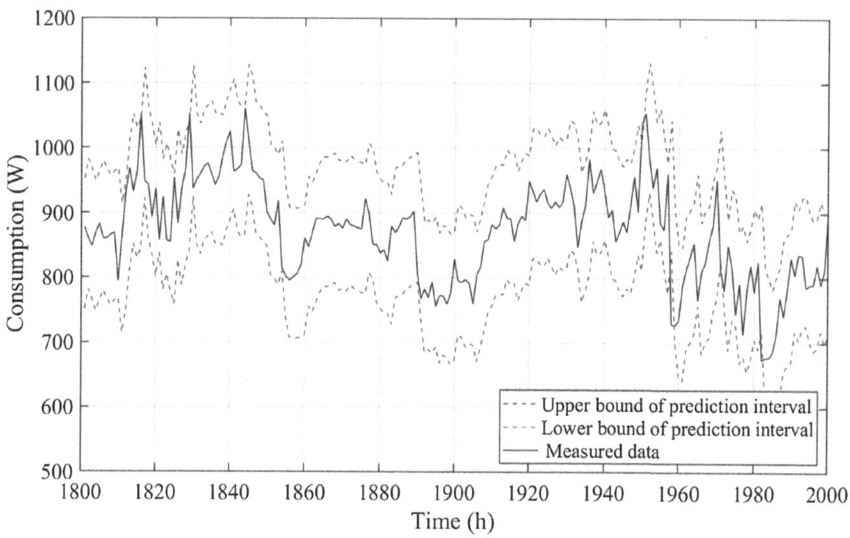

Fig. 10.19 Intervals prediction results with nominal confidence 99% obtained by the proposed GPR-TEL model with the Rational-quadratic kernel

Table 10.5 The performance evaluation indices of the proposed GPR-TEL model

Kernel	PINC	PICP	ACE	PIAW	PINAW	CWC
Exponential	0.9900	0.9800	−0.0100	201.4927	0.5218	1.3822
	0.9500	0.9000	−0.0500	153.3170	0.3971	5.2344
	0.9000	0.8550	−0.0450	128.6677	0.3332	3.4949
Squared-exponential	0.9900	0.9700	−0.0200	200.6541	0.5197	1.9323
	0.9500	0.9050	−0.0450	152.6789	0.3954	4.1471
	0.9000	0.8600	−0.0400	128.1322	0.3318	2.7839
Matern3/2	0.9900	0.9700	−0.0200	200.5622	0.5194	1.9314
	0.9500	0.9050	−0.0450	152.6090	0.3952	4.1452
	0.9000	0.8600	−0.0400	128.0735	0.3317	2.7826
Matern5/2	0.9900	0.9700	−0.0200	200.6850	0.5198	1.9326
	0.9500	0.9050	−0.0450	152.7024	0.3955	4.1477
	0.9000	0.8600	−0.0400	128.1519	0.3319	2.7843
Rational-quadratic	0.9900	0.9700	−0.0200	200.6541	0.5197	1.9323
	0.9500	0.9050	−0.0450	152.6789	0.3954	4.1471
	0.9000	0.8600	−0.0400	128.1322	0.3318	2.7839

(c) When it comes to the interval prediction results with nominal confidences 95 and 90%, the differences among the squared exponential kernel, the Matern3/2 kernel, the Matern5/2 kernel, and the Rational-quadratic are small, and the Matern3/2 kernel is slightly better than that of other models. For instance, the CWC value of the exponential kernel-based model with nominal confidences 95% and 90% are 4.1452 and 2.7826, respectively, which are slightly smaller than that of other models.

10.3.3 Gaussian Process Regression Based on Different Input Features

In the section, The GPR model based on total electric load time series and the number of hours from midnight (GPR-TEL-H), the GPR model based on total electric load time series and the day of week (GPR-TEL-W), the GPR model based on total electric load time series, the number of hours from midnight and the day of week (GPR-TEL-H-W) and the GPR model based on all information from sub-meters (GPR-SUB-A) are constructed. Figure 10.20 shows the modeling flow chart of the Gaussian process regression based on the additional time and date information as well as the information from the sub-meters. As shown in Fig. 10.20, the main steps of the modeling are as follows:

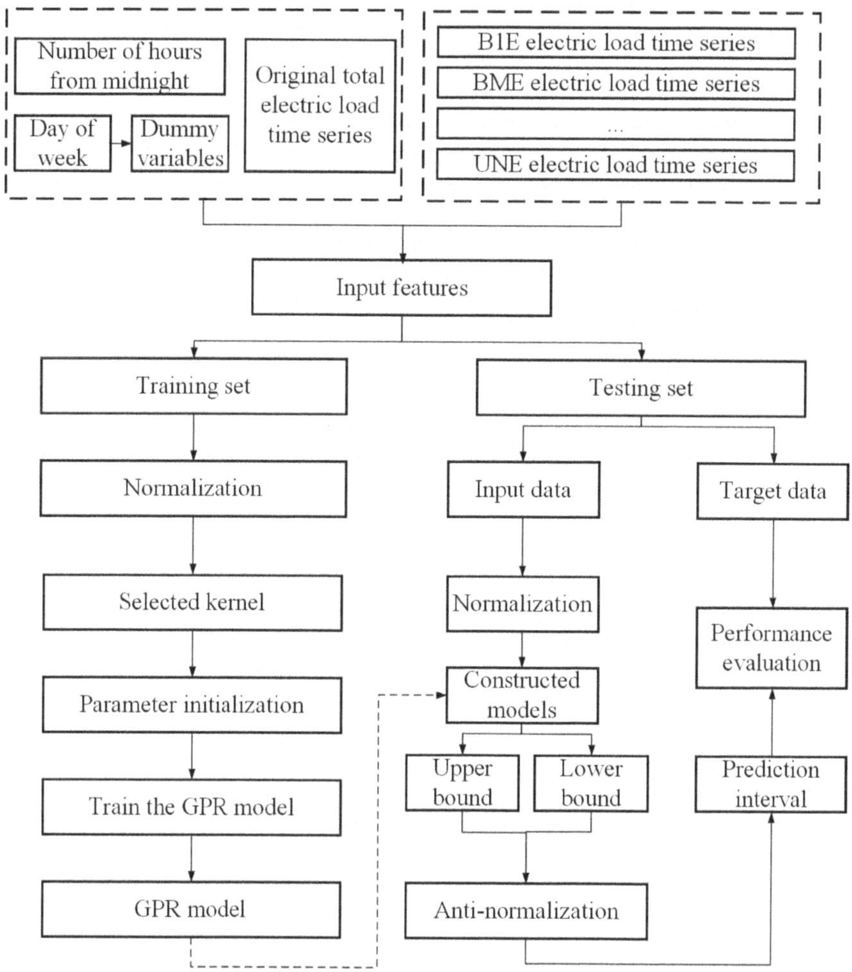

Fig. 10.20 Modeling flow chart of QR prediction model driven by multiple data

(a) Preparing the input features. The additional time and date information, as well as the information from the sub-meters, are involved. The detailed description of the features is given in Sects. 10.2.5 and 10.2.6.

(b) Getting the training data and testing data from the original total electric load time series. Normalizing all the training set as well as the input data of the testing set.

(c) Selecting the kernels and initializing the input features of the selected kernels. In the section, the σ_f is initialized as the standard deviation of the target value in the training data. The σ_l is initialized as 10.

(d) Training the GPR models under the determined parameters based on the training set.

Table 10.6 The performance evaluation indices of the proposed GPR models with different input features

Model	PINC	PICP	ACE	PIAW	PINAW	CWC
GPR-TEL-H	0.9900	0.9750	−0.0150	202.6649	0.5249	1.6361
	0.9500	0.9050	−0.0450	154.2090	0.3994	4.1886
	0.9000	0.8600	−0.0400	129.4162	0.3352	2.8118
GPR-TEL-W	0.9900	0.9700	−0.0200	202.4943	0.5244	1.9500
	0.9500	0.9200	−0.0300	154.0791	0.3990	2.1875
	0.9000	0.8300	−0.0700	129.3073	0.3349	11.4250
GPR-TEL-W-H	0.9900	0.9750	−0.0150	205.9832	0.5335	1.6628
	0.9500	0.9050	−0.0450	156.7339	0.4059	4.2572
	0.9000	0.8650	−0.0350	131.5352	0.3407	2.3010
GPR-SUB-A	0.9900	0.9750	−0.0150	226.9453	0.5878	1.8321
	0.9500	0.9300	−0.0200	172.6841	0.4472	1.6629
	0.9000	0.9100	0.0100	144.9210	0.3753	0.3753

(e) Putting the normalized input data of the testing set to the constructed model, getting the upper bound and lower bound of the interval with a certain nominal confidence, and carrying out the anti-normalization to get the final prediction intervals.

(f) Calculating performance evaluating metrics and comparing the performance of the GPR models with different kernels.

The performance evaluation indices with nominal confidence 99, 95 and 90% obtained by the proposed GPR-TEL-H model, GPR-TEL-W model, GPR-TEL-H-W model, and GPR-SUB-A model are given in Table 10.6.

As shown in Table 10.6, it can be concluded that:

(a) The GPR-TEL-H model provides the best interval prediction results with nominal confidence 99%. For instance, the ACE and the CWC value of the GPR-TEL-H model are −0.015 and 1.6361, respectively, which are significantly better than that of other models.

(b) When it comes to the interval prediction results with nominal confidences 95 and 90%, the GPR-SUB-A model provides the best performance. For instance, the CWC value of the exponential kernel-based model with nominal confidences 95% and 90% are 1.6629 and 0.3753, respectively, which are smaller than that of other models significantly.

10.3.4 Gaussian Process Regression Based on Feature Selection

In the section, the feature selection method is applied to the proposed GPR-TEL model, the GPR-TEL-H model, the GPR-TEL-W model, the GPR-TEL-W-H model, and the GPR-SUB-A model. Figure 10.21 shows the modeling flow chart in the section. As can be seen from Fig. 10.21, compared to the modeling steps described in Sect. 10.3.3, the input features are ranked and selected before dividing the training set and testing set. In the study, the mRMR algorithm is utilized to rank the importance of the features and 25% features are removed from the input features according to the ranking results. The performance evaluation results of the proposed GPR-TEL

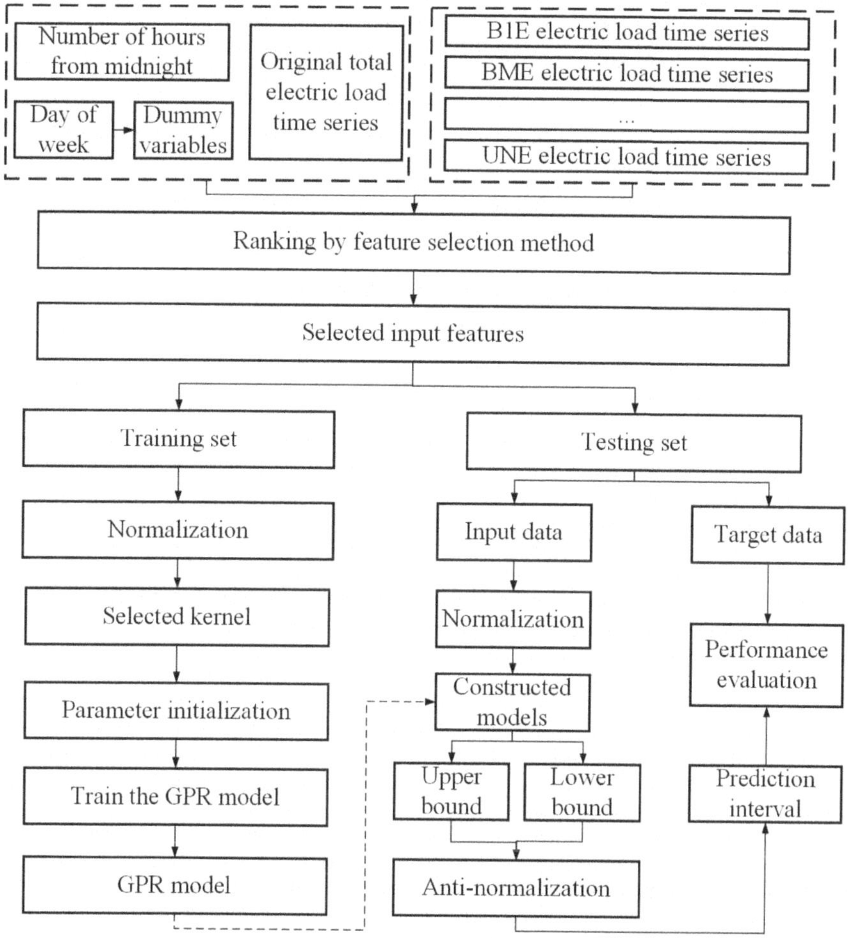

Fig. 10.21 Modeling flow chart of QR prediction model driven by multiple data

Table 10.7 The performance evaluation indices of the proposed GPR model with different input features

Models	PINC	PICP	ACE	PIAW	PINAW	CWC
GPR-TEL-F	**0.9900**	**0.9850**	**−0.0050**	**201.3769**	**0.5215**	**1.1912**
	0.9500	0.9200	−0.0300	153.2289	0.3968	2.1754
	0.9000	0.8500	−0.0500	128.5937	0.3330	4.3903
GPR-TEL-H-F	0.9900	0.9750	−0.0150	202.1398	0.5235	1.6318
	0.9500	0.9150	−0.0350	153.8094	0.3983	2.6907
	0.9000	0.8550	−0.0450	129.0809	0.3343	3.5061
GPR-TEL-W-F	0.9900	0.9650	−0.0250	202.5861	0.5247	2.3560
	0.9500	0.9100	−0.0400	154.1490	0.3992	3.3492
	0.9000	0.8600	−0.0400	129.3659	0.3350	2.8107
GPR-TEL-W-H-F	0.9900	0.9800	−0.0100	204.0801	0.5285	1.4000
	0.9500	**0.9300**	**−0.0200**	**155.2858**	**0.4022**	**1.4954**
	0.9000	0.8750	−0.0250	130.3199	0.3375	1.5156
GPR-SUB-F	0.9900	0.9800	−0.0100	225.2333	0.5833	1.5451
	0.9500	0.9300	−0.0200	171.3814	0.4439	1.6504
	0.9000	**0.9050**	**0.0050**	**143.8278**	**0.3725**	**0.3725**

model, the GPR-TEL-H model, the GPR-TEL-W model, the GPR-TEL-W-H model, and the GPR-SUB-A model are given in Table 10.7.

As shown in Table 10.7, it can be concluded that:

(a) The GPR-TEL-F model provides the best interval prediction results with nominal confidence 99%. For instance, the ACE and the CWC value of the GPR-TEL-F model are −0.005 and 1.1912, respectively, which are better than that of other models, significantly.
(b) The GPR-TEL-W-H-F model provides the best interval prediction results with nominal confidences 95%. For instance, the ACE and the CWC value of the GPR-TEL-W-H-F model are −0.02 and 1.4954, respectively, which are better than that of other models significantly.
(c) The GPR-SUB-F model provides the best interval prediction results with nominal confidences 90%. For instance, the ACE and the CWC value of the GPR-SUB-F model are 0.005 and 0.3725, respectively, which are better than that of other models significantly.

10.4 Experiment Analysis

In order to compare the performance of different forecasting methods, namely, the QR and the GPR, the performance of the two forecasting methods is compared with different input features. In order to make it easier to see the difference between the two models, some indices are defined as follows:

$$D_{\text{ACE}} = |ACE_1| - |ACE_2| \tag{10.20}$$

$$D_{\text{PINAW}} = PINAW_1 - PINAW_2 \tag{10.21}$$

$$D_{\text{CWC}} = CWC_1 - CWC_2 \tag{10.22}$$

where D_{ACE}, D_{PINAW} and D_{CWC} are the difference of *ACE*, *PINAW* and *CWC* between model 1 and model 2, respectively. If the D_{ACE}, D_{PINAW} and D_{CWC} have negative values, it means that the performance of model 1 is better than that of model 2.

The performance evaluation indices of the proposed QR models versus the GPR models with different input features are given in Table 10.8, from which it can be found that the overall performance of the QR model is better than that of the GPR model. As shown in Table 10.8, among 9 of all the experiments in the chapter, QR model provides better results than the GPR model in all evaluation metrics. There are 2 experiments in which the GPR defeat the QR model in all evaluation metrics.

The performance evaluation indices of the GPR models with selected features versus GPR models with all features are given in Table 10.9, from which it can be found that the feature selection method is effective in improving the performance of the GPR model for major experiments. As shown in Table 10.9, among 10 of all the

Table 10.8 The performance evaluation indices of the proposed QR models versus the GPR models with different input features

Features	PINC	D_{ACE}	D_{PINAW}	D_{CWC}
TEL	0.99	**−0.0050**	**−0.0003**	**−0.1910**
	0.95	**−0.0200**	**−0.0002**	**−3.0590**
	0.90	0.0050	−0.0002	0.8955
TEL-H	0.99	0.0000	**−0.0014**	**−0.0042**
	0.95	**−0.0100**	**−0.0010**	**−1.4979**
	0.90	0.0050	**−0.0009**	0.6943
TEL-W	0.99	0.0050	0.0002	0.4060
	0.95	0.0100	0.0002	1.1617
	0.90	**−0.0300**	0.0002	**−8.6143**
TEL-H-W	0.99	**−0.0050**	**−0.0049**	**−0.2629**
	0.95	**−0.0250**	**−0.0038**	**−2.7618**
	0.90	**−0.0100**	**−0.0031**	**−0.7855**
SUB-A	0.99	**−0.0050**	**−0.0044**	**−0.2870**
	0.95	**0.0000**	**−0.0034**	**−0.0125**
	0.90	**−0.0050**	**−0.0028**	**−0.0028**

Table 10.9 The performance evaluation indices of the GPR models with selected features versus GPR models with all features

Features	PINC	D_{ACE}	D_{PINAW}	D_{CWC}
TEL	0.99	**−0.0050**	**−0.0003**	**−0.1910**
	0.95	**−0.0200**	**−0.0002**	**−3.0590**
	0.90	0.0050	−0.0002	0.8955
TEL-H	0.99	**0.0000**	**−0.0014**	**−0.0042**
	0.95	**−0.0100**	**−0.0010**	**−1.4979**
	0.90	0.0050	−0.0009	0.6943
TEL-W	0.99	0.0050	0.0002	0.4060
	0.95	0.0100	0.0002	1.1617
	0.90	−0.0300	0.0002	−8.6143
TEL-H-W	0.99	**−0.0050**	**−0.0049**	**−0.2629**
	0.95	**−0.0250**	**−0.0038**	**−2.7618**
	0.90	**−0.0100**	**−0.0031**	**−0.7855**
SUB-A	0.99	**−0.0050**	**−0.0044**	**−0.2870**
	0.95	**0.0000**	**−0.0034**	**−0.0125**
	0.90	**−0.0050**	**−0.0028**	**−0.0028**

experiments, the GPR model with selected features provides better results than the GPR model with all input features in all evaluation metrics. There are 2 experiments in which the GPR model with all input features defeats the GPR model with selected features in all evaluation metrics.

References

Alobaidi MH, Chebana F, Meguid MA (2018) Robust ensemble learning framework for day-ahead forecasting of household based energy consumption. Appl Energy 212:997–1012. https://doi.org/10.1016/j.apenergy.2017.12.054

Dedinec A, Filiposka S, Dedinec A, Kocarev L (2016) Deep belief network based electricity load forecasting: an analysis of Macedonian case. Energy 115:1688–1700. https://doi.org/10.1016/j.energy.2016.07.090

Dudek G (2016) Pattern-based local linear regression models for short-term load forecasting. Electr Power Syst Res 130:139–147. https://doi.org/10.1016/j.epsr.2015.09.001

Khan N, Shahid S, Juneng L, Ahmed K, Ismail T, Nawaz N (2019) Prediction of heat waves in Pakistan using quantile regression forests. Atmos Res 221:1–11. https://doi.org/10.1016/j.atmosres.2019.01.024

Khosravi A, Nahavandi S, Creighton D, Atiya AF (2011) Lower upper bound estimation method for construction of neural network-based prediction intervals. IEEE Trans Neural Netw 22(3):337–346. https://doi.org/10.1109/TNN.2010.2096824

Koenker R, Bassett G (1978) Regression quantiles. Econometrica 46(1):33. https://doi.org/10.2307/1913643

Kong W, Dong ZY, Jia Y, Hill DJ, Xu Y, Zhang Y (2019) Short-term residential load forecasting based on LSTM recurrent neural network. IEEE Trans Smart Grid 10(1):841–851. https://doi.org/10.1109/TSG.2017.2753802

Li S, Goel L, Wang P (2016) An ensemble approach for short-term load forecasting by extreme learning machine. Appl Energy 170:22–29. https://doi.org/10.1016/j.apenergy.2016.02.114

Makonin S, Ellert B, Bajić IV, Popowich F (2016) Electricity, water, and natural gas consumption of a residential house in Canada from 2012 to 2014. Sci Data 3(1):160037. https://doi.org/10.1038/sdata.2016.37

Melkumyan A, Ramos F (2011) Multi-Kernel Gaussian processes. In: Twenty-second international joint conference on artificial intelligence, 2011/06/28

Rasmussen CE (2004) Gaussian Processes in machine learning. In: Bousquet O, von Luxburg U, Rätsch G (eds) Advanced lectures on machine learning: ML summer schools 2003. Canberra, Australia, February 2–14, 2003, Tübingen, Germany, August 4–16, 2003, Revised Lectures. Lecture notes in Computer science. Springer Berlin Heidelberg, Berlin, Heidelberg, pp 63–71

Shepero M, van der Meer D, Munkhammar J, Widén J (2018) Residential probabilistic load forecasting: a method using Gaussian process designed for electric load data. Appl Energy 218:159–172. https://doi.org/10.1016/j.apenergy.2018.02.165

Shi H, Xu M, Li R (2018) Deep learning for household load forecasting—a novel pooling deep RNN. IEEE Trans Smart Grid 9(5):5271–5280. https://doi.org/10.1109/TSG.2017.2686012

Sideratos G, Hatziargyriou ND (2012) Probabilistic wind power forecasting using radial basis function neural networks. IEEE Trans Power Syst 27(4):1788–1796. https://doi.org/10.1109/TPWRS.2012.2187803

Son H, Kim C (2017) Short-term forecasting of electricity demand for the residential sector using weather and social variables. Resour Conserv Recycl 123:200–207. https://doi.org/10.1016/j.resconrec.2016.01.016

Taylor JW, McSharry PE (2007) Short-term load forecasting methods: an evaluation based on European data. IEEE Trans Power Syst 22(4):2213–2219. https://doi.org/10.1109/TPWRS.2007.907583

Wan C, Xu Z, Pinson P, Dong ZY, Wong KP (2014) Probabilistic forecasting of wind power generation using extreme learning machine. IEEE Trans Power Syst 29(3):1033–1044. https://doi.org/10.1109/TPWRS.2013.2287871

Wilson AG, Dann C, Lucas C, Xing EP (2015) The human Kernel. Computer Sci 2854–2862

Yu C-N, Mirowski P, Ho TK (2016) A sparse coding approach to household electricity demand forecasting in smart grids. IEEE Trans Smart Grid 1–11. https://doi.org/10.1109/tsg.2015.2513900

The manufacturer's authorised representative in the EU is Springer Nature Customer Service Centre GmbH, Europaplatz 3, 69115 Heidelberg, Germany. If you have any concerns regarding our products, please contact ProductSafety@springernature.com

Printed and bound by CPI Group (UK) Ltd, Croydon, CR0 4YY
25/09/2026
02222900-0013